Critical Perspectives in Rural Development Studies

Agrarian transformations within and across countries have been significantly and dynamically altered during the past few decades compared to previous eras, provoking a variety of reactions from rural poor communities worldwide. The recent convergence of various crises – financial, food, energy and environmental – has put the nexus between 'rural development' and 'development in general' back onto the center stage of theoretical, policy and political agendas in the world today. Confronting these issues will require (re)engaging with critical theoretical perspectives, taking politics seriously, and utilizing rigorous and appropriate research methodologies. These are the common messages and implications of the various contributions to this collection in the context of a scholarship that is critical in two senses: questioning prescriptions from mainstream perspectives and interrogating popular conventions in radical thinking.

This book focuses on key perspectives, frameworks and methodologies in agrarian change and peasant studies. The contributors are leading scholars in the field of rural development studies: Henry Bernstein, Saturnino M. Borras Jr, Terence J. Byres, Marc Edelman, Cristóbal Kay, Benedict Kerkvliet, Philip McMichael, Shahra Razavi, Ian Scoones and Teodor Shanin.

This book was previously published as a special issue of the *Journal of Peasant Studies*.

Saturnino M. Borras Jr. is Canada Research Chair in International Development Studies at Saint Mary's University, Halifax, Nova Scotia. He has been deeply involved in rural social movements internationally since the early 1980s. He is a Fellow of the Amsterdam-based international think tank Transnational Institute (TNI) and of the Oakland-based Food First/Development Policy Institute. He is the Editor of the *Journal of Peasant Studies*.

Critical Agrarian Studies

Series Editor: *Saturnino M. Borras Jr.*

Critical Agrarian Studies is the new accompanying book series to the *Journal of Peasant Studies*. It publishes selected special issues of the journal and, occasionally, books that offer major contributions in the field of critical agrarian studies. The book series builds on the long and rich history of the journal and its former accompanying book series, the Library of Peasant Studies (1973-2008) which had published several important monographs and special-issues-as-books.

Available titles:

Critical Perspectives in Rural Development Studies
Edited by Saturnino M. Borras Jr.

Critical Perspectives in Rural Development Studies

Edited by Saturnino M. Borras Jr.

Routledge
Taylor & Francis Group

LONDON AND NEW YORK

Transferred to digital printing 2010
First published 2010
by Routledge
2 Park Square, Milton Park, Abingdon, Oxon, OX14 4RN

Simultaneously published in the USA and Canada
by Routledge
270 Madison Avenue, New York, NY 10016

Routledge is an imprint of the Taylor & Francis Group, an informa business

Typeset in Times by Value Chain, India

British Library Cataloguing in Publication Data
A catalogue record for this book is available from the British Library

ISBN10: 0-415-55244-3 (hbk)
ISBN13: 978-0-415-55244-8 (hbk)

ISBN10: 0-415-59177-5 (pbk)
ISBN13: 978-0-415-59177-5 (pbk)

Contents

Notes on Contributors

Saturnino M. Borras Jr. is Canada Research Chair in International Development Studies at Saint Mary's University in Halifax, Nova Scotia. He has been deeply involved in rural social movements since the early 1980s. He is a Fellow of the Amsterdam-based Transnational Institute (TNI) and of the California-based Food First/Institute for Food and Development Policy. His research interests include (trans)national agrarian movements, state–society relations in rural development, and land issues. His books include: *Pro-Poor Land Reform: A Critique* (2007) and *Transnational Agrarian Movements Confronting Globalisation* (2008) co-edited with M. Edelman and C. Kay. He is the Editor of the *Journal of Peasant Studies*.

Henry Bernstein is Professor of Development Studies in the University of London at SOAS. Having studied history and sociology at the Universities of Cambridge and London (LSE), he has been a teacher and researcher in Turkey, Tanzania, South Africa and the USA, and at a number of British universities, and is also Associate Professor at the China Agricultural University, Beijing. He has a longstanding interest in the political economy of agrarian change, as well as social theory, and more recently in globalisation and labour. He was coeditor with Terence J. Byres of *The Journal of Peasant Studies* from 1985 to 2000, and was founding editor, again with Terence J. Byres, of the *Journal of Agrarian Change* in 2001, of which he became Emeritus Editor in 2008.

Terence J. Byres is Emeritus Professor of Political Economy in the University of London, and Emeritus Editor of the *Journal of Agrarian Change*. He studied Economics and Economic History at the Universities of Aberdeen, Cambridge and Glasgow, joining SOAS as a Research Fellow in 1962. During his long tenure at the School, he was the first Head of the new Economics Department established in 1990. He is well known for his earlier work on aid and development policy and planning, and, above all, on agrarian political economy with special reference to India, on which he has published widely. More recently, he has worked on the comparative political economy of historical agrarian change. His most recent books are *Capitalism From Above and Capitalism From Below* (1996), a comparison of capitalist agrarian transition in Prussia and the United States; and *The Indian Economy: Major Debates Since Independence*, which he edited (1998). He is currently working on the relationships, in the eighteenth century, between the capitalist transformation of Scotland and the Scottish Enlightenment. *The Journal of Peasant Studies* was founded on his initiative in 1973, and he was its co-editor from 1973 to 2000. He was founding editor, with Henry Bernstein, in 2001, of the *Journal of Agrarian Change*, and the two co-edited it until 2008.

Marc Edelman (PhD in Anthropology, Columbia, 1985) is Professor of Anthropology at Hunter College and the Graduate Center of the City University of New York. His research interests include Latin American agrarian history, rural development, and social movements. Among his books are *The Logic of the Latifundio* (1992) and *Peasants Against Globalization* (1999), as well as several co-authored and co-edited volumes. He is a member of the Editorial Collective of *The Journal of Peasant Studies*.

Cristóbal Kay is Professor in Development Studies and Rural Development at the Institute of Social Studies. He is also Adjunct Professor in International Development Studies at Saint Mary's University, Halifax, Canada Environmental Sciences at the University of Birmingham. He is an editor of the *Journal of Agrarian Change* and member of the international advisory board of *The Journal of Peasant Studies*. His most recent books include *Transnational Agrarian Movements Confronting Globalisation* (edited with S. Borras and M. Edelman, 2008), *Peasants and Globalisation: Political Economy, Rural Transformation and the Agrarian Question* (edited with H. Akram-Lodhi, 2009) and *Market-Led Agrarian Reform: Critical Perspectives on Neoliberal Land Policies and the Rural Poor* (edited with S. Borras and E. Lahiff, 2008).

Benedict J. Tria Kerkvliet's teaching and research emphasise agrarian politics and political movements in Asia, especially the Philippines and Vietnam. He is an Emeritus Professor at The Australian National University and an A.liate Graduate Faculty member at the University of Hawai'i-Manoa. His recent books are *Beyond Hanoi: Local Government in Vietnam* (2004, co-edited with David Marr) and *The Power of Everyday Politics: How Vietnamese Peasants Transformed National Policy* (2005). He is a member of the International Advisory Board of the *Journal of Peasant Studies*.

Philip McMichael is Professor of Development Sociology at Cornell University. His works include *Settlers and the Agrarian Question* (1984); *The Global Restructuring of Agro-Food Systems* (1994, editor); *Food and Agrarian Systems in the World Economy* (1995, editor); *New Directions in the Sociology of Global Development* (2005, co-editor), and *Development and Social Change: A Global Perspective* (2008). Current research is on the politics of globalisation and climate change, with a focus on agrarian movements.

Shahra Razavi is Research Co-ordinator at the UN Research Institute for Social Development (UNRISD), where she directs the Institute's Gender and Development Programme. She specialises in the gender dimensions of social development, with a particular focus on livelihoods and social policy. Her publications include *Agrarian Change, Gender and Land Rights* (2003, Special Issue of *Journal of Agrarian Change*), and *Gender and Social Policy in a Global Context: Uncovering the Gendered Structure of 'The Social'* (2006, co-edited with Shireen Hassim). She is a member of the Editorial Collective of *The Journal of Peasant Studies*.

Ian Scoones is a Professorial Fellow at the Institute of Development Studies, University of Sussex. He is an agricultural ecologist by original training, although since has focused on the intersection between rural livelihoods and institutional and policy change, with a particular focus on science, technology and development questions in sub-Saharan Africa. He is currently co-director of the STEPS Centre (www.steps-centre.org) and is joint coordinator of the Future Agricultures Consortium (www.future-agricultures.org). He is a member of the Editorial Collective of *The Journal of Peasant Studies*.

Teodor Shanin OBE: PhD; Professor Emeritus University of Manchester; Fellow of the Academy of Agricultural Sciences of RF; President of Moscow School of Social and Economic Sciences. His research interests include: epistemology, historical sociology, sociology of knowledge, social economy, peasant and rural studies, theoretical roots of social work. His books include: *Peasants and Peasant Societies* (1971); *The Awkward Class* (1972); *The Rules of the Game: Models in Contemporary Scholarly Thought* (1972); *Russia as a Developing Society* (1985, 1986); *Revolution as a Moment of Truth* (1986, 1987); *Defining Peasants* (1990); *In Russian: The Great Stranger* (1987), *The Peasant Rebellion in Guberniya of Tambov 1919–1921. Antonovshchina* (with V. Danilov) (1994), *Informal Economies: Russia and the World* (1999), *Reflexive Peasantology* (2002), *Fathers and Sons: Generational History (2005), Nestor Machno: peasant movement in Ukraine* (with V. Danilov) (2006). He was among the first editors of *The Journal of Peasant Studies.*

Introduction

Saturnino M. Borras Jr.

Agrarian transformations within and across countries have been significantly and dynamically altered during the past few decades compared to previous eras, provoking a variety of reactions from rural poor communities worldwide. The changed and changing agrarian terrain has also influenced recent rethinking in critical inquiry into the nature, scope, pace and direction of agrarian transformations and development. This can be seen in terms of theorising, linking with development policy and politics, and thinking about methodologies. This collection of essays on key perspectives, frameworks and methodologies is an effort to contribute to the larger rethinking. The following paper introduces the collection.

Changed and changing agrarian terrain

Even though there are fewer people now living and working in the rural world than four decades ago, it still matters a great deal for everyone what happens there. The dynamics of social change in developing countries during the past four decades have some features distinct from those that marked the first three quarters of the twentieth century. A brief overview of changes in the global economy and politics in general, and in the agrarian world in particular, during the past four decades partly illustrates this. Henry Bernstein (2008, 247) offers a summary:

> While controversy rages, and will continue to do so, concerning the causes, mechanisms and implications, including new contradictions, of changes in the world economy, politics and culture since the 1970s... there is little doubt that important shifts with far-reaching ramifications have occurred. ... A familiar list would include: the deregulation of financial markets; shifts in the production, sourcing and sales strategies and technologies of transnational manufacturing and agribusiness corporations; the massive new possibilities attendant on information technologies, not least for mass communications, and how they are exploited by the corporate capital that controls them; the demise of the Soviet Union and finally of any plausible socialist model of development; and the ideological and political ascendancy of neoliberalism in a highly selective rolling back of the state, including the structural adjustment programmes,

economic liberalisation, and state reform/good governance agendas imposed on the countries of the South and, more recently, the former Soviet bloc. This is the context, and some of its key markers, that spelled the end of state-led development.

Michael Watts (2008, 276) adds that, 'one of the presumptions of new research focused on transnational processes and agrarian food orders is that the old or classical international division of labour within the agro-food system has been irretrievably altered in the past 25 years'. In examining the world food system, Tony Weis (2007, 5, emphasis added) observes that 'the origins of the contemporary global food economy could be traced back through a series of revolutionary changes, which once took shape over the course of millennia, then over centuries, and *which are now compressed into mere decades*'.

On the politics side of this transformation, the past four decades saw the tail-end of national liberation movements, revolutions and rebellions to which the rural poor had provided important contributions. During most of the recent period this type of peasant politics has been largely absent. Nevertheless, it was during this era when relatively newer types of agrarian movements, networks and coalitions emerged and gained political influence. Meanwhile, the recent convergence of various crises – financial, food, energy and environmental – has put the nexus between 'rural development' and 'development in general' back onto the center stage of theoretical, policy and political agendas in the world today.

Addressing these agendas requires some degree of clarification about critical theoretical perspectives and updated analytical tools. It is in this context that this collection of reflection essays on critical perspectives in agrarian change and peasant studies was put together. This essay introduces the collection. Section 1 provides an overview of persistent rural poverty and increasing inequality, agriculture and rural livelihoods, and rural politics as the context for and the object of the critical perspectives on agrarian change. An overview of the contributions is provided in Section 2. A discussion of common messages and implications of this collection is offered in Section 3, focusing on three key challenges to critical agrarian change and peasant studies: (re)engaging with critical theories, (re)engaging with the real world politics, and utilising rigorous research methodologies. Brief concluding remarks are offered in Section 4.

Agriculture and livelihoods, poverty and inequality

Although decreasing in relative terms, the absolute number of rural dwellers remains very significant. The absolute number of people living in urban centres had, in 2007, overtaken for the first time the number of people living in the countryside. By 2010, the estimate is that there will be 3.3 billion people in the rural world, with another 3.5 billion in urban communities.[1] The dramatic rural–urban demographic changes were quite recent. In 1970, the total world population was 3.7 billion, with 2.4 billion rural and 1.3 billion urban. The change in the agricultural/non-agricultural population was even more dramatic during the same period. In 1970, the agricultural population stood at 2.0 billion people and the non-agricultural population at 1.7 billion.

---

[1]Estimates by the UN Food and Agriculture Organisation Statistics (FAOSTAT). Available from: www.faostat.org [Accessed 3 November 2008]. This is the same website and the date of data downloading for all subsequent FAOSTAT sourced data cited in succeeding footnotes, unless otherwise specified.

By 2010, this will be radically reversed, at 2.6 billion agricultural population versus 4.2 billion non-agricultural.[2]

Yet even as the number of urban dwellers overtakes the number of rural population, the percentage of *poor* people in rural areas continues to be higher than that in the urban areas: three-fourths of the world's poor today live and work in the countryside (World Bank 2007). In 2008 world poverty remained a largely rural phenomenon. Often, poverty is associated with hunger. By 2006, there were 820 million hungry people. This was a marginal reduction from 1990–92's figure of 823 million. Ironically, most of the hungry people live and work in the rural areas where food is produced.[3] At the height of the recent food price crisis, the FAO (2008a) announced that in order to meet the growing global food need, food production would need to double by 2050. Much of this needed increase would have to happen in developing countries where the majority of the world's rural poor live and where 95 percent of the estimated population increase during this period is expected to occur. Alarmed, the organisation called for a 'new world agricultural order' and called on governments to 'find 30 billion dollars' of new investment in agriculture and rural development, pointing out 'the amount was modest compared to 365 billion dollars of total support to agriculture in the OECD countries in 2007 and 1,340 billion in world military expenditure the same year by developed and developing countries'.[4]

During the past four decades, amid significant rural/urban and agricultural/non-agricultural demographic changes, agricultural production and trade have witnessed dramatic growth despite marginal increase in the total size of the world's agricultural land. Some key statistics are illustrative. The world's total production of cereals was 1.6 billion tons in 1979–81, jumping to 2.3 billion tons in 2004.[5] The global production of meat nearly doubled during the same period, at 0.14 billion tons in 1979–81 to 0.26 billion in 2004.[6] Production of fruits and vegetables doubled during this period, at 0.63 billion tons in 1979–81 to 1.4 billion tons in 2004. In the midst of massive promotion of export-oriented development strategies, agricultural trade increases were most dramatic during the past four decades: the total value of all agricultural exports in 1970 was $52 billion, it increased by about 12 times in 2005, or up to $654 billion.[7] Haroon Akram-Lodhi and Cristóbal Kay (2008a, 318) observe: 'The traits of accumulation have significantly changed ... In particular, the emphasis on the expansion of the home market that previously prevailed during the mid-twentieth century has been largely, but not completely, replaced by an emphasis on the promotion of an agricultural export-led strategy as the principal means of enhancing rural accumulation.' But while there had been dramatic increases in cross-country agricultural trade during the past three to four decades, the impacts in terms of food security, household incomes, and inequality within and between countries have been varied and uneven. Moreover, as Akram-Lodhi and

---

[2]FAOSTAT data; see note 1 for the data source information.
[3]FAOSTAT data; see note 1 for data source information.
[4]FAO 2008a. FAO calls for a new world agricultural order. Available from: http://www.ipsterraviva.net/europe/article.aspx?id=6769 [Accessed 20 November 2008].
[5]The combined output of China, India and the USA took 45 percent total share during this period.
[6]The combined output of Brazil, China and the USA accounted for half of the total output in 2004.
[7]FAOSTAT data; see note 1 for the data source information.

Kay (2008a, 325) remind us: 'the impact of rural accumulation on poverty should be examined separately from the impact of rural accumulation on inequality'. Indeed the neoliberal globalisation has resulted in increasing inequality within and between countries in the world (see, e.g., Borras 2007a). Edelman and Haugerud (2005, 9) explain that, 'Global economic inequality increased dramatically between 1960 and 1990: in 1960, the wealthiest 20 percent of the world's population received 30 times the income of the poorest 20 percent; in 1997, the richest 20 percent received 74 times as much'. They add: 'By the late 20th century, the world's 200 wealthiest individuals had assets equal to more than the combined income of 41 percent of the world's population; the assets of the three richest people were more than the combined GNP of all least developed countries' (Edelman and Haugerud 2005, 9).

Although agriculture remains quite important to the livelihoods of more than three billion people, evidence suggests that rural households have increasingly diversified their ways of earning a living, as partly discussed by Ian Scoones (2009, 171–96, this collection; see also Bernstein 2007, De Haan and Zoomers 2005, Ellis 2000, Kay 2008). Labour has become more mobile and in many settings casual. Labour migration has taken multiple directions and character: rural–urban, rural–rural, urban–rural, in-country and international, permanent and cyclical. Many of these migrant jobs are casual and living conditions inhuman, both those based in urban and rural spaces (Davis 2006). Bernstein (2008, 250–51) explains that the fragmentation of the classes of labour

> signals the effects of how classes of labour in global capitalism, and especially in the South, pursue their reproduction, through insecure and oppressive – and in many places increasingly scarce – wage employment, often combined with a range of likewise precarious small-scale farming and insecure informal-sector ('survival') activity, subject to its own forms of differentiation and oppression along intersecting lines of class, gender, generation, caste and ethnicity ...

He adds that, 'many pursue their means of reproduction across different sites of the social division of labour: urban and rural, agricultural and non-agricultural, wage employment and self-employment' (see also, Davis 2006, 250–51). Furthermore, the corridors of labour flows have also brought with them multidimensional socio-cultural changes including those involving information and communication technology, resulting in previously isolated rural communities, or at least some portions of these communities, now having access to dozens of international cable channels, internet access, text messaging and audio-video conferencing free of charge between people separated from each other by thousands of miles.

Agricultural technology has continued to break new ground, some aspects of which are contested and controversial, and not so different from the previous Green Revolution package of technology and agenda (Ross 1998). Much discussion today centres on a new Green Revolution, primarily for Africa, promoted by multilateral agencies and private institutions such as the Gates Foundation. Genetically modified crops have been aggressively promoted amid increasing opposition from some high-profile organised agrarian and environmental movements (Scoones 2008, Newell 2008, Otero 2008). Satellite mapping techniques have been put to wide use, largely to expand and standardise state maps and cadastre records. There are efforts directed at harnessing the potential of information and communications technology in creating and improving rural livelihoods. Nevertheless, the same package of technology remains generally beyond the reach of poor peasants and controlled by a few transnational companies and their local distributors and retailers (Jansen and Vellema 2004).

Moreover, there are alarming environmental and climate-related problems facing the rural world today. If temperatures rise by more than three degrees, yields of major crops like maize may fall by 20–40 percent in parts of Africa, Asia and Latin America (FAO 2008b). McMichael (2009, 139, this collection) points out that 'global agriculture is responsible for between a quarter and a third of greenhouse gas (GHG) emissions'. There is a similar dilemma in confronting the energy crisis via biofuels: we would need to convert 25 percent of the current global cropland to biofuel production in order to reduce only 14 percent of our current fossil fuel consumption (FAO 2008b, 21). The question is that if global agricultural production needs to double by 2050 to feed the growing population, take nearly one billion people out of hunger, and at the same time fuel the transport and manufacturing sectors, how will this task be carried out without putting further pressure on the already fragile environment, without aggravating climate-related problems, without putting the task under the monopoly control of greedy corporate giants, and without causing massive dispossession of the rural poor?[8]

Rural politics

The dynamic changes in the agrarian world briefly described above have been politically contested by the rural poor both via 'everyday politics' and with the emergence of radical agrarian movements in different parts of the world.[9] This necessarily brings our discussion to the politics of agrarian change, and questions of agency of the rural poor. The 'rural poor' is understood here as a highly heterogeneous social category, and they include the peasantry with its various strata, landless rural labourers, migrant workers, forest dwellers, subsistence fishers, indigenous peoples, and pastoralists. This heterogeneity as well as the recent structural and institutional transformation explained above partly influence the character of rural politics. As Edelman (2008, 83), in the context of Central America, explains:

> Like the migration to which it is related, the growing 'pluriactivity' of rural households and the increasing inter-penetration of city and countryside complicate the question of *campesino* identity in ways that have ramifications for how people view their struggles and their participation in collective efforts for change.

He adds: 'The first thing to acknowledge is that the *campesino* of today is usually not the *campesino* of even 15 years ago' (Edelman 2008, 83).

The discussion of rural politics will follow the typology offered by Ben Kerkvliet (2009, 231, this collection): official politics, everyday peasant politics, and advocacy politics. Official politics 'involves authorities in organisations making,

---

[8]As of this writing, there are numerous negotiations between countries for land sales, long-term land leases or contract farming for food production: South Koreans in Madagascar, Saudi Arabia in Sudan, China in southeast Asia, Libya in Ukraine, and so on (GRAIN 2008).
[9]See, Le Mons Walker (2008) and O'Brien and Li (2006); and Petras and Veltmeyer (2001, 2003), Veltmeyer (2004), Edelman (1999), Moyo and Yeros (2005), Rosset, Patel and Courville (2006), Wright and Wolford (2003).

implementing, changing, contesting, and evading policies regarding resource allocations ... Authorities in [state and non-state] organisations are the primary actors.' 'Everyday politics occurs', according to Kerkvliet (2009, 232), 'where people live and work and involves people embracing, complying with, adjusting, and contesting norms and rules regarding authority over, production of, or allocation of resources and doing so in quiet, mundane, and subtle expressions and acts that are rarely organised or direct'. Advocacy politics 'involves direct and concerted efforts to support, criticise, oppose authorities, their policies and programs, or the entire way in which resources are produced and distributed within an organisation or a system of organisation'. It also includes actions that openly advocate 'alternative programs, procedures, and political systems'. Kerkvliet explains that 'advocates are straight-forwardly, outwardly, and deliberately aiming their actions and views about political matters to authorities and organisations, which can be governments and states but need not be'. Elaboration on each is offered below.

'Official politics' has been evolving in the neoliberal era. Taking politics seriously in critical scholarship requires consistent interrogation of this type of politics. This means confronting the issue of the state (nature, character, role, class composition, and so on) in one's analysis. Nation-states in developing countries have experienced a triple 'squeeze': globalisation, (partial) decentralisation, and the privatisation of some of their functions. Central states remain important in development processes, but have been transformed in terms of the nature, scope, level and direction of their development intervention (Evans 1997a, Ribot and Larson 2005, Keohane and Nye 2000, Gwynne and Kay 2004, Kay 2006). In this context, the recent convergence of various crises, including food, energy, climate and finance, is likely to re-emphasise, not devalue, the role played by nation-states and state authorities in the politics of agrarian transformation. International development and financial institutions continue to play a part in (re)shaping national and local policies for rural development in developing countries, despite the popular lament about the reduction of overseas development assistance for agriculture in developing countries during the past three decades.[10] Multilateral institutions, like the World Bank, continue to be instrumental in promoting neoliberal policies, such as those related to land.[11] Some of these institutions have been the target of protests by organised agrarian and environmental movements, ranging from demands for accountability (Fox and Brown 1998) to efforts at delegitimising some of these agencies. Locating one's analysis of the state in multiple levels, i.e., local, national and international, is an important challenge (Kay 2006).

'Everyday peasant politics' is the type of politics that remains almost invisible to researchers, policymakers, and agrarian movement activists, but can be very powerful in transforming national policies, as demonstrated by Kerkvliet in the case of Vietnam's agricultural policy during the past three decades (2009, see also Kerkvliet 2005). Such low-profile actions can lead to high-profile actions depending on changing political opportunities favouring peasants, as explained by Shapan Adnan (2007) in the context of Bangladesh. Some variations of the latter are also

---

[10]"The US contribution to [total overseas development assistance] fell sharply – from over 60 percent of the total in the mid-1950s to 17 percent by 1998 ... In 1947 ... US foreign aid as percentage of GDP was nearly 3 percent, while by the late 1990s it was a mere 0.1 percent.' (Edelman and Haugerud 2005, 10).
[11]See, for example, World Bank (2003), but refer also to Borras (2007b) and Rosset et al. (2006).

explained by Kathy Le Mons Walker (2008) in an analysis about 'overt everyday peasant resistance' and by Kevin O'Brien and Lianjiang Li (2006) about 'rightful resistance' in contemporary rural China, where abusive acts by local state officials usually around decisions on land use and control have been met with increasingly defiant and confrontational actions by peasants and villagers. While this type of politics has gained appropriate attention in agrarian studies during the past couple of decades or so, thanks to the compelling works by James Scott and Kerkvliet, among others, there remains a major challenge as to how to systematically integrate this perspective into the studies of 'official' and 'advocacy politics', as well as in development practice and political activist work.[12] We will return to this issue later.

Despite the usual ebb and flow from one setting to another through time, agrarian movements have been among the most vibrant sectors of civil society during the past four decades. Most of these movements are indeed rural workers' and peasants' and farmers' movements in the global south and small and part-time farmers' groups in the north. This period has also witnessed the emergence of other rural-based and rural-oriented social movements, including indigenous peoples' movements,[13] women's movements,[14] environmental movements and anti-dam movements,[15] anti-GM crop movements,[16] fishers' movements, and rural–urban alliances.[17]

The co-existence of threats and opportunities brought about by neoliberal globalisation has prompted many rural social movements to both localise further (partly in response to state decentralisation) and to 'internationalise' their actions (in response to globalisation). The seemingly contradictory social-political pressures (of globalisation and decentralisation) that are having such an impact on the nation-states, are also transforming rural social movements: more horizontal solidarity linkages, the rise of 'polycentric' rural social movement, and the emergence of radical agrarian movements that are, in a variety of ways, linked together transnationally via networks, coalitions or movements.[18]

There are distinctly new features in the current generation of agrarian movements that are important to point out as the context for and the subject of critical scholarship: (i) greater direct representation of the rural poor in (sub)national and international official and unofficial policymaking arenas,[19] (ii) more extensive scope

---

[12]See, for example, the issues raised by Malseed (2008) in the context of everyday peasant politics among the Karen people of Burma.

[13]See, for example, Yashar (2005) and Assies, van der Haar and Hoekema (1998).

[14]Refer to Deere and Royce (forthcoming) and Stephen (1997).

[15]See, for example, Peluso et al. (2008) and Baviskar (2004).

[16]Refer to Newell (2008), Scoones (2008) and Otero (2008).

[17]See Veltmeyer (2004).

[18]For an excellent discussion about the differences and possible connections between networks, coalitions and movements, refer to Fox (forthcoming).

[19]Before the existence of Via Campesina in the early 1990s, the only existing transnational agrarian movement (TAM) that had made a significant representation of the world's rural poor was the International Federation of Agricultural Producers or IFAP, which is an organisation of middle and rich farmers mainly based in the north. The politics of IFAP and most of its affiliate organisations tend to be conservative. In contrast, Via Campesina represents the solidarity of poor peasants and small farmers with class interests and politics different from those of IFAP's.

and scale of political work and issues taken up,[20] (iii) deployment of information and communication technology in movement building and collective actions to an unprecedented degree,[21] (iv) more systematic and coherent 'human rights' issue-framing and demand-making perspective and stretching citizenship rights claim-making beyond the conventional national borders,[22] (v) more assertion of the movements' autonomy from actual and potential allies.[23] Overall, contemporary radical (trans)national agrarian movements have been important actors in provoking or inspiring research agendas for critical perspectives on rural development in some ways similar to what peasant-based revolutions, national liberation movements and rebellions did during the most part of the twentieth century.

In sum, historically, the agrarian world has witnessed continuity and change in terms of general patterns of accumulation, appropriation and dispossession for capitalist development, as well as socialist construction (and later, 'de-construction') (see, e.g. Wood 2008). What has been emphasised earlier in the discussion, however, is the fact that the past four decades of the era of neoliberal globalisation have witnessed important changes in the nature, scope, pace and direction of agrarian transformations within and between countries. This can be seen in the general patterns of changes in property relations, labour, appropriation and distribution of

---

[20]The scale of Via Campesina's transnational movement or network, and the scope of its political work and influence have been unprecedented. This is despite the fact that the organisation remains absent or thin in many regions of the world (Borras, Edelman and Kay 2008, Le Mons Walker 2008, Malseed 2008).

[21]The use by TAMs of the latest information and communications technology (internet, email, electronic conferences, mobile phone, texting) is something new in the agrarian movement world. It has led to faster and relatively cheaper ways to access and exchange information, and to plan for and carry out simultaneous political actions, overcoming important traditional institutional and structural obstacles to movement building and collective actions.

[22]Radical TAMs' issue-framing and demand-making perspective is not totally new. Edelman (2005), for example, explains how most TAMs employ arguments drawn from the moral economy perspective (Scott 1976). However, TAMs have also embraced a relatively recent way of framing development discourse, invoking 'human rights' that include political, social, economic and cultural rights (Monsalve et al. 2006). In some ways it invokes a notion of 'global citizenship rights' by holding international institutions accountable, something that did not exist in any systematic way in the agrarian movement world before (Borras and Franco 2009, Monsalve et al. 2008, Patel 2006). More broadly, Via Campesina and other rural-oriented global agrarian justice movements attempt to reframe the very terms of development discourse by putting forward new (alternative) concepts such as 'food sovereignty' and 'deglobalisation' (see McMichael 2008, Bello 2003) and by developing alternative knowledge-building movements and knowledge networks such as the transnational agroecological movement in Central America (Holt-Gimenez 2006).

[23]A partly similar global advocacy on behalf of poor peasants and small farmers had actually existed and been used before. It was not carried out by agrarian movements themselves, but by intermediary NGOs. When Via Campesina was established in the early 1990s, one of the first things it did was to define its 'universal' identity (i.e., 'people of the land') and partly, and perhaps more implicitly, its class composition (i.e., poor peasants and small farmers), clarify its representation claims (i.e., 'non-mediated') and preferred form of actions (i.e., 'direct' combining confrontation and negotiation), and declared that intermediary NGOs should stop representing poor peasants and small farmers. This demand is broadly within the global civil society popular saying, 'not about us without us'. Indeed, these radical TAMs have created a distinct 'citizenship space' at the global level that did not exist before that in turn would alter the 'political opportunity structure' for their affiliate movements at the local, national and international levels (Borras and Franco 2009, see also Tarrow 2005, Borras 2008, Edelman 2003, 2008, Martinez-Torres and Rosset 2008, Desmarais 2007, Biekart and Jelsma 1994).

agrarian income and wealth, and the ways in which agrarian surpluses are disposed and invested, among others.[24] As Akram-Lodhi and Kay (2008a, 317) explain, neoliberal globalisation has 'altered the land-, labour- and capital-intensity of production, reconfiguring the rural production process in ways that may, or may not, affect processes that expand the commodification of labour and alter the purpose of production from production for use to production for exchange.' This is a highly dynamic, but uneven process from one society to the other. Meanwhile, persistent poverty and increasing inequality are among the outcomes of neoliberal globalisation, and inequality tends to be de-emphasised if not completely ignored in mainstream development discourse largely because, as O'Laughlin (2007, 42) argues, 'Inequality is difficult to conceptualise within the neoclassical language of prescriptive commodification and individual choice.' The rural poor have actively engaged such transformations in a variety of ways, ranging from quiescence to resistance. Just as the agrarian transformations themselves are politically contested, so are the interpretations of and the political strategies to influence these transformations. The development policy and academic world do not have consensus about the causes and consequences of such agrarian development processes. This has provoked recent vibrant debates and discussions within and between broad theoretical camps, e.g., materialist political economy,[25] sustainable livelihood approaches,[26] and mainstream development policy circles.[27]

Stepping back and taking a longer view, we see two broad positions that are discernible among the various important theoretical perspectives on rural development today, and according to Bernstein (2007), these are 'residual' and 'relational'. The former is based on the belief that the cause of poverty of the rural poor is their being excluded from the market and its benefits; the solution is to bring the market to the rural poor, or the rural poor to the market. The latter is founded on the belief that the cause of poverty is the very terms of poor people's insertion into particular patterns of social relations; the solutions therefore are transformative policies and political processes that restructure such social relations.[28] For *critical* perspectives in agrarian change and peasant studies, it is important to always locate one's analysis of agrarian transformation within a *relational* perspective. It is this perspective that holds together the contributions to this collection.

Introduction to the collection

The essays in this collection follow the general theme of continuity, change and challenges in critical perspectives in agrarian change and peasant studies. It is not exhaustive in terms of thematic coverage. But as the reader will soon discover, the

---

[24]See, for example, Akram-Lodhi and Kay (2008b), Ramachandran and Swaminathan (2003), Bryceson, Kay and Mooij (2000), Rigg (2006), and Spoor (2008).
[25]See, for example, the excellent volume edited by Akram-Lodhi and Kay (2008b).
[26]See, for example, De Haan and Zoomers (2005). Scoones (2009) offers an excellent critical reflection in the context of sustainable rural livelihoods approach.
[27]See, for example, the World Bank's *World development report 2008* (World Bank 2007) which is more eclectic than a usual neoliberal policy framework on this subject.
[28]In a similar fashion, Bridget O'Laughlin (2008, 199) argues that, 'Southern Africa's agrarian crisis is rooted not in what it does not have – liberal economic and political institutions – but in what it does have: a history of integration into global markets and the class relations of capitalism through violence and colonial domination.'

present collection covers substantial ground in the field. The nine contributions are authored by leading scholars in agrarian studies.

The first contribution provides us with a macro, historical perspective about national agrarian transformation, peasant differentiation and class struggle using a political economy method in the best tradition of great agrarian comparative scholars such as Barrington Moore Jr. (1967). *Terence J. Byres* examines three different paths of capitalist agrarian transition, namely, those that occurred in England, France and Prussia. These three countries represent what Byres calls 'landlord-mediated capitalism from below', 'capitalism delayed', and 'capitalism from above', respectively. He argues that 'the character of the landlord class and of class struggle have determined both the timing of each transition and the nature of the transition'. He explains that the state has always played a critical part in any transition. He also argues that 'the differentiation of the peasantry is central to transformation: it is not an outcome but a determining variable'.[29] Differentiation of the peasantry feeds into and interacts with the landlord class and class struggle, these three being critical to the eventual outcome.

After Byres' discussion about broad patterns of agrarian transition to capitalism, highlighting class struggle and peasant differentiation, among others, the collection presents an essay by *Henry Bernstein* on V.I. Lenin and A.V. Chayanov, picking up on some of the issues discussed by Byres, focusing on some aspects of the 'Lenin–Chayanov debate' on differentiation, and reflecting on the legacies of two of the most influential thinkers in agrarian studies. Bernstein outlines the differences (and similarities) between Lenin and Chayanov on a number of issues, including their works on agrarian issues, explanation of agrarian change, ideas about productivity, model of development, and legacies. He explains that these 'are offered in the hope of clarifying and stimulating consideration of patterns of agrarian change today: how they differ from, and might be illuminated by, past experiences and the ideas they generated.' Meanwhile, *Teodor Shanin's* essay focuses briefly on the key ideas of Chayanov before elaborating on what he calls the 'treble death' of Chayanov and his 'resurrection' in post-Soviet social sciences. In doing so, Shanin offers a concise discussion of the specific context of each period ('when Chayanov died'), examining the struggles between Chayanov's ideas on peasant economy and development on the one hand and those who oppose them on the other.

The first three essays deal directly with the classic debates in agrarian political economy involving ideas by Marx, Lenin and Chayanov, among others, and the ways in which these have influenced rural development in theory and practice, past and present. The next contribution, by *Cristóbal Kay*, transitions from this set of classic thought and jumps to the 1970s–80s (now classic) debate around the 'urban bias' thesis that was largely provoked by the 1970s work of Michael Lipton (1977), and criticised by several scholars; see, for example, contributions to Harriss (1982). Kay re-examines the debate between 'agriculture first' versus 'industry first' positions (see also Saith 1990). His essay 'reviews some of the main interpretations in development studies on agriculture's contribution to economic development.' The

---

[29]The social differentiation of the peasantry, as elaborately argued and explained by Lenin (Lenin 2004; original 1899), has been one of the most debated topics in agrarian studies. Despite, or perhaps because of this, there is some confusion in many of the studies and debates about this subject that Ben White (1989) has earlier pointed out. He offers a useful analytical framework for carrying out research inquiring into this question.

agrarian transformations that occurred, and the development strategies pursued, in East Asia and Latin America are examined from a comparative perspective. His main argument is that 'a development strategy which creates and enhances the synergies between agriculture and industry and goes beyond the rural–urban divide offers the best possibilities for generating a process of rural development able to eradicate rural poverty.'

Philip McMichael provides a 'food regime' analysis, which is a specific analytical framework that was originally conceptualised by Harriet Friedmann (1987) and later developed by her and McMichael. It is a perspective that explains 'the strategic role of agriculture and food in the construction of the world capitalist economy.' McMichael explains that this particular framework 'identifies stable periods of capital accumulation associated with particular configurations of geopolitical power, conditioned by forms of agricultural production and consumption relations within and across national spaces.' He argues that 'contradictory relations within food regimes produce crisis, transformation, and transition to successor regimes.' The essay 'traces the development of food regime analysis in relation to historical and intellectual trends over the past two decades, arguing that food regime analysis underlines agriculture's foundational role in political economy/ecology.'

Ian Scoones offers a reflection essay on 'sustainable rural livelihoods approaches', a more recent perspective in rural development studies that has gained currency within the international donor agency community, among others, during the past ten years or so (see also Scoones 1998, Bebbington 1999). He guides the reader back to the origins of this particular approach, and how it has developed over time. He offers an 'historical review of key moments in debates about rural livelihoods, identifying the tensions, ambiguities and challenges of such approaches.' He has identified key challenges, such as the need to bring issues of power and class, among others, into the centre of livelihoods perspectives. He thinks that 'this will enhance the capacity of livelihoods perspectives to address key lacunae in recent discussions, including questions of knowledge, politics, scale and dynamics.'

In her paper *Shahra Razavi* explains that during the past few decades, critical perspectives on rural development have 'benefited from the insights offered by feminist scholars whose intellectual project has been to bring into the political economy of agrarian change the pervasiveness of gender relations and their interconnections with broader processes of social change.' Razavi shows some of the key contributions of feminist scholarship to agrarian studies, including that in (re)conceptualising households 'and their connections to broader economic and political structures.' But she also shows 'the extent to which these have been taken up by mainstream debates', where 'the complexities of this research have been sanitised and distorted by neoclassical economists and powerful development organisations that speak the same language', or, 'alternatively ignored and sidelined by some of the political economists of agrarian change.'

Ben Kerkvliet's essay analyses the importance of politics in agrarian transformation. He explains however that 'politics in peasant societies is mostly the everyday' type, so that if we were to look only for politics 'in conventional places and forms, much would be missed about villagers' political thought and actions as well as relationships between political life in rural communities and the political systems in which they are located.' He does not dismiss the importance of the two other types of politics, namely, official politics and advocacy politics. He makes clear, however, the differences and possible overlaps between the everyday forms and the other two types. He concludes by suggesting that, 'in addition to better understanding peasant societies, the concept of everyday politics makes us – researchers – further aware of, and realise the importance of, our own everyday political behaviour' (see also Kerkvliet 1993, 2005).

Finally, *Marc Edelman* examines some approaches in 'analysing and managing relations between rural activists and academic researchers.' He argues that 'social movements engage in knowledge production practices much like those of academic and NGO-affiliated researchers' and that 'the boundaries between activists and researchers are not always as sharp as is sometimes claimed. These blurred boundaries and shared practices can create synergies in activist–academic relations' (2009, 245). He then examines tensions in the relationship. He also discusses 'the pros and cons ... of several models of activist–researcher relations, ranging from "militant" or "engaged" research to the contractual agreement between a movement and those involved in research on it' (2009, 245). He argues

> that one of the most useful contributions of academic researchers to social movements may be reporting patterns in the testimony of people in the movement's targeted constituency who are sympathetic to movement objectives but who feel alienated or marginalised by one or another aspect of movement discourse or practice.

The nine essays are very different from each other in terms of focus, but all of these touch on the themes of changes, continuities and challenges in theoretical perspectives and methodological approaches in critical agrarian change and peasant studies. In terms of period, there are two discernible clusters of essays in the collection. The first focuses on past and classic theoretical debates in agrarian political economy in explaining the dynamics of agrarian change. These are the essays by Byres, Bernstein, Shanin and Kay, covering key ideas by Marx, Engels, Kautsky, Lenin, Chayanov, Bukharin, Preobrazhensky and Stalin, on key concepts such as peasant differentiation, family farming, scale in agricultural production, vertical integration, rural–urban linkages, peasant politics and class struggles, agrarian transitions to capitalism and socialist construction. The second cluster is comprised of relatively more recent critical perspectives in agrarian studies that are, in varying ways and degrees, inspired, provoked and informed by past classic agrarian political economy debates. The second cluster includes those by Scoones on sustainable rural livelihoods approaches, McMichael on 'food regime' analytical framework, Razavi on gender perspectives, Kerkvliet on everyday politics in peasant societies, and Edelman on academic–activist research. Altogether, these essays provide us with excellent, multiple analytical handles for dealing with the difficult challenge of interpreting – and changing – current conditions in the rural world.

Common messages and implications

Having briefly situated this collection within the context of changed and changing global agrarian terrain, and having introduced the various contributions, we now turn to specifying some common messages and implications that can be drawn from this collection and that are relevant in critical perspectives on rural development. By common messages and implications we mean key themes which all, or a cluster of contributions have in common, and we deem important to critical scholarship today. While there are dozens of possible common messages and implications offered by the various contributions to this collection, we have identified three central ones in particular: engaging with critical theories in order to interpret actual conditions in the rural world, taking politics seriously in order engage on questions of how to contribute to changing existing conditions in the agrarian world, and utilising rigorous and appropriate research methodologies in order to equip us with the necessary analytical tools to carry out the first two tasks.

(i) (Re)engaging with critical theories

One key message and implication for academics, activists and development policy practitioners that can be drawn from the collection is the need to (re)engage with critical theories. Some scholars observe that 'peasant studies' have faded away from academic research agendas during the past couple of decades. In our view, it is not that researchers today are uninterested in peasants and peasant societies because many researchers are indeed interested and have ongoing studies about this theme, but perhaps they are less interested in engaging with critical theories the way past scholarship has been. For example, we will find significant extent of research interests and initiatives on rural livelihood diversification, but current research initiatives along this theme do not usually engage with theories of social differentiation of the peasantry and class (re)configuration and their implications for national development. Class and class analysis, despite their relevance and explanatory power, seem to have been dropped from (dominant) rural development studies as well as policy and political practice, as pointed out by Bernstein (2007) and Herring and Agarwala (2006), among others. Theories about the state have also become scarce in dominant development studies, including in some of the progressive currents. Kay (2008, 934) has observed that in the so-called 'new rurality approach' in rural development studies, there is an absence of 'class analysis and of the political forces which shape the State.' He concludes that 'this inability to analyse the class dynamics in society and above all to appreciate the relevance of the process of peasant differentiation leads the new ruralists astray in their policy proposals' (Kay 2008, 935). Among the most important (trans)national agrarian movements today, rigorous class analytical frameworks seem to be not an important analytical tool in political practice, as pointed out by Borras, Edelman and Kay (2008). Questions of political conflict, especially class conflict, and power have never been important features of recent dominant perspectives in rural development studies, as argued by John Harriss (2002a), as admitted by Scoones (2009), and as raised earlier by Bridget O'Laughlin (2004) in the case of sustainable rural livelihood approaches (see also O'Laughlin 2008). Two common insights around the issue of (re)engaging with theories are put forward below.

First, classic theories in agrarian political economy about agrarian transformation, peasant differentiation, peasant economy, class and class politics, class agency, family farms, 'moral economy of the peasant', and so on are foundational frameworks for critical perspectives on rural development. The various paths of agrarian transition to capitalism as explained by Byres, the key message of Chayanov in Shanin's essay, some of the unresolved aspects of the Lenin–Chayanov debate as elaborated in Bernstein's paper, the role of agriculture and food in the development of the world capitalist system in McMichael's essay, as well as the debates around 'urban bias thesis' discussed in Kay's essay, among others, offer powerful reminders that classic agrarian political economy theories remain relevant in understanding agrarian change dynamics today. For instance, a rigorous critique of the contemporary agrarian restructuring can be strengthened if it engages with some important debates in agrarian political economy, such as the 'urban bias' versus 'rural bias' debate of the 1970s–80s, as explained by Kay (2009, 103–37, this collection). Doing so can contribute to our broader understanding of the current food crisis and biofuel debates for instance, or can critically examine problematic terms such as 'rural producers' or popular development concepts such as 'entrepreneurship', as demonstrated in Oya (2007) in the context of capital accumulation dynamics in rural Senegal. The dynamics of agrarian transformation can be understood largely by having clear grasp of the notion and actual condition of the social differentiation of the peasantry. A better understanding of the impact of development policies on the rural poor can be achieved largely by having a clearer perspective on the class structure of a particular society (and other intersecting identities such as gender, race, ethnicity, caste and religion – see Bernstein 2007, 11). For example a well-intentioned 'pro-poor irrigation project' may benefit not the poorest tenants but the rich farmers and landowners; or, the socially differentiated experience and use of state law by different social classes and groups around different rural issues in the midst of current popularity of 'alternative non-state justice systems', as explained by Franco (2008a). A better understanding of today's rural politics, especially on issues of political representation and accountability in contemporary (trans)national agrarian movements can be attained partly by specifying the class composition within and between movements; otherwise it will be difficult to distinguish Via Campesina from the International Federation of Agricultural Producers (IFAP), or understand Via Campesina as both an 'arena of action' and as a 'single actor' (Borras, Edelman and Kay 2008, Borras 2004). It is relevant to ground current agrarian and rural development studies around these classic agrarian political economy theories, rather than dismiss such kinds of theorising as old-fashioned meta-narratives that are of no relevance and consequence to critical perspectives on rural development and political activism.

Second, there are relatively newer paradigms that inform, unevenly and in various ways, current rural development work and agrarian advocacy politics, such as political ecology, gender studies, 'everyday peasant politics', and sustainable rural livelihoods approaches. Some of these perspectives are more theoretically developed than others.[30] One of the challenges is how to determine actual or potential analytical links, or absence of these, to the agrarian political economy theoretical debates. An example of how relatively recent theorising is systematically grounded in classic agrarian political economy theoretical foundations is the 'food regime' framework by Friedmann (1987) and as explained by McMichael (2009, 139–69, this collection). Scoones (2009, 171–96, this collection) has explicitly raised the same point with regards to the sustainable rural livelihoods approaches.

---

[30]Compare, for example, gender studies (Razavi 2009, 197–226 (this collection), Razavi 2003, O'Laughlin 2008, Deere 2003, 1995, Deere and Leon 2001, Kabeer 1999, Tsikata and Whitehead 2003, Agarwal 1994) or the 'everyday peasant politics' (e.g., Scott 1985, 1990, Scott and Kerkvliet 1986, Kerkvliet 2009, 227–43 (this collection), 2005, 1993) with the sustainable rural livelihoods approaches (e.g., Scoones 2009, 171–96 (this collection), 1998, Ellis 2000, de Haan and Zoomer 2005, O'Laughlin 2004).

(ii) (Re)engaging with real world politics

A critical approach to rural development is one that has, in Ben White's (1987, 70, emphasis in original) words, 'a continuing concern for issues of *social and economic justice* as part of our understanding of what rural development means and as an essential part of the meaning of "development" itself, however unpopular this emphasis may be in some quarters, at some times.' This implies taking politics seriously. All contributions to this collection, in a variety of ways, raised the importance of (re)engaging the real world politics of rural development.

Development research agendas are partly provoked, inspired and shaped by those who are negatively affected by a given structural and institutional condition and who struggle to recast such relations. For instance, the peasant wars during the first three-quarters of the twentieth century have partly shaped, directly and indirectly, explicitly and implicitly, research agendas in agrarian studies during that period (Bernstein and Byres 2001, Roseberry 1983, Scott 1976, Wolf 1969, Huizer 1975, Paige 1975, Kurtz 2000). Since the early 1980s, national liberation movements, revolutions and rebellions that marked rural politics of many developing countries during that period dissipated. In its place, a few but important unarmed but radical (sub)national peasant movements and other rural-based social movements, especially indigenous peoples' movements, have emerged and become important actors in social justice-oriented struggles.[31] By the turn of the twenty-first century, the world would witness the dramatic rise of radical transnational agrarian movements (TAMs). Radical TAMs that are based among some important (sub)national agrarian movements have, in varying ways and degrees, provoked, inspired and defined contemporary research agendas in agrarian change and peasant studies today much as peasant wars did during the most of the twentieth century. This can be seen partly in recent research agendas related to global agricultural trade and the World Trade Organisation (WTO), biotechnology, the revival of land reform, as well as in growing research interests in development concepts related to new radical alternatives such as 'food sovereignty' and 'de-globalisation'.

Important works in agrarian studies have always placed agrarian power and its role in broader agrarian transformation as a key unit of analysis. We have seen this in the works of Barrington Moore Jr. (1967), Shanin (1987), Byres (1996), Scott (1976, 1985, 1990, 1998), Gaventa (1980), and more recently and in a variety of perspectives, of Hart, Turton and White (1989), Fox (1990), Nancy Peluso (1992), Marc Edelman (1999), Amita Baviskar (2004), Anna Tsing (2005) and Tania Li (2007), among others. A similar degree of importance accorded to questions of agrarian power and politics and its relevance to broader societal transformation has been raised by the contributors to this collection.

Taking politics seriously in rural development theory and practice offers a more dynamic, not static, view of agrarian change. It is the constant *political* struggles between different social classes and groups within the state and in society that largely determine the nature, scope, pace and direction of agrarian change, as forcefully argued by Byres (2009, 33–54, this collection). This has been underscored in the great works in agrarian studies cited above, as this has been the key message of more recent important works around state–society relations in rural development and democratisation such as those by Moore and Putzel (1999) and Fox (1993, 2007).

---

[31]See, for example, Petras and Veltmeyer (2001, 2003), Brass (1994), Moyo and Yeros (2005), Edelman (1999), Yashar (2005).

However, with the demise of most peasant-based, armed national liberation movements in the early 1980s, which coincided with the surge of neoliberal globalisation, politics and political analysis seem to have started to fade away from more recent mainstream scholarship and development practice. More often, politics have been (re)interpreted within 'administrative' perspectives (hence, mainstream policy promotion of so-called 'good governance', decentralisation, deconcentration, and so on). Worse, 'politics' has been interpreted only in its negative side, to simply mean 'corruption' or ineptitude of the state. Altogether, these justify the call to move away from 'politics', meaning, away from state-led development approaches, towards non-state, market-based development policy prescriptions. Many recent fashionable rural development strategies have been framed within the de-politicised approaches, e.g., the 'willing seller–willing buyer market-based land reform' and formalisation of land property rights (see, e.g., Borras, Kay and Lahiff 2008) and micro-finance.

Radical scholars in agrarian studies have consistently questioned the trend towards de-politicised development research and policy practice. For example, in an important scholarly collection that he edited in 1982, John Harriss pointed out that conventional rural development policy frameworks tend

> to focus upon the analysis of the efficiency of the use of resources in production and marketing, and to treat the social and political factors which are of central importance in the practical activity of 'Rural Development', simply as *ceteris paribus* conditions (or, in other words, they are assumed to be constant). (Harriss 1982, 16)

He would then pick this up again in a major work published as a book on 'depoliticising development' via his critical interrogation of 'social capital' as promoted by the World Bank (Harriss 2002b). In a variety of ways, all contributions to our present collection have emphasised the importance of questions of power and politics in rural development processes.

Along the same lines of taking politics seriously, some relatively recent critical scholarship has re-politicised development discourses around 'rights' and 'empowerment' that have gained currency in development studies and development policy circles more recently. For example, by asking 'how rights become real', Ben Cousins (1997; see also Franco 2008b) has brought politics into the 'rights talk'. By putting forward the notion of 'bundle of powers', as opposed to 'bundle of rights', in the context of land access, Jesse Ribot and Nancy Peluso (2003) have reminded us of the centrality of questions of politics in the struggles to control natural resources. Politics, in this context, also mean the contestations around the very meaning of land and forest resources, and 'land as territory' in the context of indigenous peoples, as earlier elaborated in Li (1996) and Peluso (1992), among others. Meanwhile, the 'rights talk' in the context of rural development has also brought with it the question of 'empowerment'. It is quite common these days to see rural development policy documents and studies invoking 'rights' and 'empowerment', but in a very de-politicised perspective. This is, for example, the case of the Hernando de Soto-inspired initiative along the lines of 'legal empowerment of the poor' where privatisation and formalisation of property rights tends to be treated as a 'magic bullet' (Nyamu-Musembi 2007). But as Fox (2007: 335) argues, the two 'good things', i.e., rights and empowerment,

> do not necessarily go together. Institutions may nominally recognise rights that actors, because of imbalances in power relations, are not able to exercise in practice. Conversely,

actors may be empowered in the sense of having the experience and capacity to exercise rights, yet they may lack institutionally recognised opportunities to do so.

One important challenge, in the context of our current discussion, is how to locate one's analysis of the dynamics of agrarian change in the *interaction* of the various institutional arenas of agrarian power or politics. The field of 'state–society relations' offers us this analytical possibility. In this field of analysis, we have to examine more closely how key actors engage each other, leading to political change within the state, in society and within state–society channels of interactions (Fox 2007). A state–society relations framework in the study of rural politics has the potential to cover more empirical and analytical grounds. This framework can accommodate the more conventional 'state-centred' perspective that gives premium to state institutions and actors (e.g., Grindle 1986). It can also accommodate 'society-centered' perspectives that put societal actors, such as social classes and social movements, as independent variables in one's analysis, such as those offered by most social movement scholars and activists (see also de Janvry 1981). Building on their strengths, but addressing some of their weaknesses, one can benefit by taking an interactive analytical approach around the notion of 'mutually transformative state–society interactions' (Fox 1993, 2007).[32] The literature on 'politics-oriented analytical frameworks' rejects a de-politicised version of state–society relations perspective which goes by many fashionable names, such as 'state–civil society partnerships', 'government–NGO collaboration', and so on, which are among the favourite frameworks currently used by mainstream development institutions. Moreover, bringing in 'everyday peasant politics' into the interactive state–society relations analytical framework can greatly increase the latter's explanatory power, as demonstrated in a number of recent studies, such as Kerkvliet (2005) and O'Brien and Li (2006).

Ultimately, however, critical scholarship is concerned not only with interpreting the world, but with changing it. Critical scholarship in agrarian change and peasant studies takes seriously the questions of 'agency' of peasants and other working classes; after all, peasants and other working classes make their own history, although as Marx already warned, they do not do it just as they please and under circumstances they choose (Marx 1968; see also McMichael 2008). It is beneficial for academics and researchers to engage with development practitioners and activists for a transformative and mutually empowering co-production of knowledge and mutually reinforcing dissemination and use of such knowledge; an 'activist scholarship' (Hale 2008) – a complex but extremely relevant issue that is the main theme of Edelman's contribution to this collection. Theorising without grounding in political realities: political relevance, political urgency, existing balance of political forces, and so on, maybe important academically, but in the end will not matter much to those who are actually suffering on a daily basis and to those who are at the forefront of struggles to change their conditions. Development practice and activism that are informed by rigorous critical theories are more effective and relevant, and are less likely to cause harm within the rural poor communities, than those that are

---

[32]The state–society relations discussion here is informed by some of the most important and relevant works, including Houtzager and Moore (2003), Evans (1997b), Migdal (2001), Migdal, Kohli and Shue (1994), Herring (1983), Wang (1997, 1999), Fung and Wright (2003), Hart (1989), Sikor and Muller (forthcoming) and Das (2007).

not. There is of course a significant difference in terms of requirements in research methods and analytical rigor between actually existing academic and activist research. But as partly argued and explained by Edelman (2009, 245–65, this collection; see also Hale 2008, Fox 2006), there are spaces for possible synergies between these camps that are likely to result in mutually reinforcing processes and outcomes.

(iii) Utilising rigorous research methodologies

Critical perspectives in agrarian change and peasant studies require complex and rigorous research methodologies. All contributions to this collection have, in varying ways and degrees, emphasised the need for rigorous and appropriate research methodologies.

One of the key messages and implications, although not always explicitly pointed out by individual contributions, is that the *interplay* between structures, institutions and actors that is a key element in agrarian change is a key unit of analysis in critical inquiry into agrarian change. However, individual contributions have different areas of emphasis, some focus on structures (e.g., Byres), others on institutions (e.g., Scoones), and still some on the actors (e.g., Kerkvliet). But each contributor examines an area of inquiry not in isolation from, but always in relation to, the others. How else can we comprehend structural processes in the absence of serious discussions about agents or actors of change, as raised in the essay of Razavi (2009, 197–226, this collection) and as asked by McMichael (2008)? How else can we comprehend the actions of key actors without slipping into a voluntaristic view if we do not embed these actors within the structures in which they are located? Norman Long (1988) himself, in advocating for actor-oriented analytical approaches in rural development studies, has argued for the importance of locating such an analysis within existing structural and institutional settings. Or, how else can we fully understand, as raised by Thelen and Steinmo (1992), the role played by institutions in the actors' confrontation with existing structures?

Moreover, all contributions have suggested that if agrarian *change* is a key area of scholarly and activist interest, then the mechanisms and processes of change are necessarily important units of analysis for critical perspectives on rural development. This implies taking seriously existing wisdom on notions such as 'social differentiation of the peasantry', as argued by Byres, but taking careful consideration of nuanced methodological issues in analysing such concepts, as explained, for example, by Ben White (1989). This method of analysis forces us to confront, not evade, the messy complex reality of the agrarian world. The explanatory power of James Scott's (1998) notion of 'state simplification' in order to render legible complex realities and its problematic consequences on humanity reminds us of the potential benefits of starting with the messy reality of the world rather than with neat theoretical grids. James Fairhead and Melissa Leach (1996) have illustrated this point in their now classic study of forests in Western Africa, as has Amita Baviskar (2004) in her study of an anti-dam movement in India. In a similar fashion, examining the political economy of land reforms in South Asia, Ron Herring (1983, 269, emphasis in original; see also Scott 1998, 49) has made a relevant observation that warrants an extended quote. He said:

> The case studies [in this book] clearly indicate change induced by land reforms, though not always in directions indicated by reform rhetoric. This structural change is of two

kinds – apparent and real. Though it seems contradictory to write of 'apparent' structural change, the usage is meaningful. Land reforms produce important alterations in the *observable* structure of agrarian systems – land records are altered, census data collected, reports are made – all presenting a picture of the rural world that is more congruent with the needs of landed elites, administrators, and ruling politicians than with reality on the ground. Landowners have strong incentives to show that they own very little land and that there are no tenants on it; reform administrators are pressured to show progress in implementation. ... The apparent change is important because it is this data-built facade which goes into planning documents, policy debates, reports of international agencies, and all too many scholarly treatments. The distortions become social facts, the primary sources for understanding the rural world for nonrural groups who are, after all, the primary movers of rural policy.

Furthermore, most of the essays in this collection have also raised a number of methodological issues that are quite relevant to critical perspectives in agrarian change and peasant studies today. These include comparative political economy, cross-disciplinary approaches, and activist scholarship.

Systematic comparative (political economy) approaches remain an important methodology in critical rural development studies. Byres, Bernstein, Shanin, Kay and McMichael demonstrate in their contributions to this collection the broad picture of the agrarian world and agrarian transformation, and in explicitly comparative framework, either cross-country or cross-regional, as in Byres' and Kay's essays, or longitudinal comparative perspective, as in Shanin's and McMichael's papers. All of these contributors have taken explicitly political economy approaches that ask at least four fundamental questions: Who owns what? Who does what? Who gets what? And what do they do with the agrarian surplus? (Bernstein 2007). In short, questions about agrarian wealth and power. Great studies about the agrarian world have taken explicit comparative approaches, longitudinal and cross-national or cross-regional, as exemplified in the works of Moore (1967), Kay (1974) and Byres (1996), to cite a few. Ben White (1987, 69–70, emphasis in original) explains that:

[T]he *comparative approach*, requiring detailed analysis of the contrasting experiences of rural development in actual societies, with recognition of the particular historical, social and political contexts at national and local level in which agrarian changes take place, in which strategies and policies have been formed and introduced and have succeeded or failed. In this way we may hope to confront and come to terms with the diversity that exists in the real world – whatever uniform tendencies some abstract theories might suggest – and to learn from it, to see the ways in which general 'tendencies' interact with specific conditions to produce particular outcomes, and to understand in this way that 'success stories' may offer valuable lessons, but not directly transferable models for other societies to follow or for external agencies to impose.

The strength of comparative inquiry is further argued by Byres (1995, 572) in the context of the agrarian question, and more generally:

Comparison is essential. It is, in part, via comparison that one might make the relevant, sensitive judgements about performance (in the context of the agrarian question, or more generally) ... It is, to a degree, in comparative terms that one might conceive of change in a suitably nuanced manner. How, otherwise, might one understand the nature, the scope, and the likely direction of change? Comparison, moreover, may point to the possibility of a substantive (non-trivial) diversity of outcome. It is in a comparative perspective that one might reach for possible lines of causality ... Comparison can clarify and make more secure the analytical judgements which we make ... It can open analytical perspectives. It can do so, when securely based theoretically, by extending our range of criteria independent of a particular context, and so allowing theory to be more nuanced in what it can reveal.

All the contributions to this collection are, to varying extents, cross-disciplinary, demonstrating the relevance and importance of cross-disciplinary approaches to rural development studies. In arguing for cross-disciplinary approaches, Harriss (2002a, 493–94) has underscored some basic arguments, including (a) contributions of disciplines other than economics to the understanding of development processes seems evident enough; (b) rigor is not the exclusive preserve of economists or of quantitative research; (c) different disciplines have different contributions to make and that it is very far from the case that all development research has to be in some way cross-disciplinary; and (d) there is a much-to-be-desired tension between 'discipline' and 'anti-discipline'. 'Discipline' in research is productive. Without it we cannot distinguish science or knowledge from opinion and are left floundering in a sea of relativism. Harriss (2002a, 494) concludes that cross-disciplinary approaches in international development are relevant 'because research priorities should be set by the practical problems that development involves, more than by the puzzles that are generated out of theoretical speculation'.

Following the last quotation from Harriss, critical perspectives on rural development may also benefit from collaborative academic–activist research approaches. This is demonstrated in the contribution by Edelman. If, as Harriss said, 'research priorities should be set by the practical problems that development involves', then collaborative research between academics, development practitioners and activists becomes an important undertaking. Co-production of knowledge and a mutually reinforcing dissemination and use of such knowledge among academics, development practitioners and activists are likely to address some of the key weaknesses of a purely theoretical research detached from the real world, or of a too practice-oriented initiative without theoretical and methodological rigor. Edelman (2009, 245–65, this collection) has specified important areas of tensions and synergies in academic–activist research, much as has been pointed out recently by other scholars such as the several of the contributors to Hale (2008) and by Fox (2006).

Concluding remarks

The entire world is in crisis, a crisis with multiple dimensions. There is a food crisis, an energy crisis, a climate crisis and a financial crisis. The solutions put forth ... – more free trade, more GMOs, etc. – purposefully ignore the fact that the crisis is a product of the capitalist system and of neoliberalism, and they will only worsen its impacts ... To find real solutions we need ... getting speculative finance capital out of our food system, and re-nationalising food production and reserves offer us the only real way out of the food crisis. Only peasant and family farm agriculture feed people, while agribusiness grows export crops and agrofuels to feed cars instead of human beings ... Industrial agriculture warms the planet, and peasant agriculture cools the planet ... Genuine integral agrarian reform and the defense of the territories of indigenous peoples are essential steps to roll back the evictions and displacement in the countryside, and to use our farm land to grow food instead of exports and fuels ... Only agroecological peasant and family farming can de-link food prices from petroleum prices, recover degraded soils, and produce healthy local food for our peoples. (Via Campesina 2008)

Summarised in the excerpts of Via Campesina's Fifth Congress Declaration was the global movement's reading of the current global crisis confronting the (rural) world, as well as their proposed solutions to the crisis. Academics, development practitioners and movement activists have been, and are likely to be inspired to rally around,[33] or provoked to raise some valid critical questions on the issues raised by Via Campesina,[34] either in their entirety or at least by some components of this set of issues; or, indeed by the 'silences' in Via Campesina's official discourse,[35] as well as about what Via Campesina represents (or claims to represent) and does not represent. The reason why the excerpts of Via Campesina's recent world assembly declaration were offered as part of the concluding discussion of this introductory essay is to remind us of at least some of the 'messy and complex' realities in the agrarian world – again coming back to the point raised by James Scott (1998) – that ought to be the starting point of critical scholarship in agrarian change and peasant studies.

Whatever position one takes with regard to the official discourse and advocacy politics of Via Campesina and other contemporary (trans)national agrarian movements (TAMs), it can hardly be denied that the issues raised by these movements are compelling and their advocacy strategies are relatively effective – enough to inspire or provoke varied reactions from different quarters, conservative or radical. One thing that has been highlighted by the Via Campesina declaration is that, today, perhaps more than ever, what happens in and in relation to the rural world is critical to our understanding of the broader world and the very future of human society. Confronting the issues identified earlier in this paper, as well as the question put onto the research agendas by TAMs like Via Campesina, regardless of one's standpoint on issues at hand, will require (re)engaging with critical theories, taking politics seriously, and utilising rigorous and appropriate research methodologies. These are the common messages and implications of the various contributions to this collection in the context of a scholarship that is critical in two senses: questioning prescriptions from mainstream perspectives and interrogating popular conventions in radical thinking.

References

Adnan, S. 2007. Departures from everyday resistance and flexible strategies of domination: the making and unmaking of a poor peasant mobilisation in Bangladesh. *Journal of Agrarian Change*, 7(2), 183–224.

Agarwal, B. 1994. *A field of one's own: gender and land rights in South Asia*. Cambridge: Cambridge University Press.

Akram-Lodhi, H. and C. Kay. 2008a. Neoliberal globalisation, the traits of rural accumulation and rural politics: the agrarian question in the twentieth century. *In:* H. Akram Lodhi and C. Kay, eds. *Peasants and globalisation: political economy, rural transformation and the agrarian question.* London: Routledge, pp. 315–38.

Akram-Lodhi, H. and C. Kay, eds. 2008b. *Peasants and globalisation: political economy, rural transformation and the agrarian question.* London: Routledge.

Araghi, F. 2008. The invisible hand and the visible foot: peasants, dispossession and globalisation. *In:* H. Akram Lodhi and C. Kay, eds. *Peasants and globalisation: political economy, rural transformation and the agrarian question.* London: Routledge, pp. 111–47.

---

[33]As Araghi (2008,138) declares: 'It is the outcome of this struggle that will resolve not only the peasant question, but indeed all our questions.' Refer also to McMichael (2008).

[34]Refer, for example, to some of the issues raised by Bernstein (2008, and 2009, 55–81, this collection).

[35]Refer, for example, to the relevant discussions in Borras *et al.* (2008).

Assies, W., G. van der Haar and A.J. Hoekma. 1998. *The challenge of diversity: indigenous peoples and reform of the state in Latin America*. Amsterdam: CEDLA.

Baviskar, A. 2004. *In the belly of the river: tribal conflicts over development in the Narmada Valley*. Delhi: Oxford University Press.

Bebbington, A. 1999. Capitals and capabilities: a framework for analysing peasant viability, rural livelihoods and poverty. *World Development*, 27(12), 2021–44.

Bello, W. 2003. *Deglobalisation: ideas for a new world economy*. London: Zed.

Bernstein, H. 2007. Rural livelihoods in a globalising world: bringing class back in. *Conference on policy intervention and rural transformation: towards a comparative sociology of development*, 10–16 September 2007, China Agricultural University, Beijing.

Bernstein, H. 2008. Agrarian questions from transition to globalisation. *In:* H. Akram Lodhi and C. Kay, eds. *Peasants and globalisation: political economy, rural transformation and the agrarian question*. London: Routledge, pp. 239–61.

Bernstein, H. 2009. V.I. Lenin and A.V. Chayanov: looking back, looking forward. *The Journal of Peasant Studies*, 36(1), 55–81.

Bernstein, H. and T.J. Byres. 2001. From peasant studies to agrarian change. *Journal of Agrarian Change*, 1(1), 1–56.

Biekart, K. and M. Jelsma, eds. 1994. *Peasants beyond protest in Central America*. Amsterdam: Transnational Institute.

Borras, S.M. Jr. 2004. La Via Campesina: an evolving transnational social movement. *Transnational Institute (TNI) Briefing Paper*, November 2004. Amsterdam: Transnational Institute (TNI).

Borras, S.M. Jr. 2007a. 'Free market', export-led development strategy and its impact on rural livelihoods, poverty and inequality: the Philippine experience seen from a Southeast Asian perspective. *Review of International Political Economy*, 14(1), 143–75.

Borras, S.M. Jr. 2007b. *Pro-poor land reform: a critique*. Ottawa: University of Ottawa Press.

Borras, S.M. Jr. 2008. Re-examing the agrarian movement–NGO solidarity relations discourse. *Dialectical Anthropology*, 32, 203–9.

Borras, S.M. Jr., M. Edelman and C. Kay. 2008. Transnational agrarian movements: origins and politics, campaigns and impact. *In:* S.M. Borras Jr., M. Edelman and C. Kay, eds. *Transnational agrarian movements confronting globalisation*. Oxford: Wiley-Blackwell.

Borras, S.M. Jr. and J.C. Franco. 2009. Transnational campaigns for land and citizenship rights. *IDS working paper series*. Brighton: IDS, University of Sussex.

Borras, S.M. Jr., C. Kay and E. Lahiff, eds. 2008. *Market-led agrarian reform: critical perspectives on neoliberal land policies and the rural poor*. London: Routledge.

Brass, T., ed. 1994. *The new farmers' movements in India*. London: Frank Cass.

Bryceson, D., C. Kay and J. Mooij, eds. 2000. *Disappearing peasantries? Rural labour in Africa, Asia and Latin America*. London: Intermediate Technology Publications.

Byres, T.J. 1995. Political economy, the agrarian question and the comparative method. *The Journal of Peasant Studies*, 22(4), 561–80.

Byres, T.J. 1996. *Capitalism from above and capitalism from below: essays in comparative political economy*. London: Palgrave Macmillan.

Byres, T.J. 2009. The landlord class, peasant differentiation, class struggle and the transition to capitalism: England, France and Prussia compared. *The Journal of Peasant Studies*, 36(1), 33–54.

Cousins, B. 1997. How do rights become real? Formal and institutions in South Africa's land reform. *IDS Bulletin*, 28(4), 59–67.

Das, R., ed. 2007. Peasant, state and class. *The Journal of Peasant Studies*, 34(2–3), special issue.

Davis, M. 2006. *Planet of slums*. London: Verso.

Deere, C.D. 1995. What difference does gender make? Rethinking peasant studies. *Feminist Economics*, 1(1), 53–72.

Deere, C.D. 2003. Women's land rights and social movements in the Brazilian agrarian reform. *Journal of Agrarian Change*, 3(1–2), 257–88.

Deere, C.D. and M. León. 2001. *Empowering women: land and property rights in Latin America*. Pittsburgh, PA: University of Pittsburgh Press.

Deere, C.D. and F. Royce, eds. Forthcoming. *Alternative visions of development: rural social movements in Latin America*. Gainesville, FL: University Press of Florida.

De Haan, L. and A. Zoomers. 2005. Exploring the frontier of livelihood research. *Development and Change*, 36(1), 27–47.

De Janvry, A. 1981. *The agrarian question and reformism in Latin America*. Baltimore, MD: The Johns Hopkins University Press.

Desmarais, A. 2007. *La Via Campesina: globalisation and the power of peasants*. Halifax, Nova Scotia: Fernwood.

Edelman, M. 1999. *Peasants against globalisation: rural social movements in Costa Rica*. Stanford, CA: Stanford University Press.

Edelman, M. 2003. Transnational peasant and farmer movements and networks. *In:* M. Kaldor, H. Anheier and M. Glasius, eds. *Global civil society 2003*. Oxford: Oxford University Press, pp. 185–220.

Edelman, M. 2005. Bringing the moral economy back in ... to the study of 21st century transnational peasant movements. *American Anthropologist*, 7(3), 331–45.

Edelman, M. 2008. Transnational organising in agrarian Central America: histories, challenges, prospects. *Journal of Agrarian Change*, 8(2–3), 229–57.

Edelman, M. 2009. Synergies and tensions between rural social movements and professional researchers. *The Journal of Peasant Studies*, 36(1), 245–65.

Edelman, M. and A. Haugerud. 2005. Introduction. *In:* M. Edelman and A. Haugerud, eds. *The anthropology of development and globalisation: from classical political economy to contemporary neoliberalism*. Oxford: Blackwell, pp. 1–74.

Ellis, F. 2000. *Rural livelihoods and diversity in developing countries*. Oxford: Oxford University Press.

Evans, P. 1997a. The eclipse of the state? Reflections on stateness in an era of globalisation. *World Politics*, 50(1), 62–87.

Evans, P., ed. 1997b. *State–society synergy: government and social capital in development*. Berkeley, CA: University of California.

Fairhead, J. and M. Leach. 1996. *Misreading the African landscape*. Cambridge: Cambridge University Press.

FAO 2008a. Hunger on the rise due to soaring food prices, 3 July 2008. Available from: http://www.fao.org/newsroom/EN/news/2008/1000866/index.html [Accessed 17 November 2008].

FAO 2008b. *The state of food and agriculture – biofuels: prospects, risks and opportunities*. Rome: FAO.

Fox, J., ed. 1990. *The challenges of rural democratisation: perspectives from Latin America and the Philippines*. London: Frank Cass.

Fox, J. 1993. *The politics of food in Mexico: state power and social mobilisation*. Ithaca, NY: Cornell University Press.

Fox, J. 2006. Lessons from action–research partnerships: LASA/Oxfam America 2004 Martin Diskin memorial lecture. *Development in Practice*, 16(1), 27–38.

Fox, J. 2007. *Accountability politics: power and voice in rural Mexico*. New York, NY: Oxford University Press.

Fox, J. 2009 (forthcoming). Coalitions and networks. *In:* H. Anheier and S. Toepler, eds. *International encyclopedia of civil society*. New York, NY: Springer.

Fox, J. and L.D. Brown. 1998. *The struggle for accountability: the World Bank, NGOs and grassroots movements*. Cambridge, MA: MIT Press.

Franco, J. 2008a. Peripheral justice?: rethinking 'non-state justice' systems in the Philippine countryside. *World Development*, 36(10), 1858–73.

Franco, J. 2008b. Making land rights accessible: social movement innovation and political-legal strategies in the Philippines. *Journal of Development Studies*, 44(7), 991–1022.

Friedmann, H. 1987. International regimes of food and agriculture since 1870. *In:* T. Shanin, ed. *Peasants and peasant societies*. Oxford: Basil Blackwell, pp. 258–76.

Fung, A. and E. Olin Wright, eds. 2003. *Deepening democracy: institutional innovations in empowered participatory governance*. London: Verso.

Gaventa, J. 1980. *Power and powerlessness: quiescence and rebellion in an Appalachian Valley*. Urbana, IL: University of Illinois Press.

GRAIN 2008. Seized: the 2008 landgrab for food and financial security. Available from: http://www.grain.org/briefings/?id=212 [Accessed 18 December 2008].

Grindle, M. 1986. *State and countryside: development policy and agrarian politics in Latin America*. Baltimore, MD: The Johns Hopkins University Press.

Gwynne, R. and C. Kay. 2004. *Latin America transformed: globalisation and modernity*, second edition. London: Arnold.

Hale, C., ed. 2008. *Engaging contradictions: theory, politics, and methods of activist scholarship*. Berkeley, CA: University of California Press.

Harriss, J., ed. 1982. *Rural development: theories of peasant economy and agrarian change*. London: Hutchinson.

Harriss, J. 2002a. *Depoliticising development: the World Bank and social capital*. London: Anthem Press.

Harriss, J. 2002b. The case for cross-disciplinary approaches in international development. *World Development*, 30(3), 487–96.

Hart, G. 1989. Agrarian change in the context of state patronage. *In:* G. Hart, A. Turton and B. White, eds. *Agrarian transformations: local processes and the state in Southeast Asia*. Berkeley, CA: University of California Press, pp. 31–49.

Hart, G., A. Turton and B. White, eds. 1989. *Agrarian transformations: local processes and the state in Southeast Asia*. Berkeley, CA: University of California Press.

Herring, R. 1983. *Land to the tiller: the political economy of agrarian reform in South Asia*. New Haven, CT: Yale University Press.

Herring, R. and R. Agarwala. 2006. Restoring agency to class: puzzles from the subcontinent. *Critical Asian Studies*, 38(4), 323–56.

Holt-Gimenez, E. 2006. *Campesino a campesino: voices from Latin America's farmer to farmer movement for sustainable agriculture*. Oakland, CA: Food First Books.

Houtzager, P. and M. Moore, eds. 2003. *Changing paths: international development and the new politics of inclusion*. Ann Arbor, MI: University of Michigan Press.

Huizer, G. 1975. How peasants become revolutionaries: some cases from Latin America and Southeast Asia. *Development and Change*, 6(3), 27–56.

Jansen, K. and S. Vellema, eds. 2004. *Agribusiness and society: corporate responses to environmentalism, market opportunities and public regulation*. London: Zed Books.

Kabeer, N. 1999. Resources, agency, achievement: reflections on the measurement of women's empowerment. *Development and Change*, 30(3), 435–64.

Kay, C. 1974. Comparative development of the European manorial system and the Latin American hacienda system. *The Journal of Peasant Studies*, 2(1), 69–98.

Kay, C. 2006. Rural poverty and development strategies in Latin America. *Journal of Agrarian Change*, 6(4), 455–508.

Kay, C. 2008. Reflections on Latin American rural studies in the neoliberal globalisation period: a new rurality. *Development and Change*, 39(6), 915–43.

Kay, C. 2009. Development strategies and rural development: exploring synergies, eradicating poverty. *The Journal of Peasant Studies*, 36(1), 103–38.

Keohane, R. and S. Nye Jr. 2000. Governing in a globalising world. *In:* S. Nye Jr. and J. Donahue, eds. *Visions of governance for the 21st century*. Cambridge: Cambridge University Press, pp. 1–41.

Kerkvliet, B.J. Tria. 1993. Claiming the land: take-overs by villagers in the Philippines with comparisons to Indonesia, Peru, Portugal, and Russia. *The Journal of Peasant Studies*, 20(3), 459–93.

Kerkvliet, B.J. Tria. 2005. *The power of everyday politics: how Vietnamese peasants transformed national policy*. Ithaca, NY: Cornell University Press.

Kerkvliet, B.J. Tria. 2009. Everyday politics in peasant studies (and ours). *The Journal of Peasant Studies*, 36(1), 227–43.

Kurtz, M. 2000. Understanding peasant revolution: from concept to theory and case. *Theory and Society*, 29(10), 93–124.

Le Mons Walker, K. 2008. From covert to overt: everyday peasant politics in China and the implications for transnational agrarian movements. *Journal of Agrarian Change*, 8(2–3), 462–88.

Lenin, V.I. 2004. *Development of capitalism in Russia*. Honolulu, Hawaii: University Press of the Pacific.

Li, T.M. 1996. Images of community: discourse and strategy in property relations. *Development and Change*, 27(3), 501–27.

Li, T.M. 2007. *The will to improve: governmentality, development, and the practice of politics*. Durham, NC: Duke University Press.

Lipton, M. 1977. *Why poor people stay poor: urban bias in world development*. London: Temple Smith.

Long, N. 1988. Sociological perspectives on agrarian development and state intervention. *In:* A. Hall and J. Midgley, eds. *Development policies: sociological perspectives*. Manchester: Manchester University Press, pp. 108–33.

Malseed, K. 2008. Where there is no movement: local resistance and the potential for solidarity. *Journal of Agrarian Change*, 8(2–3), 489–514.

Martinez-Torres, M.E. and P.M. Rosset. 2008. La Vía Campesina: transnationalising peasant struggle and hope. *In:* R. Stahler-Sholk, H.E. Vanden and G. Kuecker, eds. *Latin American social movements in the twenty-first century: resistance, power, and democracy*. Lanham, MD: Rowman & Littlefield, pp. 307–22.

Marx, K. 1968. The eighteenth Brumaire of Louis Bonaparte. *In: Marx/Engels selected works in one volume*. London: Lawrence and Wishart, pp. 96–179.

McMichael, P. 2008. Peasants make their own history, but not just as they please.... *In:* S.M. Borras Jr., M. Edelman and C. Kay, eds. *Transnational agrarian movements confronting globalisation*. Oxford: Wiley-Blackwell, pp. 61–89.

McMichael, P. 2009. A food regime genealogy. *The Journal of Peasant Studies*, 36(1), 139–69.

Migdal, J. 2001. *State in society: studying how states and societies transform and constitute one another*. Cambridge: Cambridge University Press.

Migdal, J., A. Kohli and V. Shue, eds. 1994. *State power and social forces: domination and transformation in the Third World*. Cambridge: Cambridge University Press.

Monsalve, S., *et al.* 2006. Agrarian reform in the context of food sovereignty, the right to food and cultural diversity: land, territory and dignity. A civil society article presented at the International Conference on Agrarian Reform and Rural Development, March 2006, Porto Alegre, Brazil.

Monsalve, S., U. Bickel, F. Garbers and L. Goldfarb. 2008. *Agrofuels in Brazil*. Heidelberg: FIAN International. Available from: http://www.fian.org/resources/documents/others/agrofuels-in-brazil/pdf [Accessed 16 December 2008].

Moore, B. Jr. 1967. *Social origins of dictatorship and democracy: lord and peasant in the modern world*. Harmondsworth: Penguin.

Moore, M. and J. Putzel. 1999. Thinking strategically about politics and poverty. *IDS working article series 101*. Brighton: IDS, University of Sussex.

Moyo, S. and P. Yeros. 2005. The resurgence of rural movements under neoliberalism. *In:* S. Moyo and P. Yeros, eds. *Reclaiming the land: the resurgence of rural movements in Africa, Asia and Latin America*. London: Zed, pp. 8–66.

Newell, P. 2008. Trade and biotechnology in Latin America: democratisation, contestation and the politics of mobilisation. *Journal of Agrarian Change*, 8(2&3), 345–76.

Nyamu-Musembi, C. 2007. De Soto and land relations in rural Africa: breathing life into dead theories about property rights. *Third World Quarterly*, 28(8), 1457–78.

O'Brien, K. and L. Lianjiang. 2006. *Rightful resistance in China*. Cambridge: Cambridge University Press.

O'Laughlin, B. 2004. Book review article of several books on rural livelihoods. *Development and Change*, 35(2), 385–92.

O'Laughlin, B. 2007. A bigger piece of a very small pie: intrahousehold resource allocation and poverty reduction in Africa. *Development and Change*, 38(1), 21–44.

O'Laughlin, B. 2008. Gender justice, land and the agrarian question in southern Africa. *In:* H. Akram-Lodhi and C. Kay, eds. *Peasants and globalisation: political economy, rural transformation and the agrarian question*. London: Routledge, pp. 190–213.

Otero, G., ed. 2008. *Food for the few: neoliberal globalism and biotechnology in Latin America*. Austin, TX: University of Texas Press.

Oya, C. 2007. Stories of rural accumulation in Africa: trajectories and transitions among rural capitalists in Senegal. *Journal of Agrarian Change*, 7(4), 453–93.

Paige, J. 1975. *Agrarian revolution: social movements and export agriculture in the underdeveloped world*. New York, NY: Free Press.

Patel, R. 2006. International agrarian restructuring and the practical ethics of peasant movement solidarity. *Journal of Asian and African Studies*, 41(1–2), 71–93.

Peluso, N. 1992. *Rich forests, poor people: resource control and resistance in Java*. Berkeley, CA: University of California Press.

Peluso, N., S. Affif and N. Fauzi. 2008. Claiming the grounds for reform: agrarian and environmental movements in Indonesia. *Journal of Agrarian Change*, 8(2–3), 377–407.

Petras, J. and H. Veltmeyer. 2001. Are Latin American peasant movements still a force for change?. *The Journal of Peasant Studies*, 28(2), 83–118.

Petras, J. and H. Veltmeyer. 2003. The peasantry and the state in Latin America: a troubled past, an uncertain future. *In:* T. Brass, ed. *Latin American peasants*. London: Frank Cass, pp. 41–82.

Ramachandran, V.K. and M. Swaminathan, eds. 2003. *Agrarian studies: essays on agrarian relations in less-developed countries*. London: Zed.

Razavi, S., ed. 2003. Agrarian change, gender and land rights. *Journal of Agrarian Change*, 3(1–2), 2–32.

Razavi, S. 2009. Engendering the political economy of agrarian change. *The Journal of Peasant Studies*, 36(1), 197–226.

Ribot, J. and A. Larson, eds. 2005. *Democratic decentralisation through a natural resource lens*. London: Routledge.

Ribot, J. and N. Peluso. 2003. A theory of access. *Rural Sociology*, 68(2), 153–81.

Rigg, J. 2006. Land, farming, livelihoods, and poverty: rethinking the links in the rural south. *World Development*, 34(1), 180–202.

Roseberry, W. 1983. From peasant studies to proletarianisation studies. *Studies in Comparative International Development*, 18(1–2).

Ross, E. 1998. *The malthus factor: poverty, politics and population in capitalist development*. London: Zed.

Rosset, P., R. Patel and M. Courville, eds. 2006. *Promised land: competing visions of agrarian reform*. Berkeley, CA: Food First Books.

Saith, A. 1990. Development strategies and the rural poor. *The Journal of Peasant Studies*, 17(2), 171–244.

Scoones, I. 1998. Sustainable rural livelihoods: a framework for analysis. *IDS working article*, 72. Brighton: IDS, Sussex.

Scoones, I. 2008. Mobilising against GM crops in India, South Africa and Brazil. *Journal of Agrarian Change*, 8(2–3), 315–44.

Scoones, I. 2009. Livelihoods perspectives and rural development. *The Journal of Peasant Studies*, 36(1), 171–96.

Scott, J. 1976. *The moral economy of the peasant: rebellion and subsistence in Southeast Asia*. New Haven, CT: Yale University Press.

Scott, J. 1985. *Weapons of the weak*. New Haven, CT: Yale University Press.

Scott, J. 1990. *Domination and the arts of resistance: hidden transcripts*. New Haven, CT: Yale University Press.

Scott, J. 1998. *Seeing like a state: how certain schemes to improve the human condition have failed*. New Haven, CT: Yale University Press.

Scott, J. and B. Kerkvliet. 1986. Everyday forms of peasant resistance in Southeast Asia. *The Journal of Peasant Studies*, 13(2), special issue.

Shanin, T., ed. 1987. *Peasants and peasant societies*, new edition. London: Penguin in association with Basil Blackwell.

Shanin, T. 2009. Chayanov's treble death and tenuous resurrection: an essay about understanding, about roots of plausibility and about rural Russia. *The Journal of Peasant Studies*, 36(1), 83–101.

Sikor, T. and D. Muller, eds. Forthcoming. The limits of state-led land reform. *World Development*, special issue.

Spoor, M., ed. 2008. *The political economy of rural livelihoods in transition economies. Land, peasants and rural poverty in transition*. London: Routledge.

Stephen, L. 1997. *Women and social movements in Latin America*. Austin, TX: University of Texas Press.

Tarrow, S. 2005. *The new transnational activism*. Cambridge: Cambridge University Press.

Thelen, K. and S. Steinmo. 1992. Historical institutionalism in comparative politics. *In:* S. Steinmo, K. Thelen and F. Longstreth, eds. *Structuring politics: historical institutionalism in comparative analysis*. Cambridge: Cambridge University Press, pp. 1–32.

Tsikata, D. and A. Whitehead. 2003. Policy discourse on women's land rights in Sub-Saharan Africa: the implications of the re-turn to the customary. *Journal of Agrarian Change*, 3(1–2), 67–112.

Tsing, A. 2005. *Friction: an ethnography of global connection.* Princeton, NJ: Princeton University Press.

Veltmeyer, H. 2004. Civil society and social movements: the dynamics of intersectoral alliances and urban–rural linkages in Latin America. *Civils society and social movements programme article no. 10.* Geneva: UNRISD.

Via Campesina. 2008. Peasant agriculture and food sovereignty are solutions to the global crisis, 19–22 October 2008. Maputo, Mozambique. Available from: http://www.via campesina.org [Accessed 23 December 2008].

Wang, X. 1997. Mutual empowerment of state and peasantry: grassroots democracy in rural China. *World Development,* 25(9), 1431–42.

Wang, X. 1999. Mutual empowerment of state and society: its nature, conditions, mechanisms, and limits. *Comparative Politics,* 31(2), 231–49.

Watts, M. 2008. The southern question: agrarian questions of labour and capital. *In:* H. Akram Lodhi and C. Kay, eds. *Peasants and globalisation: political economy, rural transformation and the agrarian question.* London: Routledge, pp. 262–87.

Weis, T. 2007. *The global food economy. The battle for the future of farming.* London: Zed Books.

White, B. 1987. Rural development: rhetoric and reality. *Journal für Entwicklungspolitik,* (1), 54–72.

White, B. 1989. Problems in the empirical analysis of agrarian differentiation. *In:* G. Hart, A. Turton and B. White, eds. *Agrarian transformations: local processes and the state in Southeast Asia.* Berkeley, CA: University of California Press, pp. 15–30.

Wright, A. and W. Wolford. 2003. *To inherit the earth: the landless movement and the struggle for a new Brazil.* Oakland, CA: Food First Books.

Wolf, E. 1969. *Peasant wars of the twentieth century.* New York, NY: Harper and Row.

Wood, E.M. 2008. Peasants and the market imperative: the origins of capitalism. *In:* H. Akram Lodhi and C. Kay, eds. *Peasants and globalisation: political economy, rural transformation and the agrarian question.* London: Routledge, pp. 37–56.

World Bank 2003. *Land policies for growth and poverty reduction.* A World Bank policy research report prepared by Klaus Deininger. Washington, DC: World Bank.

World Bank 2007. *World development report 2008: agriculture for development.* New York, NY: Oxford University Press for the World Bank.

Yashar, D. 2005. *Contesting citizenship in Latin America: the rise of indigenous movements and the post-liberal challenge.* Cambridge: Cambridge University Press.

The landlord class, peasant differentiation, class struggle and the transition to capitalism: England, France and Prussia compared

Terence J. Byres

The three examples considered – England, France and Prussia – are all very important instances of capitalist agrarian transformation. They illustrate, moreover, strikingly different paths of agrarian transition. These are termed, respectively, landlord-mediated capitalism from below, capitalism delayed, and capitalism from above, and an explanation is offered of how these different outcomes came to pass and of why there was such a marked divergence in the nature of agrarian transition. It is argued that the character of the landlord class and of class struggle have determined both the timing of each transition and the nature of the transition. Both the quality of the landlord class and the manner and outcome of the class struggle have sometimes delayed, perhaps for prolonged periods, and sometimes hastened transition; and have had profound implications for the nature and quality of the transformation and how reactionary or progressive it has been. In this the state has always played a prominent part. It is further argued that differentiation of the peasantry is central to transformation: it is not an outcome but a determining variable, a *causa causans* rather than a *causa causata*. Differentiation of the peasantry feeds into and interacts with the landlord class and class struggle, these three being critical to the eventual outcome. The distinctly varying trajectories in the three crucial instances are explained in these terms.

Introduction

The three examples I have chosen to consider are all very important instances of agrarian transformation. They illustrate, moreover, strikingly different experiences of transformation: differing paths of agrarian transition.

England, the first historical example of such transformation, of 'capitalism triumphant', one might describe as landlord-mediated capitalism from below (that, at least, is how I would choose to describe it). The former feudal landlord class became a capitalist landlord class, letting its land to capitalist tenant farmers on fixed-term leases at 'competitive' rents: and 'English farming came to be dominated by the triple division into landlords, [capitalist] tenant farmers and hired labourers' (Hobsbawm and Rudé 1973, 6). The transition to capitalist agriculture proceeded vigorously in Tudor England, in the sixteenth century, and was completed during the seventeenth. Not everyone would accept such a characterisation. Brenner, for example, would dispute both its historical priority, arguing (2001) that this, in fact, belongs to the Low Countries; or any suggestion of its representing capitalism from below.

In Prussia, by contrast with England, the class of feudal landlords – the Junkers – had ceased, by the beginning of the nineteenth century, to be feudal landlords. Prussian feudalism gave way to a distinctive form of agrarian capitalism, in the wake of the 'freeing' of Prussian serfs in 1807: a transition to capitalism that came some three centuries after the English transition. Prussian feudal landlords ceased to be a landlord class and became a class of capitalist farmers, working the land with an oppressed force of wage labour, who had formerly been serfs. Here was Lenin's celebrated 'capitalism from above': the impulse was an exclusively landlord one, accompanied by 'the degradation of the peasant masses' (Lenin 1964, 33). There was no question of capitalism from below.

France embodies, one might say, 'capitalism delayed'. Here, the capitalist impulse in the countryside was frustrated, or, at least, significantly delayed. At the end of the nineteenth century, France could still be portrayed as 'the classical land of small peasant economy' (Engels 1970, 460) – a land, one might say, dominated by poor and middle peasants. Here we have a stubbornly enduring peasantry: a peasantry that refused to go. Sharecropping was still widespread, and persisted well into the twentieth century. The French landlord class at no point, either before or after 1789, had shown any significant move of either the English kind or the Prussian. There was no dominant 'capitalism from above' and no broad progressive landlord role; and there was no 'capitalism from below', from within the ranks of the peasantry.

In this paper, I offer an explanation of how these different outcomes came to pass and of why there was such a marked divergence in the nature of agrarian transition. I do so in terms of the kind of landlord class, the kind of class struggle and the kind of peasant differentiation that were integral to 'agrarian transformation'. I argue that the character of the landlord class and of class struggle have determined both the *timing* of each transition and the *nature* of the transition. Both the quality of the landlord class and the manner and outcome of the class struggle have sometimes delayed, perhaps for prolonged periods, and sometimes hastened transition; and have had profound implications for the nature and quality of the transformation and how reactionary or progressive it has been. In this the state has always played a prominent part. I further argue that differentiation of the peasantry is central to transformation: it is not an outcome but a determining variable, a *causa causans* rather than a *causa causata*. My argument is that differentiation of the peasantry feeds into and interacts with the landlord class and class struggle, these being critical to the eventual outcome. Such is the theme of the present paper. Differentiation is no mere outcome. The distinctly varying trajectories in the three crucial instances are explained in these terms.

Some preliminary analytical observations

In each instance I consider a transition from feudalism to agrarian capitalism. I start with some preliminary observations concerning feudal social formations, to help to clear our analytical path. I will then proceed to take each of the cases separately.

Contra many social historians, conflict, rather than harmony, was the principal underlying feature of the relationship between the main classes of feudal society in Europe. As Marc Bloch, the great historian of European feudalism (1961a, 1961b) and of the French countryside (1966), stressed of European feudalism: 'agrarian revolt is as natural to the seigneurial regime as strikes, let us say, are to large-scale capitalism' (1966, 170). Rodney Hilton, too, the outstanding Marxist historian of medieval England, and a formidable comparativist, emphasised the importance of conflict – *class* conflict – in feudal Europe 'between peasants and ruling groups over the disposal of the surplus (disputes about rents and services) and over the sanctions used to enforce its appropriation (serfdom, private jurisdiction)' (Hilton 1974, 207). Indeed, the nature, manifestations and implications of this lord/peasant class conflict is a recurring theme in all of his work: 'the conflict between the peasants as a whole on the one hand and the landowning class and its institutions on the other' (1974, 210). This was true of each of my three examples.

Peasant resistance in medieval Europe, in fact, falls into different sub-genres. First, 'some movements were obviously direct confrontations between lords and peasants over the proportion of the surplus product of peasant labour which should go in rents, services and taxes' (Hilton 1973, 62). Very broadly, 'on the whole, the[se] more elemental movements with the simplest demands were at the village level' (1973, 64). The demands in question may have been the 'simplest', but they were quite fundamental. They occurred at the points of production and of distribution. These existed throughout the medieval period. They may be sub-divided into those that proceeded via individuals or groups of individuals within a village; and those that involved whole peasant communities seeking village enfranchisement. Secondly, there were those that appear as 'movements of social, religious or political protest' (1973, 62). These were a feature of the later medieval epoch. By contrast, 'the movements affected by the new developments in medieval society tended to be regional in scope, and generally to have wide horizons, which were extended not merely beyond the village but beyond purely social aspirations' (loc. cit.). They may have sought religious goals, or attempted to confront and moderate the increasing encroachments of the state. Both of the sub-genres were obvious in England and France. It is the direct confrontation over surplus that interests me most here. It is clearly visible in feudal Prussia.

Under feudalism, the peasantry is to be viewed as a single class. Kosminsky, the author of the classic Marxist treatment of thirteenth-century England's agrarian history, insists: 'And yet whatever distinctions and contradictions may have existed within the peasantry, they do not preclude our seeing in the peasantry of the epoch a single class, occupying a definite place in the feudal mode of production, and characterised by the anti-feudal direction of its interests and its class struggle.' (1956, 198). Hilton takes the same position, with respect to the final era of English feudalism, 1350–1450, as he does for the whole of the feudal era in Europe (1975, ch. 1, 3–19, entitled 'The Peasantry as a Class'). What *all sections of the peasantry* had in common was their servile condition: tied to the land, subject to an array of feudal restrictions, with surplus appropriated via extra-economic coercion – in Chris Wickham's incisively reductive phrase, feudalism is 'coercive rent-taking' (1985, 170). Moreover, as the feudal era proceeded, they were subject to attempts at increased exploitation and seigneurial onslaught. That bound them together in hostile conflict against feudal lords.

Despite seeing the peasantry as a class, it is crucial to note that these peasantries are socially differentiated. In Europe, 'the peasant community was not a community of equals' and 'the stratification of peasant communities, moreover, was at least as old as the earliest records which we have of them' (Hilton 1973, 32). In analysis of differentiation of the peasantry, it is common to refer to three strata: a rich peasantry, a middle peasantry and a poor peasantry. It is appropriate to do so with respect to medieval Europe, and historians like Kosminsky and Hilton proceed thus (see, for example, Hilton 1978, 272, 280, Kosminsky 1956, 354). Hilton stresses that 'the internal stratification of the peasantry, during the medieval period, was strictly limited' (1973, 34). Rich peasants hired wage labour, especially at peak seasons: whether from the ranks of poor peasants or from that of a class of completely landless labourers. That this might generate conflict seems possible. Yet, there was remarkably little conflict of this kind (Hilton 1966, 166, 1975, 53). Nor was there a social gulf, or a 'competitive element', between rich and poor peasant (1975, 53): 'The social gulf that was still the most important was that between the peasant and the lord.' Certainly, the conflict between peasant and lord was far more important than conflict within peasant communities. There was a clear absence of struggle *within* the feudal peasantry. Yet, differentiation might deepen within feudal limits. Such deepening needs close attention. These observations are valid for all three of my case studies.

I have posited a servile feudal peasantry. Yet there was the possibility of a *relatively free peasantry* under feudalism. From the eleventh century onwards there was, in different parts of Europe, land hunger and the possibility of alleviating it through the settling of uncultivated land (Hilton 1973, 43, 92). This proceeded as 'the consequence of increasing population, increased production for the market, and firmer and more ambitious political organisation by the aristocracy and the ruling kings and princes' and 'it involved a response by landowners to the search by peasants for more land, which took the form of attempts to direct this land hunger towards the colonisation of forest, scrubland and marsh' (1973, 43). Hilton distinguishes two distinct movements in this respect. The first was in West Germany, France, England, and Italy, where unsettled areas of such land 'were filled up by the overflow from crowded villages in old settled areas' (1973, 43). The 'best-known' (1973, 43) and most important historically was, however, the second, which involved the settlement of empty lands to the east, in Slav territories. Peasants needed inducements to cultivate such land, and so was created a free peasantry in Prussia: a peasantry that remained among the freest in Europe until the sixteenth century. That was not so in any general way in England and France. This Hilton describes as 'the expansion of German colonisation in central and Eastern Europe' (1973, 43). It included, most importantly, Prussia east of the Elbe.

England

Differentiation and the emergence of a stratum of powerful rich peasants

Differentiation of the English peasantry existed from earliest times. It was a structural feature of feudalism at its very inception (let us say, by the sixth century AD). Hilton (1978, 272) cites Kosminsky (1956, 207) to the effect that 'the deep-seated causes of peasant differentiation probably lie as far back as the disintegration of the pre-feudal lands into the ownership of single families'.[1] Hilton stresses that 'the contours of a peasantry divided between a wealthy minority, a solid middle

---

[1] Kosminsky also observes: 'The formation of an upper layer among the free peasantry may be partially connected with processes taking place even in pre-feudal society, with the advance of early property differentiation and may represent, as it were, certain elements of incompleteness in the feudalisation of English society' (1956, 225–6). Significant aspects of this may be seen in Rosamond Faith's book *The English Peasantry and the Growth of Lordship* (1997). She notes: 'Differentiation begins to appear in the sixth century' (1997, 5).

peasantry and a significant proportion of smallholders can easily be seen in the Domesday Book, 1086' (1978, 272, see also 1973, 33). That it developed and deepened, in the wake of the Norman Conquest, as the medieval era proceeded, is also clear. We have a superb treatment for the thirteenth century and for the century 1350–1450 by Hilton (1949 and 1978) and Hilton and Fagan (1950) and valuable accounts of the thirteenth century by non-Marxist historians (Postan 1966b, 617–32, Miller and Hatcher 1978, 149ff). I have considered this in detail elsewhere (Byres 2006b, 32–53). Here, with the transition to capitalism in mind, I would note the following of the rich peasantry.

In the thirteenth century, a lower stratum of rich peasants held more than 30 acres, while a select few worked more than 60. Hilton stresses 'the growth of a rich upper stratum among the peasants'. Thus: 'Whether we look at peasant life in the south-east, in the Thames valley, in East Anglia or in the Midlands, we find standing out from the ordinary run of tenants with their fifteen or twenty-acre holdings, a small group of families sometimes free, more often serf, holding a hundred acres or more' (Hilton 1949, 130). These were 'the village aristocrats', some of whom 'were climbing towards yeoman status' (Miller and Hatcher 1978, 149). The rich peasants 'controlled the commons, declared local custom, and maintained order' (Hilton 1978, 278). Here, then, was clear 'space' for rich peasants to operate within. That they took advantage of this, to maintain and further their interests, is clear. Motivated by the existence of a sizeable market for agricultural products, that they wanted to accumulate more land is also clear. But there were strict limits on such accumulation (Hilton 1978, 278).

In the century 1350–1450, 'the village community was dominated by the richer peasant families, who ran the manorial court in its jurisdictional, punitive and land-registration functions.' (Hilton 1978, 281). Rich peasants strengthened their position, and did accumulate more land, in the shape of 'the abandoned demesnes of the aristocracy' (Hilton 1978, 282), although, within the village community, the limits upon the accumulation of land continued. Thus, 'by the end of the fourteenth century an upper class of peasants [was clearly in evidence]' and 'four or five families in the village were now cultivating sixty or a hundred acres of arable land, and tending several hundred head of livestock' (Hilton and Fagan 1950, 29–30). This should not tempt one into seeing the rich peasantry of this era as stronger than it actually was. To be sure, 60–100 acres of arable land and several hundred head of livestock represented a sizeable, and unprecedented, holding and considerable village standing. 'But if the upper peasants were able to take advantage of the economic embarrassment of both nobles and poorer peasants, they still remained socially and politically a subject group in society' (Hilton and Fagan 1950, 30). Here, indeed, was a class in the making 'chafing at . . . [the] feudal restrictions on . . . [its] economic enterprise' (Hilton and Fagan 1950, 31), without the class autonomy that would ensure the accumulation it sought to pursue. Here was a stratum of 'capitalist farmers in embryo', and the rich peasant constituted, in this sense, a 'revolutionary . . . figure' (Hilton and Fagan 1950, 31). Hilton identifies this era as a '"marking time" phase' (Hilton 1978, 281). This notion represents 'the natural tendency to focus on the minority of rich peasants, the "yeomen", because they seemed to be the group which would evolve into the capitalist tenant farmers of a later period.' (Hilton 1978, 281). They would so evolve, but were still constrained in this respect: were still, to use Ellen Wood's phrase, in the grip of 'a precapitalist logic taken to its absolute limits' (Wood 2002, 59). The era of vigorous transition would come in the sixteenth century. What is crucial is that a long

urge to accumulate might be broken. The great bulk of its land was rented. The rental form would be critical.

Until the mid-fifteenth century it was an antagonistic relationship, with rich peasant pitted against lord in the social conflicts waged in the English countryside. Hilton argues that, if, up to 1450, the *peasantry* achieved a victory, those who gained most were the rich peasants (the 'villein sokemen'), who were often the leaders in collective action (Hilton 1973, 89). Here, then, was a significant stratum of rich peasants, some of whom were truly substantial: before 1450, not yet detached, as a class, from the rest of the peasantry; still powerfully constrained, in myriad ways, in its urge towards accumulation, and sharing, with other segments of the peasantry, feudal restrictions; and sharing common enemies, along with those other segments of the peasantry, in the landlord class and the state. But its economic interests were very different from those of other strata of the peasantry. Here was a class of rich peasants ripe for transformation into a class of capitalist farmers. The English transition to capitalist agriculture would not have been possible without that. The fifteenth century saw the end of feudalism in England. This ushered in the possibility of the formerly antagonistic relationship between rich peasant and feudal landlord being transformed into one of mutual advantage. But this was not so for the rest of the peasantry. Their antagonism increased, and with that came bitter class struggle, now with a transformed landlord class, or at least with one whose transformation was under way. The sixteenth century was an era of 'fierce agrarian struggles' (Hilton and Fagan 1950, 192).

Already, by the middle of the fifteenth century, landlords had rented out most of their demesne land, largely on leasehold, at competitive rents (Tawney 1912, 202–3). That was not sufficient to secure the rental income they sought. What we now see is their transformation from a *feudal* into a *capitalist* landlord class. The former feudal landlord class responded vigorously and ruthlessly to its predicament. At the end of the feudal era, the landlord class was determined to renew its fortunes and was reconstituted as a capitalist landlord class. This entailed a concerted effort to let land already let on copyhold, i.e. customary, terms, and protected by custom from increases in rent, as leasehold land, with the new capitalist tenants paying competitive rents. Landlords, in search of far higher rents, sought to dispossess small, customary tenants of their rented land, which was farmed in strips, and let it out in large units on 'economic' leases rather than customary copyhold; to appropriate part of the commons and similarly let it out. Such land, to be suitable for capitalist farming, whether arable or pasture, had to be enclosed. It might be enclosed either by the landlord himself, or by the new capitalist tenant, usually the latter. This was bitterly resisted.

Tawney gives a detailed account of peasant resistance. Now the rich peasantry, or its most powerful stratum, had been detached from the peasantry, and no longer provided strength and leadership in its struggle. He analyses at length the deeply felt grievances which 'at one time or another in the sixteenth century, set half the English counties ablaze' (1912, 304). Thus: 'Sometimes the discontent swelled to a small civil war, as it did in Lincolnshire and Yorkshire in 1536, and in the eastern and southern counties in 1549' (Tawney 1912, 318). Between 1547 and 1549 there was 'violent agitation and ... drastic expedients' (Tawney 1912, 362). Tawney draws attention to disturbances and sometimes violent unrest – induced by high rents, seizing of land for pasture, enclosure, the taking of the commons – in 1550, 1552, 1554, 1569, 1595, and in 1607 (Tawney 1912, 319–20). The riots in the Midlands in 1607, a reaction to a preceding decade of enclosure and depopulation, were the 'last serious agrarian uprising in England' (Tawney 1912, 320).[3] By then the peasantry had been defeated. By the end of the Tudor era the transition had been completed, the new class structure was in place, and the way was set for the stark opposition, in the countryside, of agricultural proletariat and capitalist employer in alliance with a powerful capitalist landlord class.

France

Differentiation, the rich peasantry and the absence of a clear movement towards capitalism

As in England, a differentiated peasantry was a feature of the French countryside from the very dawn of feudalism. This was so at least as early as the ninth century, sometimes 'with immense differences in the sizes of peasant holdings' (Hilton 1973, 33): Hilton cites evidence for the Paris region (Perrin 1945), and Picardy (Coopland 1914). Likewise, Bloch points to 'marked differences in size' of the *manse* and 'glaring inequalities' 'in the early Middle Ages', the earliest he notes also being in the ninth century: these inequalities indicating clear social stratification, the critical dividing line being 'the possession or lack of a plough team' (Bloch 1966, 190–1, 152–3). He argues that 'the indications seem to be that these small rural groups were at all periods divided into quite well-defined classes' (1966, 190). Such social stratification continued 'in later centuries' (Hilton 1973, 33). Bloch further notes considerable stratification in the twelfth and thirteenth centuries (1966, 191). When, by the end of the fifteenth century, serfdom had collapsed, as it had in England, a differentiated peasantry continued, but there were 'no clear movements in the direction of agrarian capitalism' (Hilton 1978, 282). A differentiated peasantry continued to be reproduced thereafter. We have an illuminating account of its nature in the seventeenth century, by Goubert, first for the last quarter of that century for the region of Beauvaisis (1956), and subsequently for all of France for the whole century (1986). Soboul provides an incisive treatment for the eighteenth century up to 1789, and thereafter until the end of the nineteenth century (1956).

Goubert tells us that 'were we to erect a social pyramid of peasant property, it would have a very broad base and an absurdly slender apex' (1956, 57). At the base was 'a swollen mass of dwarf peasants and labourers' (Soboul 1956, 87). This 'host of *manouvriers* constituted, in nearly every village, the majority – the overwhelming majority – of the inhabitants' (Goubert 1956, 59, 1986, 98): the *manouvrier* typically owning a few acres, a cottage, and a small garden; half without a cow, rarely keeping a pig, and without horses (Goubert 1956, 58–60,

---

[3]The fascinating Kett's Rebellion of 1549, in Norfolk, which was directed against enclosure, has given rise to a particularly large body of scholarship. Russell (1859) has become something of a *locus classicus*. Later scholars include, for example, Bindoff (1949), Land (1977), Cornwall (1977, 1981), MacCulloch (1979, 1981), Fletcher and MacCulloch (2004), Walter (2004). For them the focus is not upon the kind of questions addressed here: i.e. questions of class struggle, the transition to capitalism, etc. Among writers who do have such a focus see Tawney (1912, 149, 326, 331–7), Hilton and Fagan (1950, 193), and Whittle (2000, 1, 45, 69, 76, 287, 312–3, 329). Whittle points to the complex nature of the rebellion, which took place in a region that was in the vanguard of agrarian capitalism, in the era of transition. Kett's Rebellion bore the contradictions of transition at a time when those contradictions had yet to work themselves out fully.

1986, 98–9); 'forced to hire himself out to *laboureurs* and big farmers and, perhaps, to try and take up a secondary occupation (cooper, wheelwright, tailor, weaver)' (Goubert 1956, 60, and see 1986, 100–5). In the eighteenth century, the indebtedness of small peasants to larger ones increased, there was an increasing tendency towards the hiring of small peasants as wage-labourers, and there emerged a growing category of landless day-labourers, who worked for wages (Soboul 1956, 87).

Above them were differing categories of rich peasant. At 'the very peak of the peasant social pyramid' (Goubert 1956, 58) was a tiny group of *very rich* peasants. These were the large and enterprising tenants in areas of large-scale farming: either the *laboureurs-fermiers*, the substantial tenant-farmers; or the *receveurs de seignurie*, the receivers for the lords of the manors, or *fermiers-receveurs* (Goubert 1956, 58, 64–5). These tenants leased lands concentrated in large units (anything from 80 to 150 hectares and more), especially church land (Goubert 1956, 548); they employed wage labour; they produced large surpluses; and they were extremely important as creditors. But they existed only where land was 'owned [as] a large compact domain', which was by no means common (Goubert 1956, 64); and, other than in the north, around Paris and Lille 'they were only to be found in the villages in ones or twos; in some villages they were not to be found at all' (Goubert 1956, 58, 1986, 111–14).

There were, secondly, wealthier *laboureurs*. These were *rich peasants*. A *laboureur* 'was, almost by definition, a man who owned a plough and a pair of horses' (Goubert 1956, 63); who was not a sharecropper, but owned land; 'ploughing ... his own land with his own horses'; perhaps taking in other land on lease; using his horses for carting; hiring his horses out to peasants who owned no horses; the *laboureurs* becoming 'the creditors of the mass of small peasants and, when occasion demanded, their employers at low wages' (Goubert 1956, 63, 1986, 114–16). They had considerable reserves and produced regular surpluses. Some *laboureurs* 'farmed no more than 35–40 acres'. But the larger ones – the true rich peasants – farmed up to 100 acres (Goubert 1956, 64). Like the previous category, however, their significance in the total social formation should not be exaggerated: 'the *laboureur* is a fairly rare social specimen' (Goubert 1956, 4, and see 1986, 114–16).

In France, however, by 1789, no transition to capitalism had taken place. Soboul insists that at the end of the Old Régime the differences within the French peasantry 'were in no sense fundamental' (1956, 84). Differentiation of the peasantry there assuredly was, but it was, still, *quantitative* rather than *qualitative* in nature (Soboul 1956, 88). Thus, while 'mere quantitative differentiation between those possessing more or less real or movable property, land or money' had always existed, this 'did not entail a modification of the relations of production' (Soboul 1956, 88). For Soboul, 'the structure of landed property ... remained feudal' (1956, 84), and 'social differences within the [peasant] community were much less important than the antagonism between the peasantry as a whole and the landed aristocracy' (1956, 84). A rich peasantry had yet to detach itself.

Contenders for a capitalist primum mobile *role*

The full capitalist transformation of the French countryside was not finally effected until the very end of the nineteenth century. Eugen Weber suggests that a critical juncture can be located in the 1840s: but only for the areas of large-scale farming, found in the north (1979, 117–18). There, capitalist transformation had already proceeded, largely via rich tenant-farmers. For the great bulk of France – the centre, the south and the west, areas of small peasant farming, where sharecropping was often prevalent – a turning-point may be found in the 1890s or 1900s: and not before (Weber 1979, 118–29). Part of the explanation for this may be sought in the nature of the class struggle fought in the French countryside. Before considering that, however, it is instructive to consider the possible contenders for a *primum mobile* role in the securing of such a transition before 1789.

In England, the landlord class played a critical role: transforming itself into a capitalist landlord class, letting the land as leasehold at competitive rents, to rich peasants who became capitalist tenants, and ensuring that the land was enclosed. No such role was played by the French landlord class. As Lefebvre noted, 'the most important feature of France's rural physiognomy, the one characteristic whose underlying influence may well have had the most profound consequences' was that 'priests, nobles, and bourgeois almost never managed their properties directly; their domains were extremely fragmented and rented-out as middle-sized farms, even as individual fields' (1977, 34). As Meek observes in his classic work on Physiocracy, in France of the mid-eighteenth century:

> The main feature which distinguished French agriculture from that of England at this time was the relative lack of enclosures and the consequent survival of very large numbers of small peasant proprietors, who, although they were normally subject to heavy seigneurial dues, had the right to transfer their property or pass it on to their heirs. (Meek 1962, 23)

The French landlord class was, in this respect, thoroughly unprogressive.

If the French landlord class was thus defective, what of those who administered its estates? These were the *fermiers-généraux*. Here was a particular form of agrarian concentration, that of estate administration, which gave access to the marketable surplus. Crucially, 'numerous *fermiers-généraux* also traded in agricultural produce' (Soboul 1956, 86). They would most usually take the form of 'a businessman, a notary, a large shopkeeper' (Soboul 1956, 86). They stood between proprietor and share-cropper: leasing in sharecropping units (the *métairies*) from one or perhaps more than one proprietor and sub-letting them; perhaps assigned by the proprietor (if the proprietor were a lord) to collect seigneurial revenues; and possibly entrusted with feudal rights of usage. The surplus they acquired from their sharecropping tenants enabled them to pursue a lucrative business as traders. They might, presumably, have been transformed, ultimately, into large capitalist farmers. Soboul rejects them, however, as a likely destroyer, or transformer, of 'the old mode of production'. The *fermier-général* represented 'the application of commercial capital to agricultural production', and was caught in the contradictions characteristic of merchant's capital: that is to say, 'being integrated into the old system and profiting from it, generally himself the receiver of seigneurial revenues, the *fermier-général* had a vested interest in the maintenance of the old system which guaranteed his own position' (Soboul 1956, 86). That logic is powerful. He was a possible, but not the most likely, agent of agrarian transition.

Soboul argues that the large tenant-farmers had to be discounted. They 'were indeed active agents in the evolution of agriculture inasmuch as they combined small farms into large ones and strove for a scientific and intensive form of agriculture' (Soboul 1956, 86–7). But their potential revolutionary role was limited. Soboul insists that 'they brought into agriculture capital of commercial origin ... [so that] their productive activity was subordinate to their commercial function'. Moreover, they had deep involvement 'in the old productive system', since they were often 'receivers of seigneurial revenues and ecclesiastical tithes' (Soboul 1956, 87).

It is to the *laboureurs* that Soboul attributes the potential revolutionary role. It was they, certainly, who proved to be so in the north. They had no binding commitment to 'the old productive system'. It was they who 'tended to destroy and not to preserve feudal production' (Soboul 1956, 87). If there were signs of the rural community beginning to disintegrate, this was 'because of the progress of capitalist production, which involved both the creation of a market for commodities and a market for labour ... [which] took place within the community essentially through the action of the *laboureurs*' (Soboul 1956, 88). It would be a transition, if it were to proceed, via the agency of the rich peasantry: not the aforementioned very rich peasantry, as represented by the substantial tenants, but the *laboureurs*. Their capacity to accumulate was, however, constrained by the siphoning off of much of their surplus. From the seventeenth century they, along with the rest of the peasantry, were taxed very heavily by the state: in Richelieus's phrase, the peasantry became the 'donkey of the state' (Goubert 1986, 189). Royal taxes, along with others (those of the church, the seigneur and sometimes the peasant community itself), which, for the first time, they now outstripped, constituted a heavy burden (see Goubert 1986, ch. 15, on 'The Peasants and Taxation').

Peasant resistance: class struggle and the frustrating of the capitalist impulse

While in England in the thirteenth century it was landlords who took the initiative, to which peasants then responded, in France participation in wider economic activity in the twelfth and thirteenth centuries enabled *peasant communities* – whole villages – to take aggressive action (Hilton 1973, 87). This was the major form taken by peasant movements in France in the period from the late twelfth to the middle of fourteenth century (Hilton 1973, 83). There was a powerful movement for village enfranchisement (Hilton 1973, 74–85): with demands very similar to those made on an *individual* basis in England – for example, for fixed judicial fines, abolition or regularisation of the *taille* (the seigneurial tax), eradication of *mainmorte* (death duty), a fixed rather than arbitrary marriage tax, fixed payments to the lord on alienation of property (Hilton 1973, 80–1). Thus: 'on the whole these were movements which aimed to obtain from the lord of the village a charter granting, at least, exemption from various exaction and fixed rather than arbitrary obligations, and at most an element of autonomy in the running of the village community' (Hilton 1973, 75). The demands also included 'freedom of personal status' (Hilton 1973, 75). The struggle was often prolonged, violent and bitter (Hilton 1973, 81–3). The peasants were often defeated by a combination of the nobility and royal power; and where they made gains they often paid heavily (Hilton 1973, 85). Its achievements were tangible but partial and gained at considerable cost (Hilton 1973, 84–5). Yet there was some success: chartered privileges were enjoyed and there was abolition of certain tolls and certain services (Hilton 1973, 81–5).

Within such communities, it was *rich peasants* who were the driving force and the major beneficiaries:

> The struggle for village charters in the earlier period was part of the peasant reaction to the economic expansion of the period when the development of production for the market made the wealthy peasants who benefited from it socially and politically ambitious, and their demands and achievements were a straightforward reflection of the potentialities of the situation, primarily economic. (Hilton 1973, 74)

But, if rich peasants gained most, by the end of the fifteenth century, as we have seen, they did not constitute a presence comparable to that of the English rich peasantry. Nor, thereafter, were they a force sufficient to spearhead capitalist transformation. And, if serfdom had collapsed by then, the French peasantry continued to be subject to a host of feudal restrictions.

A prolonged struggle was pursued, which often erupted into violence, against the *seigneurie* (the feudal lords) until the end of the old régime: a struggle punctuated by peasant uprisings and agrarian revolt, by riots, disturbances and *jacqueries*. The 'great insurrections were altogether too disorganised to achieve any lasting result' and 'almost invariably doomed to defeat and eventual massacre' (Bloch 1966, 170).[4] At another level, 'the patient, silent struggles stubbornly carried out by rural communities over the years would accomplish more than these flashes in the pan' (Bloch 1966, 170). The more or less cohesive 'French rural community' was unsuccessful in ridding itself of feudal obligations before 1789. But, united around struggles against the *seignurie* for communal land and involved in a 'constant struggle for its economic existence and administrative autonomy' (Soboul 1956, 80), it did succeed in maintaining 'an economic and social system founded on the interplay of communal pressures, the limitation of private property and the existence of collectively exploited lands' (Soboul 1956, 80). The rich peasantry, the *laboureurs*, had not detached itself from the peasantry. By 1789 there had been no transition to capitalism, spearheaded by the rich peasantry or any other rural class.

The French Revolution cleared the ground for a possible unleashing of capitalism. It removed the massive barrier inherent in the feudal relationships which had persisted despite an apparently 'free' peasantry ('free' inasmuch as they were not serfs). It 'destroyed the seigneurial regime and abolished feudal rights. It proclaimed the total right to property; hence the freedom to enclose and cultivate and the restriction of collective rights' (Soboul 1956, 88). It modified the distribution of land and proprietary rights in land, as church land and the land of *émigré* nobles were sold. The major beneficiaries were the urban middle class and the rich peasantry (Soboul 1956, 88). Indeed, 'the French rural community' was destroyed. But the

---

[4]Ladurie, in his *The French Peasantry 1450–1660*, provides a cogent account of peasant revolts over the two centuries from 1450 to 1660 (1987, ch. 5, 359–99), with particular stress upon the capacity of 'taxation ... to arouse the resentments of the peasants' (1987, 359). In view of what has been said already about the increased burden of taxation, this is not, perhaps, surprising. The seventeenth century, indeed, the *grande siècle*, has attracted detailed treatment of peasant revolts and uprisings, by, for example, the Soviet scholar Porchnev (1963), in a meticulously researched and exciting work; and by Goubert (1986, ch. 16, 205–19), Bercé (1990), Mousnier (1971), Mollat and Wolff (1973), and Foisil (1970). We cannot here consider that rich literature and the considerable debate embodied in it.

attempt to secure a compulsory division of communal lands failed (Soboul 1956, 89). On the one hand, 'the *laboureurs* ... finally constituted themselves [as] a class', and clearly became the 'dominant class in the countryside' (Soboul 1956, 89). With the changes, moreover, 'the hitherto latent antagonism between the *laboureurs* ... and the mass of peasants ... became overt' (Soboul 1956, 89). On the other, the 'mass of peasants' – a poor and middle peasantry – 'clung desperately to the traditional forms of production and stubbornly called for the maintenance of the limitations which collective constraint imposed on private property' (Soboul 1956, 89). From the very early post-1789 years they demanded the retention of gleaning and stubble rights, of common grazing land, of the requirement to reap with sickles, of the communal herd, and opposed attempts to enclose lands and meadows (Soboul 1956, 91–2), and a bitter struggle lasted throughout the nineteenth century. The resistance was most marked in the southeast, the southwest and the centre of France; while in the countryside of the north capitalist transformation did not meet such opposition (Soboul 1956, 93). This prolonged resistance 'considerably inhibited the capitalist transformation of French agriculture' (Soboul 1956, 91). It was sufficient to continue to stifle the capitalist impulse, and prevent the unleashing of capitalism throughout the French countryside, until the very end of the nineteenth century.

Prussia

From the 'free peasants east of the Elbe' to subjugation as serfs: class struggle red in tooth and claw

Engels refers, for the period up to the sixteenth century, to 'the free peasants east of the Elbe' (1965, 156), and Brenner to 'what had been, until then, one of Europe's freest peasantries' (1976, 43). Here was a relatively free peasantry within feudalism, or, as an outpost of feudalism. This arose as follows. From the eleventh and twelfth centuries: 'lay and church magnates obtained grants of land, or took part in movements of conquest in sparsely settled wooded areas', and 'they put into the hands of special agents the recruitment of peasants from the Rhineland and the Low Countries to take up new holdings and, in effect, to create new village settlements' (Hilton 1973, 43). These

> colonising German and Polish landowners, in the eastern territories beyond the Elbe, deliberately created villages whose inhabitants were offered, as bait, freer terms and conditions of life than in the western ones. The landlords' agents were authorised to offer holdings on free and heritable terms, to be held often for no more than money rents and church tithes. Superior jurisdictions and fiscal pressures were avoided and the agent himself, provided with a holding three or four time the size of the peasants', became in effect the immediate lord, presiding over the village court and taking a proportion of the fines. (Hilton 1973, 92)

These peasant colonists had tenure on free status, with 'low fixed money rents and with no labour services, together with a degree of local autonomy' (Hilton 1973, 44). It was a peasantry in which the seeds of differentiation were planted, and differentiation must have proceeded to a not insubstantial extent, remaining so until the sixteenth century (Byres 1996, 54–5, 67). Hilton stresses: 'Although such villages might to some extent reflect the peasant ideal, they came into existence *as a*

result of seigneurial, not peasant initiative' (1973, 92, emphasis added). This had
nothing to do with class struggle waged by peasants.[5]

But class struggle between Prussian lords and this relatively free peasantry would
come red in tooth and claw. In the wake of the Black Death and other later
visitations of the fourteenth century, came widespread flight from the land and, as
the fifteenth century wore on, depopulation, totally deserted villages, and a serious
shortage of labour brought an ominous seigneurial offensive. The Junkers
themselves began to acquire and to farm deserted peasant land. This, at first, was
something of an emergency measure, until new peasants might be found. In the
sixteenth century, however, when corn prices began to rise it became more
permanent. The Junkers confronted a serious labour shortage, whether in the form
of the free wage labourers employed hitherto or that of labour supplied via labour
services. They reacted: with increasing curtailment of peasants' freedom to move, the
imposition of mandatory maximum wages, an assault on fixed money rents, and,
finally, a concerted move to extend mandatory labour services. In the sixteenth
century, the latter multiplied and became the norm and a free peasantry disappeared
completely. In Prussia 'the peasants did not accept the deterioration of their rights
and conditions without resistance' (Carsten 1989, 13–14): i.e. there was class struggle
between peasants and lords. They appealed to princely authority through the judicial
process, and there were peasant uprisings. Whatever the apparent judicial victories,
they were a very limited deflection of the powerful forces engulfing the peasantry;
and peasant uprisings were cruelly suppressed with no great difficulty.[6] The tide of
enserfment would not be stemmed. In this class struggle, the Junkers won a crushing
victory. Where, in England, the peasantry had successfully resisted the seigneurial
reaction, and so sounded feudalism's death knell, in Prussia the opposite was true.
There, 'the "manorial reaction" shattered the free institutions of East Elbia and
wrought a radical shift in class relationships' (Rosenberg 1958, 29). By the end of the
sixteenth century, the Prussian Junkers had succeeded, with the aid of state power, in
enserfing to themselves 'the formerly free peasants of the German East ... [in] an
almost complete subjugation of large segments of the peasantry' (Gerschenkron
1966, viii). It was the case that: 'Everywhere in north-east Germany, and equally in
neighbouring Poland, there developed in the course of the sixteenth century the
system of *Gutsherrschaft* consisting of demesne farming and serf labour, which was
entirely different from the agrarian system of central, western, and southern
Germany [*Grundherrschaft*]' (Carsten 1989, 19). The outcome was 'the classic Junker
estate economy ... as a form of seigneurial market production (*Teilbetrieb*) in which,

---

[5]I have cited Hilton to maintain comparative perspective. I have considered the creation of a
free peasantry in Prussia in my book (Byres 1996, 48–52). On this see Engels (1965, 154–5),
Rosenberg (1943, 2, 1944, 228, 1958, 30), Carsten (1954, 80–1, 111, 1989, 3), Leyser *et al.*
(1978, 75–6), Hagen (1985, 3–4).
[6]For a brief treatment of peasant resistance see Byres (1996, 60–1). It has been suggested by
Rosenberg (1944, 233) that 1525 – i.e. the German Peasant War – was a probable turning-
point: that 'a long period of peasant unrest had come to an end with the crushing defeat
suffered by the rebellious Prussian peasants in the uprising of 1525'. In fact, as Rosenberg
himself suggests (1944, 233), the peasants east of the Elbe were 'relatively docile'. Engels,
indeed, points out that in only one region of East Elbia was there an active peasant movement
in 1525 – in Eastern Prussia (Engels 1965, 156). That is where they were crushingly defeated.
Elsewhere east of the Elbe, the peasants 'left their insurgent brethren in the lurch, and were
served their just deserts' (Engels 1965, 156). If that was the case, they paid a harsh penalty.

by means of extra-economic coercion, the landlords forced the peasantry to shoulder the cost of the labour, horsepower and tools necessary to demesne farming' (Hagen 1985, 111). When serfdom had broken down irretrievably in England and France, in Prussia it was established with a vengeance. It would remain in place until the early nineteenth century.[7]

Peasant differentiation in late feudal Prussia and the unlikelihood of capitalism from below[8]

As Prussian junkerdom mounted a feudal offensive that changed fundamentally agrarian relationships east of the Elbe, processes of peasant differentiation first would be brought to a severe halt, and then would be reversed and would have a tight rein placed upon them. Between the sixteenth and the early nineteenth century, the Junkers expanded their demesne at the expense of the peasantry. They, further, imposed ever new burdens on an enserfed peasantry. Yet, by the second half of the eighteenth century the East Elbian peasantry was by no means a homogeneous one.

There was, first, a line of division between an overwhelming majority of unfree peasants and a tiny minority of free peasants. The free peasants were 'large peasants with especially favourable conditions' (Harnisch 1986, 41). They were analogous to the very rich peasants identified in France by Goubert (the *laboureurs-fermiers* and the *fermiers-receveurs*). They were 'completely free of obligations or subservience to the noble estates' (Berdhal 1988, 29); and frequently served the Junkers' interests, working closely with the Junkers, serving as chief administrative and police officers and directing the village's labour force. Like their French counterparts, they may be discounted as likely candidates for a capitalist transforming role. Their relationship with the Junker was close, so that a possible independent role was unlikely. Any transformation would be likely to be in concert with, rather than in competition with, the Junkers. Conceivably, that might have been as capitalist tenant farmers renting from a transformed feudal landlord class. But they 'comprised a very small percentage of the rural population' (Berdhal 1988, 29), and alone could not have become a class of capitalist farmers. In terms of sheer control of a sufficient quantum of resources within the village their transforming significance was very limited. As Harnisch observes, 'we can more or less discount the comparatively small number' of free peasants (Harnisch 1986, 41).

The unfree peasantry were divided into 'true *Bauern*', described by Harnisch as 'the middle and large peasants' (1986, 56) and those who were not 'true *Bauern*'. The class of peasantry beneath the Bauern – the great majority of the rural population – had far less favourable property rights (if they had any at all), seldom had strips of land on the estate, and were unable to support a team of animals. They were divided into a stratum of smallholders (*Kossaten*), with between 5 and 10 hectares (12 to 25 acres) and at most 15 hectares (37 acres), but without teams of draught animals; cottagers with small plots (*Budner*); garden cottagers (*Hausler*); and a multitude of day-labourers (*Tagelohner*). For the purposes of this essay I concentrate on the 'true *Bauern*'.

---

[7]This I consider at length in my book (Byres 1996, 53–68). Here I simply state the barest of outlines.
[8]This is treated in detail in Byres (1996, 81–90), of which the following is a highly condensed version. I there draw especially upon the unerringly excellent Harnisch (1986) and the valuable Berdhal (1988); as well as Blum (1978) and Hagen (1985).

The 'true *Bauern*' had property, inheritance and contractual rights; were able to support a team, or teams, of draught animals; and had holdings of between 20 and 70 hectares (i.e. 50 to 170 acres). They might be sub-divided into full-*Bauern*, with at least four teams (*Spannen*) of two horses or oxen each; half-*Bauern* (*Halbbauern*), with two teams; and quarter-*Bauern* (*Viertelbauern*), with one team. Their ownership of draught animals needs to be qualified by the reality that 'the labour obligation required peasants to keep far more draft animals than they needed for their own operation' (Blum 1978, 150). A full peasant, for example, with heavy labour obligations, and with 12 horses, might need 8 for his labour obligations and 4 for his own needs (Blum 1978, 150). Or a peasant with two teams would have one of the teams to meet his labour services. Such peasants 'had to maintain one team, with a farm servant (*Knecht*) and quite often also a maid, merely to be able to meet their dues' (Blum 1978, 150). They were a feudal rich peasantry, heavily constrained by feudal obligations. Surplus appropriation was heavy and stripped them bare. Labour rent varied regionally. It might be two–three days per week, but the lighter the burden of labour obligations the heavier the feudal dues in kind (with instances of dues in grain constituting 20 percent of an average harvest where there was only one day of service). They did participate in production for the market, but their marketed surplus was not a true commercial surplus of the kind that market-oriented rich peasants in a non-feudal situation might set out to market regularly. The outcome was not a growing source of accumulation. Rather, it was that 'their net proceeds were minimal, which meant that they could only keep their farmsteads going through the utmost exertions' (Harnisch 1986, 47). There simply did not exist a class of rich peasants with the size, strength and resources of the English rich peasantry. There was, certainly, peasant resistance – clear class struggle – centring upon the need for the abolition of feudal dues especially labour services: increasing in intensity, and with a growing determination and resilience, in the second half of the eighteenth century.[9] The most prominent proponents of such struggle were rich and substantial peasants – the full- and half-*Bauern* – who were particularly irked by labour services, and from whom the Junkers were seeking to extract even heavier services. This now posed a threat to the feudal order.

The agrarian traverse in Prussia in the nineteenth century: capitalism from above[10]

By the end of the eighteenth century, increasingly aware of peasant pressure and fearful of revolutionary activity such as had erupted in France, 'an influential group among the Prussian leaders of state now realised that [abolition of feudal dues] ... had become an urgent necessity' (Harnisch 1986, 64). This led to reforms in 1799, on the royal demesnes, and that action was surely induced, in part, by peasant pressure – class struggle – over previous decades. Then, in 1807, in the wake of crushing defeat by Napoleon's armies in 1806, feudalism was abolished, with the emancipation of the serfs – some 350 years after its demise in England. There followed a period of transition. If the sixteenth century was the era of the transition to capitalist agriculture in England, the nineteenth century was so for Prussia. But in Prussia it was the erstwhile feudal landlords who became capitalist farmers. It was 'capitalism from above'. It was the first such agrarian traverse in history.

---

[9]On 'Peasant Struggle: Its Nature and Implications' see Byres (1996, 90–6).
[10]I consider the Prussian transition in detail in my book (Byres 1996, ch. 4, 104–58).

As Harnisch points out, 'the landowning families stayed in full possession of their large estates as well as their often extensive forests' (Harnisch 1986, 37), just as the English landlord class retained ownership of their land. The Junkers took over and enclosed, in the teeth of opposition, the land of both poor and rich peasants – the land they worked and common land. But a change took place in the composition of Prussian landowners. By the late eighteenth century, the Prussian nobility had accumulated much debt. In the 1820s there was a severe depression and the market for grains virtually collapsed. It was then that decisive changes took place. A large number of noble estates were sold to commoners. The new estate owners, equipped with fresh capital, 'led the way in the transformation of Prussian agriculture' (Berdhal 1988, 282). By the 1850s, the proportion of Junker estates owned by those without title – commoners – had tripled or quadrupled since 1807, and in 1856 the figure was 56 percent. But it was not only the new owners who took to capitalist farming. The old were similarly receptive to new ways.

The Prussian landlord class retained ownership of most of the land, engrossing large quantities of peasant land. But, unlike English landlords, they ceased to be landlords. We note a particular characteristic of this dominant landlord class which was important to their transformation – or those of them who had been feudal landlords. Of critical importance was their character as takers of labour rent: which meant that they took decisions with respect to the form that production would take (which crops would be grown, etc.); and that, therefore, before 1807 they were not totally divorced from the process of production. Such a landlord class is more likely to be poised for possible transformation to hirers of wage labour than is one which appropriates surplus via kind or money rent. The transition to kind, and, even more, money rent, as was the case in England, constitutes an important change for a landlord class. Such a transition represents a severing of links with production. It is not inconceivable that such a landlord class might be transformed into a class of capitalist farmers: hiring wage labour, and appropriating surplus thus. But such a transformation is more likely where the landlord class has a direct relationship with labour (through labour rent) and has links with the process of production.

With the disappearance of obligatory labour services, Prussian landlords had lost their captive labour supply. Former serfs, indeed, were unwilling to work on Junker holdings. The relationship with the new forms of labour, however, was not immediately fully capitalist. It involved, initially and for some time, transitional forms. The Junkers, then, did not spring fully-caparisoned as capitalist farmers from the belly of feudalism. They would take time to slough off their feudal skins. At first, 'peasant labour services and the compulsory farm service of peasant youth on Junker farms were replaced by contractually hired farm servants and the cottager system ... the latter [involving] the exchange of labour for an allocation of the land' (Perkins 1984, 5). While farm servants were technically free, there were restrictions on their movement. This was followed by the system of confined labourers, hired on written short-term contracts. In each case, there was an absence of the money wage. Living standards were pitifully low. Ultimately, the Junkers were forced to employ day labourers, or 'free labourers' (*frei Arbeiter*), paid a money wage. It was wage labour, free in Marx's double sense, but not without the vestigial traces of feudalism. By 1871 the transition was complete. By then the Junkers were, in every useful sense, fully capitalist. It was a capitalism marked deeply by Prussia's immediate feudal past and the powerful subjugation of the peasantry which it entailed.

Some rich peasants did emerge – from the suggested sources, of free peasants and true-*bauern* – to be transformed, eventually, into capitalist farmers. But they were exceptional, 'a small minority of *Grossbauern* ("big peasants")' (Lenin 1962, 239) in close alliance with Junker capitalist farmers. The dominant class in the Prussian countryside, the 'masters of the countryside', were the latter.

Conclusion

In England, a feudal landlord class, rendered obsolete by class struggle waged by a united peasantry, was transformed into a progressive, capitalist landlord class, which let its land, which was enclosed, in leases at 'competitive' rents; and a rich peasantry (or, at least, its upper stratum), emerging from prior feudal differentiation, was metamorphosed into a class of capitalist tenant farmers able to pay 'competitive' rents and earn the average rate of profit. Integral to that outcome was the victory of the reconstituted landlord class over the peasantry during the sixteenth century, the era of transition. In France, a clearly unprogressive landlord class displayed no evidence of either transformation into a capitalist landlord class or a class of capitalist farmers; while the rich peasantry, the *laboureurs*, a potential class of capitalists, was constrained, in part, by its surplus being effectively appropriated by the state and by landlords. This continued until 1789, while capitalist transformation was further postponed, until the end of the nineteenth century, by a relentless struggle waged by poor and middle peasants. In Prussia, a powerful feudal landlord class, the Junkers, having crushed, in the sixteenth century, the free peasantry that had existed east of the Elbe, and enserfed it comprehensively, eventually, in the nineteenth century, was transformed into a class of capitalist farmers; while the possibility of capitalism from below, via a rich peasantry, was wholly pre-empted by the absence of a rich peasantry of sufficient strength. The Junkers' decisive victory over the 'free' peasantry in the sixteenth century, which ushered in Prussian feudalism, was repeated in the nineteenth century over the formerly enserfed peasantry, in an equally crushing conquest, which was the prelude to a capitalism from above.

I have elsewhere sought to consider something of the relevance of the historical experience for contemporary developing countries (Byres 2002). I do not wish to repeat that here or seek to draw any detailed conclusions. Clearly, however, one must proceed with the utmost caution.

I can do no better, perhaps, than quote the late Rodney Hilton, the outstanding Marxist historian of feudal England, who had a formidable knowledge, too, of medieval Europe, and a keen interest in the contemporary world (for an appreciation of Hilton, see Byres 2006a):

> historians and sociologists are engaged in comparative studies of peasant societies in different epochs. It would be very risky to transfer any generalisations about peasant societies of medieval Europe to any other time. For example, the capitalist farmers who were to be an important element in the history of early European capitalism emerged in a general environment of small-scale enterprise. What could the fate of peasant societies in the present world of almost world-wide commercial and industrial monopoly capitalism have in common with that of peasant societies of the late medieval world? Clearly, the tasks of leadership in contemporary peasant society have nothing in common with the tasks of the past, except in the recognition that conflict is part of existence and that nothing is gained without struggle. (Hilton 1973, 236)

To that I might add that when dealing with peasantries, in the past or the present, and however different the one is from the other, it is always important to consider the nature, the extent and the progress of the social differentiation that characterises such peasantries, and the nature of the landlord class.

References

Bercé, Y.-M. 1990. *History of peasant revolts*, trans. A. Whitmore. Ithaca, NY: Cornell University Press.

Berdhal, R.M. 1988. *The politics of the Prussian nobility. The development of a conservative ideology*. Princeton, NJ: Princeton University Press.

Bindoff, S.T. 1949. *Ket's rebellion 1549*. London: George Philip and Son for the Historical Association.

Bloch, M. 1961a. *Feudal society, volume 1, the growth of ties of dependence*, trans. L.A. Manyon. London: Routledge and Kegan Paul.

Bloch, M. 1961b. *Feudal society, volume 2, social classes and political organisation*, trans. L.A. Manyon. London: Routledge and Kegan Paul.

Bloch, M. 1966. *French rural history. An essay on its basic characteristics*, trans. J. Sondheimer. London: Routledge and Kegan Paul.

Blum, J. 1978. *The end of the old order in rural Europe*. Princeton, NJ: Princeton University Press.

Brenner, R. 1976. Agrarian class structure and economic development in pre-industrial Europe. *Past and Present*, February, 70, 30–75.

Brenner, R. 2001. The low countries in the transition to capitalism. *Journal of Agrarian Change*, 1(2), 169–242.

Byres, T.J. 1996. *Capitalism from above and capitalism from below. An essay in comparative political economy*. Basingstoke: Macmillan.

Byres, T.J. 2002. Paths of capitalist agrarian transition in the past and in the contemporary world. *In:* V.K. Ramachandran and M. Swaminathan, eds. *Agrarian studies. Essays on agrarian relations in less-developed countries*. New Delhi: Tulika Books, pp. 54–83.

Byres, T.J. 2006a. Rodney Hilton (1916–2002): in memoriam. *Journal of Agrarian Change*, 6(1), 1–16.

Byres, T.J. 2006b. Differentiation of the peasantry under feudalism and the transition to capitalism: in defence of Rodney Hilton. *Journal of Agrarian Change*, 6(1), 17–68.

Carsten, F.L. 1954. *The origins of Prussia*. London: Oxford University Press.

Carsten, F.L. 1985. *Essays in German history*. London: The Hambledon Press.

Carsten, F.L. 1989. *A history of the Prussian Junker*. Aldershot: Scolar Press.

Coopland, W.G. 1914. The abbey of Saint-Bertin and its neighbourhood, 900–1350. *In:* P. Vinogradoff, ed. *Oxford studies in social and legal history*, vol. IV. Oxford: Clarendon Press.

Cornwall, J. 1977. *Revolt of the peasantry 1549*. London: Routledge and Kegan Paul.

Cornwall, J. 1981. Kett's rebellion in context. *Past and Present*, November, 93, 160–64.

Engels, F. 1965. On the history of the Prussian peasantry. *In: The peasant war in Germany*. Moscow: Progress Publishers, pp. 154–65.

Engels, F. 1970. The peasant question in France and Germany. *In: Selected works of Karl Marx and Frederick Engels*, vol. 3. Moscow: Progress Publishers, pp. 457–76.

Faith, R. 1997. *The English peasantry and the growth of lordship*. London: Leicester University Press.

Fletcher, A. and D. MacCulloch. 2004. *Tudor rebellions*, fifth edition. Harlow: Pearson Education (Longman).

Foisil, M. 1970. *La révolte des nu-pieds et les révoltes normandes de 1639*. Paris.

Forster, R. and O. Ranum, eds. 1977. *Rural society in France. Selections from the annales. Economies, sociétés, civilisations*. Baltimore, MD: The Johns Hopkins University Press.

Gerschenkron, A. 1966. *Bread and democracy in Germany*, new edition. New York, NY: Howard Fertig.

Goubert, P. 1956. The French peasantry of the seventeenth century: a regional example. *Past and Present*, November, 10, 55–77.

48   *Terence J. Byres*

Goubert, P. 1986. *The French peasantry in the seventeenth century*, trans. I. Patterson. Cambridge: Cambridge University Press.

Hagen, W. 1985. How mighty the Junkers? Peasant rents and seigneurial profits in sixteenth-century Brandenburg. *Past and Present*, August, 108, 80–116.

Harnisch, H. 1986. Peasants and markets: the background to the agrarian reforms in feudal Prussia east of the Elbe, 1760–1807. *In:* R. Evans and W.R. Lee, eds. *The German peasantry*. London: Croom Helm, pp. 37–70.

Hilton, R. 1949. Peasant movements in England before 1381. *Economic History Review*, second series, ii, 117–36.

Hilton, R. 1957. A study in the pre-history of English enclosure in the fifteenth century. *In: Studi in onore di Armando Sapori*, 2 volumes, vol. 1. Milan, pp. 674–85.

Hilton, R. 1966. *A medieval society. The west Midlands at the end of the thirteenth century*. London: Weidenfeld and Nicolson.

Hilton, R. 1973. *Bond men made free. Medieval peasant movements and the English rising of 1381*. London: Temple Smith.

Hilton, R. 1974. Medieval peasants – any lessons?. *The Journal of Peasant Studies*, 1(2), 207–19.

Hilton, R. 1975. *The English peasantry in the later middle ages*. London: Oxford University Press.

Hilton, R. 1978. Reasons for inequality among medieval peasants. *The Journal of Peasant Studies*, 5(3), 271–84.

Hilton, R. 1980. Feodalité and seigneurie in France and England. *In:* D.F. Johnson, F. Bedarida and F. Crouzet, eds. *Britain and France: ten centuries*. Folkestone: Wm. Dawson & Son, pp. 37–50.

Hilton, R. 1990. *Class conflict and the crisis of feudalism*. London: Verso.

Hilton, R. and H. Fagan. 1950. *The English rising of 1381*. London: Lawrence and Wishart.

Hobsbawm, E.J. and G. Rude. 1973. *Captain swing*. Harmondsworth: Penguin University Books.

Kosminsky, E.A. 1956. *Studies in the agrarian history of England in the thirteenth century*, trans. R. Kisch. Oxford: Basil Blackwell.

Ladurie, E.L. 1987. *The French peasantry 1450–1660*, trans. A. Sheridan. Aldershot: Scolar Press.

Land, S.K. 1977. *Kett's rebellion. The Norfolk rising of 1549*. Ipswich: The Boydell Press.

Lefebvre, G. 1977. The place of the revolution in the agrarian history of France. *In:* R. Forster and O. Ranum, eds. *Rural society in France. Selections from the annales. Economies, sociétés, civilisations*. Baltimore, MD: The John Hopkins University Press, pp. 31–49.

Lenin, V.I. 1962. The agrarian programme of social-democracy in the first Russian revolution, 1905–1907. *Collected works*, vol. 13. Moscow: Foreign Languages Publishing House.

Lenin, V.I. 1964. The development of capitalism in Russia. *Collected works*, vol. 3. Moscow: Progress Publishers.

Leyser, K.J., *et al.* 1978. Germany, history of. *In: Encyclopaedia Britannica*, fifteenth edition. Vol. 8. Chicago, IL: Encyclopaedia Britannica.

MacCulloch, D. 1979. Kett's rebellion in context. *Past and Present*, August, 84, 36–59.

MacCulloch, D. 1981. A rejoinder. *Past and Present*, November, 93, 165–73.

Meek, R.L. 1962. *The economics of physiocracy. Essays and translations*. Cambridge, MA: Harvard University Press.

Miller, E. and J. Hatcher. 1978. *Medieval England – rural society and economic change 1086–1348*. London: Longman.

Mollat, M. and P. Wolff. 1973. *Popular revolutions of the late middle ages*. London: Allen and Unwin.

Mousnier, R. 1971. *Peasant uprisings in seventeenth century France, Russia and China*, trans. B. Pearce. London: Allen and Unwin.

Perkins, J.A. 1984. The German agricultural worker, 1815–1914. *The Journal of Peasant Studies*, 11(3), 3–27.

Perrin, C.E. 1945. Observations sur le manse dans la région parisienne au début du IXe siècle. *Annales d'Histoire Sociale*.

Porchnev, B. 1963. *Les soulèvements populaires en France de 1623 à 1648*. Paris.

Postan, M.M., ed. 1966a. *The Cambridge economic history of Europe. Volume 1. The agrarian life of the middle ages*, second edition. Cambridge: Cambridge University Press.

Table 1. Lenin and Chayanov: some 'markers'.

'Markers'	V.I. Lenin	A.V. Chayanov
Dates	1870–1924	1888–1937
Career	Professional revolutionary and Marxist intellectual	Agricultural economist, applied researcher and policy analyst
Key works (on agrarian issues)	• *Development of Capitalism in Russia* (1899) • Other work on capitalism and agriculture, and political strategy in relation to Russian peasantry, especially between Revolutions of 1905 and 1917 • (little between 1917–1924: Revolution, war communism 1918–1921, New Economic Policy 1921 on)	• Many empirical and then also theoretical studies from 1909, published in Russian and later German, culminating in *On the Theory of Non-capitalist Economic Systems* (Germany 1924; English 1966), *Peasant Farm Organisation* (Moscow 1925; English 1966), *The Theory of Peasant Co-operatives* (Moscow 1927; English 1991)
Key ideas	• 'Prussian' and 'American' paths of development of capitalism in agriculture, the latter via class differentiation of the peasantry into agrarian capital and labour	• 'Theory of peasant economy' (as species of genus 'family economy') • Centred on household reproduction (demographic cycle), generating demographic vs class differentiation • 'Self-exploitation' of peasant households
Model of development	• Capitalist transition (as above): changes in social relations of production as condition of development of productive forces (with growing economies of scale) • Transition to socialism in the countryside, i.e. large-scale 'scientific' farming; need to gradually 'remould the small farmer' towards this end • (Contributions of agriculture to industrialisation)	• 'the development of agriculture on the basis of cooperative peasant households, a peasantry organised cooperatively as an independent class and technically superior to all other forms of agricultural organisation' • Cooperation to achieve economies of scale suited to different purposes/activities in different branches of production ('differential optima')
Legacies	• Wide availability of his writings, distributed by the USSR and the Comintern in many languages. • *The Development of Capitalism in Russia* a key text for subsequent agrarian political economy	• 'treble death' (see Shanin 2009), with work known to wider international audience only from 1966 taken up by variants of neo-populist analysis and policies to promote small-farm(er) development

and also informed criticism of later currents of agrarian populism or neo-populism often viewed as the legacy of Chayanov by some of their champions and (especially?) their critics.

Shanin's question of what 'alternatives' those critics were able to offer points to a crucial disjuncture: the definitive opposition of his essay is *not* Chayanov vs Lenin but Chayanov vs Stalin as the principal architect of the forced dispossession and collectivisation of the Russian peasantry from 1929. In effect, it was Stalin's abrupt *action* that cast its long shadow over subsequent history – as powerful an example as any of the force of events in this context, of twists and turns that shape the lives and deaths of texts. Stalin's action rendered Lenin and Chayanov (and many others) irrelevant to what actually happened in the moment of 1929. Whether their ideas are later brought back to life, and when, why and how – for what purposes and with what effects – is considered for Chayanov in Shanin's bravura essay, and provides the backdrop to what follows.

Looking back I: Russia

I begin by noting some other significant disjunctures, or at least asymmetries, in how Lenin and Chayanov may be contrasted.

Careers, intellectual and practical

(1) Lenin was a Marxist intellectual and professional revolutionary. His central concern was to analyse the conditions and prospects of political revolution in Russia, in order to inform strategy and tactics, programmes and positions, for the Bolshevik party. This entailed intense (often polemical) engagement with other currents of Russian Marxism and radicalism, and by extension with those of communist and socialist parties elsewhere. His work sought to connect socioeconomic analysis (the development of capitalism) with a political sociology of class forces and interests, and how they were manifested in particular conjunctures, events, parties and indeed personalities, during a period of massive upheaval (the 1905 revolution, the Stolypin reforms, the First World War, the circumstances, course and aftermath of the 1917 revolution).

Chayanov was a professional agricultural economist who became the leading figure in the Russian Organisation and Production School. He did not produce a political sociology of the peasantry or of policy making, and might best be regarded as a kind of scholar-technocrat of exceptional intellectual culture and originality, commitment and immersion in practical activity. As Stalin's collectivisation was launched, Chayanov was dismissed from his post as Director of the Research Institute for Agricultural Economics, arrested and eventually executed in 1937.

(2) Lenin's principal work in agrarian political economy came early in his career. His study of *The Development of Capitalism in Russia* (1899) was the only comprehensive account of the development of capitalism in a backward country, peripheral to the centres of industrial capitalism, in the corpus of classic Marxism. Subsequently he wrote a sequence of important articles that demonstrate his continuing interest in agrarian questions, especially (but not exclusively) in relation to the need for a Bolshevik *political* analysis of, and strategy towards, the Russian peasantry after the 1905 revolution and the Stolypin reforms that followed it (Kingston-Mann 1980). However, he was able to contribute little on the peasantry and agricultural strategy in the few years between the end of the revolutionary war and his death in January 1924.

Chayanov published a massive corpus of empirical and theoretical studies from his precocious early writing to his two most elaborated works (written when he was still in his thirties, and works in progress, as he emphasised): *Peasant Farm Organisation* and *The Theory of Peasant Co-operatives* published in Moscow in 1925 and 1927 respectively, that is, after Chayanov's return to Russia to resume his work there in the period of the New Economic Policy (NEP) (and after Lenin's death).

(3) Lenin's iconic status after 1917 as the greatest revolutionary of his time, and the availability of his writings, distributed by the USSR and the Comintern in many languages, meant that his ideas were widely influential, especially among the emergent and important communist parties of Asia. For the project of agrarian political economy pioneered by *JPS*, *The Development of Capitalism in Russia* was a key text, and especially its analysis of peasant class differentiation.

Chayanov's ideas, of course, were largely lost in the years of Stalinism, as Shanin's account shows so vividly, and became available to Anglophone scholars only with the English translation in 1966 of *Peasant Farm Organisation* (together with the key essay *On the Theory of Non-Capitalist Economic Systems*), a founding text of the interest in peasant studies that gathered from the 1960s. Following long delays, *The Theory of Peasant Co-operatives* (which Shanin considers Chayanov's more important work) was published in English translation in 1991, and remains much less widely known than *Peasant Farm Organisation*. Two special issues of *JPS* made their own valuable contributions to the recovery and dissemination of Chayanov's work. His novella *The Journey of my Brother Alexei to the Land of Peasant Utopia* was the centrepiece of R.E.F. Smith's *The Russian Peasant 1920 and 1984* (1976). And as new archival sources became available after the end of the Soviet Union, Frank Bourgholtzer (1999) produced another historical treasure in *Aleksandr Chayanov and Russian Berlin*, a collection of Chayanov's letters from his stay in Berlin and more briefly England in 1922–23, with some additional material and a biographical essay on this hitherto largely obscure moment of Chayanov's life.

I turn next to other differences (and an unremarked similarity) between Lenin and Chayanov.

Key ideas

The Development of Capitalism in Russia was theoretically framed by Lenin's reading of *Capital* in the context of intense polemic against the economic and political arguments of the Narodniks of the time (Lenin 1967a, Ch. 1), and drew on and analysed a comprehensive range of up-to-date empirical material. Its second chapter, with its extensive tables of *zemstvo* statistics, argued the case for class formation among the Russian peasantry as both expression and driver of the development of capitalism in the countryside.[1] Lenin provided a model of three basic peasant classes – rich, middle and poor peasants – which anticipated their (eventual) transformation into classes of agrarian capital (rich peasants) and proletarian labour (poor peasants), with a minority of middle peasants joining the ranks of the former and the majority joining the ranks of the latter. It is important to distinguish peasant

---

[1]Chapters 3 and 4 considered respectively 'The Landowners' Transition from Corvée to Capitalist Economy' and 'The Growth of Commercial Agriculture'. The *zemstvo* statistics on farming were produced by organs of provincial (*guberniya*) and district (*uezd*) government established after the abolition of serfdom in Russia in 1861. 'In the decades from the 1880s onward, Russia's leading economists, statisticians, sociologists and agricultural experts assessed, analysed and fought over the materials furnished by the successive *zemstvo* inquiries. Their articles and books provided the richest analytical literature we have on the on the peasant economy of any country in the period since the Industrial Revolution.' (Thorner 1966, xii).

class differentiation as a *tendency* and as an observable *trend* in any given place and time. Lenin used Marx's theoretical concepts and method to derive the fundamental tendencies of a social dynamic from available empirical evidence, an approach that he sometimes termed (necessary) 'exaggeration' and that required considerable dialectical deftness, signalled in many of his observations, for example, that 'infinitely diverse combinations of the elements of capitalist evolution are possible'.[2] Whether Lenin got the trend of peasant differentiation right from the *zemstvo* statistics he drew on was, of course, contested by his Narodnik opponents and by Chayanov (below), as well as questioned by later scholars (e.g., and from different positions, Banaji 1976a, Kingston-Mann 1980, Lehmann 1982).

The Development of Capitalism in Russia was Lenin's principal contribution, following from which – together with his studies of Germany and the USA – he formulated his conception of paths of transition to capitalism in farming 'from above', the 'Prussian path' of the 'internal metamorphosis' of landed property, and 'from below', the 'American path' of class differentiation in the absence of (pre-capitalist) landed property, and characterised them as respectively reactionary and progressive in terms of their social and political effects.[3]

By contrast with Lenin, Chayanov argued that indices of apparent inequality among Russian peasants – in particular size of land farmed and stock of instruments of labour (draft animals and equipment) – were not due primarily to class formation but reflected the locations of households in the demographic cycle, traced in the 'labour–consumer balance' or ratio of producers (working adults) to consumers (working adults plus dependants: children and the old) at different moments in the recurrent process of generational reproduction. This links with another fundamental element of Chayanov's 'theory of peasant economy': that the aim – or 'motivation' (a term he used) – of peasant households is to meet the needs of (simple) reproduction while minimising 'drudgery' (of labour). This can have both virtuous and vicious effects, as it were. On one hand, the mode of economic calculation of peasant households distinguishes them from the conventional capitalist enterprise which costs all 'factors of production' in its drive for profit maximisation and accumulation. Indeed, for Chayanov 'peasant economy' was an instance of a broader and generic 'family economy' centred on the organisation of 'family *labour*'.[4] On the other hand, the imperatives of reproduction in family labour enterprises mean that labour costs (drudgery) are discounted in adverse conditions, generating peasant 'self-exploitation'. In effect, peasants tend to farm more intensively than capitalists, albeit at lower levels of labour productivity; similarly they are often constrained to buy or rent land at higher prices, and to sell their product at lower prices, than capitalist farmers are prepared to do. Chayanov

---

[2] From the Preface to the second edition of *The Development of Capitalism in Russia* in 1907, and continuing: 'only hopeless pedants could set about solving the peculiar and complex problems arising merely by quoting this or that opinion of Marx about a different historical epoch' (Lenin 1967a, 33).

[3] On Germany the key work was Kautsky's *Die Agrarfrage*, also published in 1899, which Lenin (1967a, 27) considered 'after Vol. III of *Capital*, the most noteworthy contribution to recent economic literature'. A full English version of Kautsky appeared only in 1988, although Jairus Banaji (1976b) had earlier published an influential translation of extracts from Kautsky.

[4] See *On the Theory of Non-capitalist Economic Systems* (in Chayanov 1966) – and a 'non-capitalist economic system' that also 'resists' capitalism as many populists claim, or simply assume?

devoted a great deal of attention to the integration of peasant households in capitalist commodity markets, touching on a variety of complex factors and processes, including issues of the capitalisation of peasant farming and the sale (and purchase) of labour power. Much of what he said continues to be of great interest and utility, even if he failed to theorise the social relations of such processes (see further below).[5]

Models of development

As implied above, there is a sharp disjuncture between Lenin's profound analyses of the development of capitalism in agriculture, and more generally, and what he may have considered an appropriate model of development in the novel and testing circumstances of constructing socialism in a 'backward' country where the transition to capitalism was incomplete and the economy ravaged by years of war and foreign invasion. Lenin's brief statements between 1917 and 1923 were addressed to practical issues of immediate urgency, above all the supply of grain during 'war communism' (that is, no kind of communism at all) and in the subsequent shift to the NEP. Of most interest are his short but strategic contributions to the Tenth Congress of the Bolshevik Party (RCP(B)) in March 1921 that introduced the NEP. Its centrepiece was the substitution of a (lower) tax in kind on peasant farmers for the harsh requisitioning of ostensibly 'surplus' grain under the 'war communism' of 1918–21. In his addresses to the Congress on this vital matter, Lenin concluded that in the conditions then prevailing 'It is our duty to do all we can to encourage small farming' (Lenin 1967c, 238) while 'it will take generations to remould the small farmer' (Lenin 1967b, 216).

Chayanov's model of development was encapsulated in the definition of neo-Narodism he was ordered to provide by his interrogator from OGPU (Joint State Political Directorate, in effect the secret police apparatus), after his arrest in 1930: 'the development of agriculture on the basis of cooperative peasant households, a peasantry organised cooperatively as an independent class and technically superior to all other forms of agricultural organisation' (Bourgholtzer 1999, 16).[6] We can read the former as meaning 'independent' of (i) predatory landed property and its exactions, (ii) capital and its imperative of accumulation, and (iii) state socialism and collectivisation as the 'proletarian line' in agriculture – respectively the burden of the past, the pressures of the present, and the threat of the future. 'Technical superiority' refers to the optimal scale of a farm that can be managed and worked by *family* labour (which will vary with the technologies at its disposal), informed by the inimitable knowledge of its natural environment that it accumulates.

The peasant household/farm economy thus remains the basic cell of Chayanov's model of agricultural development but it requires cooperation to achieve its technical superiority. His intense interest and practical engagement in cooperatives in Russia[7] – for input supply, machinery, technical services and, of strategic

---

[5]Of less interest is his use of marginalist analysis to model decision-making in the peasant household. Chayanov's marginalist economics is considered by Sivakumar (2001).

[6]Eerily, the name of the interrogator, chief of the Secret Department of the OGPU, was Yakov Agranov.

[7]And elsewhere; his first publication (1909) was on cooperatives in Italian agriculture (Chayanov 1966, 279). Viktor Danilov (1991) provides a useful sketch of the intense Russian interest in cooperatives before and during Chayanov's time.

importance, credit and processing/marketing – focussed on how they can satisfy the 'differential optima' of various organisational economies of scale appropriate to different activities/functions in different branches of agricultural production inserted in commodity markets (including those of 'state capitalism' under NEP (Chayanov 1991)). Effective cooperation in turn requires state support and regulation that replaces or substitutes for the kinds of 'vertical' concentration/integration characteristic of capitalist commodity markets and organised by large(r)-scale capitals, and thereby excises the 'self-exploitation' of peasant farmers generated by such adverse circumstances as shortages of land and credit and the market power of large (often locally monopolistic) merchants and processors. In effect, Chayanov's work on cooperatives also served to describe the agricultural sector as a technically and functionally differentiated whole, and by extension its place within 'national economy', building on his model of the farm household/enterprise as the individual cell of agrarian economy. How successfully it did so remains another matter.[8]

And the similarity: Chayanov was no less committed than Lenin to technical progress that raises the productivity of labour in (peasant) farming, including the use of machinery and chemicals, and with it the incomes and security of farm households. In short, Chayanov was no less a moderniser, which distinguishes him radically from many agrarian populists, on whom more below. Before that, it is necessary to return to the central theoretical issue of peasant class differentiation, and incidentally a historical matter: whether Lenin and Chayanov actually 'debated' with each other.

Peasant differentiation: dynamics, extent, implications

This marks the fundamental difference between Lenin and Chayanov. For Lenin differentiation of the peasantry (as of other petty commodity production/producers) was intrinsic and central to the development of capitalism and the class dynamic of its 'laws of motion' theorised, with unique power, by Marx. In emphasising differentiation, Lenin added significantly to Marx's model of agrarian transition, based as it was principally on British experience and what can be called an 'enclosure' model of primitive accumulation.

Chayanov contested any such strong argument about class differentiation of the Russian peasantry on two grounds: theoretically that the logic of peasant economy (simple reproduction) excludes the capitalist imperative of accumulation for its own sake (expanded reproduction), and empirically according to the *zemstvo* reports and surveys he and his teams of researchers conducted.[9] In effect then, Chayanov could not provide a theoretical explanation of any dynamic or tendency of class differentiation in the countryside, while having to recognise its existence to at least some degree (see below).

These observations also point to issues of the *extent* (as well as forms) of class differentiation in different areas of the Russian countryside at different times, and their implications. Thus Lenin, introducing the 'tax in kind' of the NEP (above),

---

[8]Sivakumar (2001, 38) suggests that 'Chayanov and his colleagues had neither a sound theory of value nor a sound macroeconomic theory; eventually this became the Achilles heel of the Organisation and Production School'.

[9]The accumulation of capital is different from the importance of investment both to meet household needs at improved levels of income/consumption, and to develop effective cooperatives.

suggested that the extent of peasant class differentiation had diminished by 1921 as a result of land redistribution: 'Everything has become more equable, the peasantry in general has acquired the status of the middle peasant. Can we satisfy this middle peasantry as such, with its economic peculiarities ... ?' (Lenin 1967b, 216).[10]

Whether peasant differentiation then increased again as a result of the market liberalisation of NEP – as Lenin had predicted it would (1967b, 225) – and if so its consequences, was a preoccupation of the Agrarian Marxists who followed in Lenin's footsteps during the 1920s, while developing a more nuanced methodology for identifying and measuring class differences (Cox 1979, Cox and Littlejohn 1984). Chayanov had touched on class differentiation briefly towards the end of *Peasant Farm Organisation* (1966, Ch. 7), and *The Theory of Peasant Cooperatives* includes an analysis of a survey conducted by his research institute in 1925. He distinguished six 'basic types of peasant household', while insisting that the vast majority of Russian farmers are 'middle peasants' of types three, four and five who form the social base for developing cooperation (Chayanov 1991, Ch. 2). In these instances, we can assume that Chayanov is 'debating' or contesting Lenin's approach and findings (without naming him) and those of the contemporary Agrarian Marxists.

Once again, all this was swept aside by the decision to launch collectivisation in 1929, justified by the assertion that it was the rise of '*kulak* power' under NEP that produced the grain supply crisis of the late 1920s (Lewin 1968). Stalin's collectivisation generated many difficult questions at the time and since, the object of large literatures and debates. For example, was it in some sense the logical outcome of Leninist or Bolshevik or generically Marxist conceptions of the objective nature of progress *qua* development of the productive forces? Of their justifications of the subjective will and force necessary to develop the productive forces in the historically unprecedented, and unanticipated, circumstances of socialist revolution in a backward country?[11] Was its 'nastiness', in terms of both the massive violence and suffering of its execution and its fatal consequences for remaining hopes of socialist politics in the USSR, nonetheless 'necessary': whether to resolve the immediate grain supply crisis of 1927–29, to establish the conditions of a productively superior large-scale agriculture in the longer term, and/or to contribute to the accumulation fund for industrialisation? And did it achieve those objectives?

This, then, is to return to the central disjuncture noted earlier, and highlighted by Shanin's essay in this collection. On one hand, there is Lenin's work on the development of capitalism in agriculture and the absence of any similarly rich consideration by him (or any other major Marxist thinker?) of basic questions of socialist construction, and especially in countries where 'the peasant is a very essential factor of the population, production and political power', as Engels (1951) remarked of France and Germany in the 1890s. On the other hand, there is Chayanov's account of the economic life of Russia's peasants, rejected by subsequent materialist political economy, and his 'alternative' model of

---

[10]Does 'economic peculiarities' contain an indirect reference to the ideas of Chayanov and his Organisation and Production School of agricultural economics? Lenin's principal argument for the tax in kind was to incentivise middle peasant production, primarily through liberalising exchange, and he also emphasised the importance of local circuits of trade and a role for (consumer) cooperatives.
[11]Shanin (1986, 11) remarks that collectivisation 'was not a natural deduction from Marxism or from Lenin but a fairly arbitrary result of the 1926–28 failure of rural policies and of interparty factional struggle'.

development – peasants/small farmers + cooperatives + a supportive state – that was either ignored or subsumed into the more generalised arena of dispute between Marxism and agrarian populism. In short, there is a kind of vacuum that exerts its own force in the later career of agrarian political economy, that I turn to next.

Looking back II: legacies

The founding of *JPS* manifested the intense intellectual and political currents and concerns of its own historical moment, characterised by continuing struggles against imperialism in which peasants were major actors (notably the Vietnamese struggle for national liberation); the prospects for economic and social development in the recently independent, and mostly agrarian, former colonies of Asia and Africa, as well as in Latin America; and the ways in which those prospects could be explored through investigating practices, experiences and theories of socialist transformation of the countryside. In *JPS* these concerns were typically (but not exclusively) informed by a materialist political economy – with all its variants and internal debates – and by a strongly historical approach, applied to agrarian social formations before capitalism; transitions to capitalism in its original heartlands and subsequently; agrarian change in colonial conditions and in the revolutionary circumstances of the USSR and China, Vietnam and Cuba; processes of agrarian change in the independent countries of the Third World in the contexts of their various projects of 'national development'; and then the unravelling of those projects (as of actually existing state socialisms) in a new period of 'globalisation' and its neo-liberal hegemony.[12]

How did the ghosts of the Russian debates and experiences haunt agrarian studies from the 1960s, especially in the pages of *JPS*? As the agenda of *JPS* developed in its early years its commitment to agrarian political economy became clearer, which meant a closer connection with the approach exemplified by Lenin in *The Development of Capitalism in Russia* than with that of Chayanov. This was not an *a priori* nor dogmatic intent; indeed *JPS* published two special issues presenting some of Chayanov's writing for the first time in English (above), as well as articles sympathetic to Chayanov. Nonetheless, 'Lenin' here stands as shorthand for the interest in and commitment to more generally Marxist investigation and debate in the 1960s and 1970s, not least of the highly topical concerns noted in the previous paragraph. Moreover, Lenin's approach to the development of capitalism in agriculture – like that of Marx, Engels, Kautsky, Luxemburg, Bukharin, Preobrazhensky, Kritsman, Gramsci, and others – extended beyond the logics and paths, problems and prospects, of peasant farming. It encompassed different types of capitalist agriculture – their origins, paths of development, modalities of accumulation, labour regimes, location in social divisions of labour, relations with other forms of capital and with the state – while Chayanov restricted himself to the capitalist farm *qua* commercial enterprise familiar from any standard economics textbook (the model against which the distinctive principles of the peasant farm household were defined).

---

[12]I draw here on elements of Bernstein and Byres (2001) which considers the conjuncture and concerns of the founding moment of *The Journal of Peasant Studies* and provides a thematic survey of the work it published from 1973–2000.

The development of capitalism

First, there was considerable interest in the Russian experience of agrarian change from the 1890s to the 1920s analysed (at different moments) by Lenin and Chayanov as well as by Kritsman and the Agrarian Marxists.[13] Chayanov's work was revisited in four notable articles in *JPS* by Mark Harrison (1975, 1977a, 1977b, 1979), which also engaged with other key texts, notably Shanin's *The Awkward Class* (1972), as did Cox (1979). Shanin had counterposed a model of social mobility in peasant Russia to both Lenin's analysis of class differentiation and Chayanov's demographic differentiation which he termed 'biological determinism' (1972, 101–9).[14] Harrison and Cox supported the argument for class differentiation, albeit with qualifications that require more subtle methodologies and sensitivity to the nuances of rural dynamics. Kritsman was an innovator in both these respects, according to Cox, enabling him to identify counter-tendencies to class differentiation, while Harrison (1977a) proposed that the particular effects of patriarchy in patterns of commodification and class differentiation in Russia contribute to explaining practices of household partition more fruitfully than Shanin's argument from social mobility.

Beyond this, there was a plethora of studies of agrarian change *qua* the development of capitalism that both drew on and extended the range of the Russian debates historically, geographically, and in terms of themes, concepts and methods. Historically, the new peasant studies opened up explorations of agrarian change in colonial conditions which also extended the geographical range beyond Europe and Russia, the principal referents for Lenin, as for Marx, Kautsky and others, including Chayanov. Seminal texts included Barrington Moore's magisterial comparison (1966) of the agrarian conditions of transitions to modernity in England and France, the USA, Japan, China and India (further informed by his knowledge of the histories of Russia/Germany and Russia/USSR); Eric Wolf's *Peasant Wars of the Twentieth Century* (1969) with its case studies of Mexico, Russia, China, Vietnam, Algeria and Cuba; and James Scott's study of peasant rebellion in colonial southeast Asia, *The Moral Economy of the Peasant* (1976).

Thematically, a central question from the start of peasant studies, and one pursued extensively in *JPS*, was whether 'peasants' constitute a distinct and coherent object of study, whether as economic form or mode of production, 'class', type of society, 'community' and/or culture, or some other entity, that can be usefully identified and analysed across different historical circumstances and periods. Chayanov provided an economic model which others might then seek to elaborate in social and cultural terms, as in Scott's formulation (1976) of a 'subsistence ethic' inspired by, or grafted on to, Chayanov's economics.[15]

---

[13]In another notable special issue of *JPS*, Terry Cox and Gary Littlejohn (1984) provided extensive essays on the Agrarian Marxists of the 1920s together with an abridged translation by Littlejohn of a key work by their leading figure L.N. Kritsman on *Class Stratification of the Soviet Countryside*.

[14]Shanin has been a key figure in the revival of interest in Chayanov's work with a preference for *The Theory of Peasant Cooperatives* over *Peasant Farm Organisation*, more evidently the economists' Chayanov, as it were. Together with Shanin (1972), another key text of the 'Russian influence' on the founding moment of peasant studies was Moshe Lewin's *Russian Peasants and Soviet Power* (1968).

[15]As well as inspired by E.P. Thompson's celebrated essay (1971) although Thompson's militants of moral economy were not peasants but eighteenth-century English plebeians rioting against rising bread prices.

There are several major and connected problems attached to notions of a generic, and trans-historical, figure of 'the peasant' in the line of Chayanov.[16] First, Chayanov lacked any significant theorisation of *social relations*. For the basic cell of the peasant household, the key relation is the internal producer–consumer ratio and its shifts over the cycle of household (demographic) reproduction. While his work on the insertion of peasant households in wider economic systems and circuits, especially those of capitalist commodity exchange, contains rich empirical analysis and many fruitful insights, these are *not theorised*. What they 'do' for Chayanov is to delineate the conditions *external* to peasant households and that impose greater or lesser constraints on their pursuit of (simple) reproduction ('subsistence' in Scott's term), which points to the second problem.

This is that the theory of the peasant household is fundamentally one of its 'internal' logic (and organisational 'machine' as Chayanov often put it) which gives it its trans-historical character: it applies to a variety of historical circumstances ('historical epochs', note 2 above) in which a variety of 'external' forces bear down on peasants – landlords, merchant capital, and states (representing the exactions of rent, commercial profit, and taxation respectively) – thereby intensifying the response of 'self-exploitation' to meet the pressures of simple reproduction.[17] Is it possible to maintain the utility of such a trans-historical construct without slipping into the ahistorical? As sympathetic a commentator as Daniel Thorner remarked that 'Although it encompassed a very wide range of possibilities, Chayanov's theory of peasant farming remained essentially a static one' (1966, xxii). This is an acute observation. Chayanov does indeed specify 'a very wide range of possibilities' concerning 'external' conditions and how peasants respond to them, but their responses represent only so many 'adaptations' derived from the unchanging logic of household reproduction and its mode of calculation.

A third kind of problem follows: if many 'peasant' households are driven by (deepening) insertion in commodity relations to reproduce themselves increasingly through non-farming activities, at what point do they remain 'peasants' in any meaningful sense? Chayanov (1991, 27) acknowledges the salience of this question in his sixth type of household 'whose main income is derived from the sale of their labour-power ... (but which) nonetheless have their own farming activities, usually on a very small scale and nearly always for their own consumption'.[18] And he was clear that agricultural cooperatives are not for rural households of this type (a small minority at the time, in his view).

The question indicated of the meaning of 'peasant-ness' or 'peasant-hood', and appropriate indicators to specify (and perhaps measure) it, is not intended to invite a

---

[16]From this point, due to limits of space, my brief summaries become increasingly anonymous as well as selective. Interested readers who want to follow up references to work in *JPS* on the issues and debates sketched can refer to the survey by Bernstein and Byres (2001), and the thematic index of *JPS* from 1973–93 (Bernstein *et al.* 1994). There is also, in a polemical and personalised register, Tom Brass's survey (2005) of the 'third decade' of *JPS*, and a later thematic index for 1993–2003 (*The Journal of Peasant Studies* 2005).

[17]Unusually (exceptionally?) among Marxists, Banaji (1976a) commended the very abstraction of Chayanov's theoretical model of the family labour farm as providing something lacking in Lenin.

[18]Almost an echo of Lenin's remarks on 'allotment-holding wage-workers', and his warning against 'too stereotyped an understanding of the theoretical proposition that capitalism requires the free, landless worker' (1967a, 181).

resurgence of the syndrome of determining and attaching the 'right' class (or other) labels to categories of social agents. Indeed, many of the contributions to *JPS* (and elsewhere) on peasantries and their differentiation in the development of capitalism illuminated the great fluidity, as well as range of variation, of their forms in different places at different times. The Chayanovian take is that such fluidity and variation manifest only the 'very wide range of possibilities' that attach to the basic, and unchanging, cell of the peasant household; the Leninist take is that they represent how, in different sets of historical conditions, the dynamic of the development of capitalist commodity relations shapes the conditions, practices and fates of petty producers – and indeed is *internalised* within their enterprises and circuits of reproduction.

The materialist inspiration of much of the political economy featured in *JPS* thus illuminated both general analytical concepts and specific historical experiences across the range of forms evident (and less evident) in the development of agrarian capitalism. Lenin's perspective and approach were assessed, reassessed, and advanced, not least in the elaboration and testing of increasingly sophisticated methods for investigating peasant class differentiation, informed by awareness of the difference between tendencies and trends noted earlier, itself an antidote to simplistic understandings of Lenin's schema by friend and foe alike.[19] Moreover, that schema had to be recast in key respects by the impact of feminism, and the gendered nature of peasant households (absent in Lenin and Chayanov) as of class formation in the countryside, was explored in a series of articles in *JPS*, especially from the 1980s. Gender analysis added further, and necessary, complexities to investigating and understanding the fluidity of the social boundaries between peasant capitalism and other forms of capital (and the state), and between peasants and classes of rural labour. Moreover, some notable essays not only acknowledged that fluidity but explored the determinants of its intricate patterns in specific historical and social conditions.

These kinds of analytically and empirically precise studies also engaged with wider theoretical debates. One important example was the theorisation of agrarian petty commodity production (PCP) or simple petty production (SCP) and its constitutive social relations, in capitalism and transitions to capitalism. This was of particular interest for several reasons. First, it can be viewed in part as an attempt, if often implicitly so, both to provide a theoretical approach alternative to Chayanov's model of the peasant household/farm and to fill the gap of a fully articulated alternative model in Lenin's work on the peasantry (see note 17). Second, it brought together, in a common debate with similar theoretical preoccupations, work on both peasants and the 'family farm' in developed capitalist countries. Third, the consideration of PCP/SCP also addressed questions of peasant differentiation from another angle: whether and how peasant production and differentiation are inflected, and in some circumstances constrained, by specific forms of pre-capitalist social relations and practices and their reproduction (and reconfiguration). This, in turn, often connected with vigorous debates concerning the 'articulation of modes of production'.

---

[19]Lenin's two paths of development of agrarian capitalism (above) were explored and extended in the seminal work of Terence J. Byres (1991, 1996); see also Byres in this collection (2009, 33–54).

The political economy of *JPS* connected the socioeconomic study of peasant production and reproduction with wider themes: the political role of peasants in different times and places of the turbulent upheavals that created modernity; the nature and forms of rural proletarianisation and labour regimes; the size and forms of the peasant (and more generally agricultural) 'surplus' and its contributions (or otherwise) to an accumulation fund for industrialisation; the class character and role of states in managing such transfers of the peasant/agricultural 'surplus' (or failing to do so); and analysis and critique of 'development' regimes, policies and practices, both in colonial conditions and following political independence (to which Shanin alludes).

There was not much to be found in Chayanov to inform exploration of such central themes, compared with Lenin and the Marxist tradition more generally, with, perhaps, one exception of considerable resonance. This concerns widespread debate of the relatively slow and uneven development of capitalism in farming, at least in the form of the predominance of large capitalist farms, for which Chayanovian concepts might be part of the explanation, and especially his emphasis on how self-exploitation underlies the apparent staying power ('resistance' in more heroic versions) of small-scale/peasant farming throughout the era of modern capitalism. 'Self-exploitation' is a profoundly ambiguous notion. On one hand, for poor peasants (and many middle peasants, according to circumstances) it involves back-breaking drudgery at very low levels of labour productivity and income – as Chayanov (1991, 40) emphasised: 'No one, of course, can welcome peasant hunger, but one cannot fail to recognise that in the course of the most ferocious economic struggle for existence, the one who knows how to starve is the one who is best adapted'.[20] On the other hand, the capacity of peasants to intensify their labour is central to a neo-populist path of agricultural development (see below). This would have alarmed Chayanov for whom the mundane realities of 'self-exploitation' derive from how peasants are incorporated in capitalist markets; the passage just quoted continues: 'All that matters is that those who shape economic strategy should make this latter method superfluous.'

Transitions to socialism

Between 1973 and 2000, *JPS* published nothing on Soviet agriculture from the 1930s to the 1980s. This is all the more striking given the rich contributions to, and debates within, *JPS* (and on the left more generally) that skirt this profound silence: on one side of it, the Russian debates and experiences from the 1880s to 1920s, already noted; on the other side, the extensive topical contributions on farming on Russia, and other regions of the former USSR, in the *aftermath* of the collapse of the Soviet Union (including the reinsertion of 'peasant' themes and their inevitable controversies). Did that silence imply a tacit, if reluctant, acceptance that Stalinist collectivisation was a 'nasty but necessary' condition of subsequent Soviet

---

[20]Relevant here is Kitching's vivid observation that peasants represent 'the historically classical and demographically dominant example of people who are poor *because* they work so hard' (2001, 147). Some explanations for the 'persistence' of peasant farming under capitalism, including its 'functions' for capital, rest on notions of (intense) 'self-exploitation'. There are resonances of this in Kautsky (1988); analyses from value theory of the peculiar trajectories of capitalism in agriculture include Mann and Dickinson (1978) and Djurfeldt (1981).

industrialisation?[21] Similarly the rich material on peasant movements and rebellions, and the places of peasants in the social revolutions of the twentieth century – Eric Wolf's 'peasant wars' – tended to stop at that moment when communist or social parties achieved the 'capture' of state power (Wolf 1969).[22]

This silence meant that the articles that filled many of the pages of *JPS* on the 'Lenin–Chayanov' debate looked to Lenin's earlier polemics with the Narodniks while ignoring the immense issues attached to collectivisation, let alone the subsequent performance of Soviet agriculture and its contradictions. Consequently the principal focus was on the places, prospects and fates of peasantries in historic and contemporary processes of transition to capitalism (outlined above), with questions of the forms and functioning of socialist agriculture attached to consideration of China and Vietnam as explicit 'alternatives' to capitalism and its 'development' policies in the Third World, for some as 'alternatives', explicit or implicit, to the once hegemonic claims of the 'Soviet model' – and also as 'alternatives' to various currents of agrarian populism ('taking the part of peasants'), which I touch on next.

The critique of populism/neo-populism

As Gavin Kitching (1982) showed to such effect, populist ideas are a response to the massive social upheavals that mark the development of capitalism in the modern world. Advocacy of the intrinsic value and interests of the small producer, both artisan and 'peasant', as emblematic of 'the people', arises time and again as an ideology, and movement, of opposition to the changes wrought by the accumulation of capital. This is the case in both the original epicentres of such accumulation (north-western Europe, North America) and those other zones exposed to the effects of capitalist development through their integration in its expanding and intensifying world economy, from nineteenth-century Russia to the 'South' of today. Agrarian populism, in particular, is the defence of the small 'family' farmer (or 'peasant') against the pressures exerted by the class agents of a developing capitalism – merchants, banks, larger-scale capitalist landed property and agrarian capital – and indeed, by projects of state-led 'national development' in all their capitalist, nationalist and socialist variants, of which the Soviet collectivisation of agriculture was the most potent landmark.

There are many varieties of populism, and of agrarian populism, that should be distinguished not only by their specific discursive elements and intellectual forms, but also by the particular historical circumstances in which they emerge and their varying

---

[21]And if its 'necessity' is less demonstrable than its nastiness? Soviet collectivisation was considered in only two articles before 2001: Nirmal Chandra (1992) started from the question whether it was the only possible response to the grain supply crisis of 1927–29, and then identified and ('counterfactually') assessed an alternative industrialisation strategy formulated by Bukharin; while Peter Nolan (1976) contributed a seminal comparison of collectivisation in the USSR and China, including the contrasting relations of their communist parties with the peasantries of their countries, a factor also highlighted as a critical political deficit of the Bolsheviks by Cox (1979) and Harrison (1979). After 2000 there were special issues of *JPS* on the contemporary Chinese countryside (30(3–4) 2003), presenting rather mainstream analyses, and on Russia over the past century (31(3–4) 2004), with an emphasis on 'peasant adaptation'.

[22]With the partial exception of some articles on China and Vietnam, and on agricultural policy in the very different circumstances of nationalisation of large export-oriented plantations and estates in, for example, Cuba, Mozambique and Nicaragua.

political strength and salience. Not surprisingly, the moral dimension of agrarian populism – as defence of a threatened (and idealised) way of life – often encompasses strong elements of anti-industrialism and anti-urbanism. Such ideologies are often explicitly anti-proletarian too, as new classes of wage labour represent the same threatening urban–industrial milieu as classes of capital and 'modernising' regimes of different political complexions. Much agrarian populist ideology, then, is backward-looking and explicitly reactionary, but this cannot be said of Chayanov.

For present purposes it is useful, if hardly sufficient, to differentiate at least two currents of populism that permeate agrarian debates, not least in relation to the problems and prospects of peasants or family farmers. The first can be termed 'political' populism, the second 'technicist' populism or, better, neo-populism. I come back to the former in the next section, after saying something about neo-populism defined here as based in (conventional) economics and its associated policy discourses, with claims to Chayanov as intellectual ancestor. It champions an equitable agrarian structure of small farms as most conducive to efficiency and growth. The economic case for efficiency incorporates arguments about the intrinsic advantages of the deployment of family labour in farming (lower supervision and transaction costs) and the factor endowments of poorer countries (plentiful labour, scarce capital), and combines them with arguments about equity (the employment and income distribution effects of small-scale farming). The neo-populist case thus rests on belief in the 'inverse relationship': that smaller farms manifest higher productivities of land – output per area – than larger farms, as well as generating higher net employment (albeit at necessarily lower levels of labour productivity). This remains a central plank in continuing populist economic arguments for redistributive land reform, encapsulated concisely and seductively in the notion of 'efficiency *and* equity' (Lipton 1977): the philosopher's stone of policy discourses that seek to reconcile the contradictions of capitalist development.[23]

A successful small farmer path of development also requires conducive market institutions, and a supportive state – in effect, removing the counterproductive oppressions of usury and merchant profit, and of taxation, in the same way that redistributive land reform removes the burden of rent (and/or of 'functional landlessness'). The obstacle to achieving these conditions, in another central term in the vocabulary of contemporary neo-populism, is the power of 'urban bias': the notion that policies in the South in the period of statist developmentalism (1950s–70s) favoured cheap food policies in the interests of strong urban constituencies and a (mistaken) emphasis on industrial development, at the expense of smaller and poorer farmers. This was a notable component of the World Bank's encompassing assault, from the 1980s, on state-led development strategies and their outcomes, with the added argument that poorer countries would do best to remove policy 'distortions' that impede the contributions of agriculture to their export performance (on the principle of comparative advantage) as well as supplying their domestic markets. Most recently, more technicist neo-populist approaches have adapted, more and less easily, to new conceptions of market-led land and agrarian reform,

---

[23]The neo-populist answer to the central question of land reform – to whom should land be redistributed and why? – is: to those who are both able to use it best (small farmers) *and* who need it most (the 'rural poor'). It is one of the profound ironies of modern agrarian history that the land reforms that have come closest to this ideal, an equitable distribution of 'land to the tiller', have been the outcomes of moments of violent social turmoil that brought together massive peasant mobilisation with revolutionary political parties.

closely associated with the World Bank and its hegemonic grip on development discourse in this conjuncture of 'globalisation'.

These instances of neo-populism as model of agricultural development – small is beautiful in farming, redistributive land reform, the removal of 'urban bias' – are often shared with the 'political' populism of rural movements and the radical intellectuals who identify with them. And they continue to be subjected to criticism critique from positions that might be considered 'Leninist', at least in the broad sense that they are rooted in the analysis of class relations in the countryside and beyond. That criticism need not be rehearsed here (see, for example, Byres 2004), but one can ask: what this has to do with Chayanov, and the legacies associated with him? Certainly Chayanov would have recognised, and embraced, some elements or contemporary expressions of the 'technicist' version of small farmer development outlined, if not others.[24] At the same time, in the recent trajectory of neo-populism Chayanov's expansive vision – for example, his emphasis on organisational economies of scale ('differential optima') to be achieved through the cooperative pooling of resources and efforts by peasant households with state support – has been increasingly reduced to a set of arguments from neo-classical economics that can be accommodated to the dominant neo-liberal paradigm. The effect is that household enterprises ('family' farms) should be constituted on the basis of *individualised* property rights and production in properly competitive markets for land, as well as other factor and product markets, a position that Chayanov would not have endorsed any more than contemporary 'political' populism does. And indeed he would have been entirely sceptical of the view that an equitable structure of small farming could be 'sustainable' in such conditions – itself an extraordinary article of faith.

Why were populism and neo-populism, often personified in the figure of Chayanov, so strong a polemical axis in the founding moment and subsequent career of *JPS*? Here are several suggestions. First, the revival of Marxism – and especially the centrality of Lenin's *Development of Capitalism in Russia* to the new wave of agrarian political economy – needed to distance/distinguish itself from other types of 'radical' critique of capitalism, including longstanding varieties of populism and 'peasantism'. Some of the theoretical bases and consequences of this have been sketched, with special reference to class analysis and class formation in the countryside. Second, this acquired a particular focus in contesting versions of neo-populism applied to agrarian questions in newly independent Asia and Africa (as well as in Latin America) and manifested in development policies prescribed for those great zones of the world. This contestation centred, once more, on issues of class dynamics with particular reference to two 'classic' issues: belief in economies of scale in farming (as in manufacturing) as a necessary condition of the development of the productive forces, and the effects of agrarian class structure for industrialisation, not least the transfer of the agricultural 'surplus' through some or other form of taxation, be it of landed property, agrarian capital and/or rich peasants (or all classes of the peasantry?). Third, some authors identified with, and were sometimes members of, existing communist and socialist parties which

---

[24]The most potent explicit critique of Chayanov in this respect, constructed through the contrast with Lenin's *Development of Capitalism in Russia*, was by Utsa Patnaik (1979): 'Neo-populism and Marxism: The Chayanovian View of the Agrarian Question and Its Fundamental Fallacy'. Patnaik based herself on Chayanov (1966); his *Theory of Peasant Cooperatives* was not then available in English translation.

confronted political formations with an agrarian populist ideology, programme and, indeed, appeal in the countryside. This leads to the next section.

Politics

I begin by returning to the vacuum suggested earlier at the core of much of the 'Lenin–Chayanov' debate, an empty space between, on one hand, the lack of any political sociology in Chayanov,[25] and, on the other hand, the tensions and strains in the long and tortuous history of Marxism as it has sought to connect its intellectual claims and analyses with its political ambitions (and realistic 'alternatives' to capitalism) – and over which Stalinism cast so long a shadow for its followers and opponents alike. The most useful commentary that I know of in this context is the last of Mark Harrison's sequence of four articles cited above. In it Harrison (1979) reaffirmed his respect for Chayanov by way of identifying the problems of a (politically) 'subordinate Marxism' restricted to 'reactive theoretical critique' and unable to advocate 'practical theory' as Chayanov had done. Harrison illustrated this 'with reference to three (connected) themes of great significance for Soviet history (and beyond): the lack of Bolshevik political work, experience and organisation in the countryside; a tradition of Bolshevik ultra-leftism towards the peasantry; and the failure to transcend these problems after 1917, despite some 'fresh and creative impulses' shown by Lenin and Bukharin in the early and late 1920s, respectively.' (Bernstein and Byres 2001, 13, n. 20).

This is not to disregard the large, often innovative and valuable, body of work on peasant politics published in *JPS* and elsewhere, across the range from 'everyday forms of resistance' (Scott 1985, Scott and Kerkvliet 1986, Kerkvliet 2009, 227–43, this collection) through 'new farmers' movements' in India (Brass 1994) to 'peasant wars' (Wolf 1969) – and often framed as debates between 'political' populism and Marxism. However, from the latter viewpoint, as noted earlier, analysis tended to stop at the moment when communist or social parties achieved the 'capture' of state power, with the two exceptions indicated: the 'lost' Soviet 1920s, and a brief moment of enthusiasm for the communes of Mao's China before their subsequent dismantling. The symptomatic quality of these exceptions carries forward to the present moment, when Marxism is (even more) prone to its own inverse relationship between political marginality and 'reactive theoretical critique' with its seductions of class purism and ultra-leftism. And in the current conjuncture, this registers its vulnerability to the force of a revived and vigorous 'political' agrarian populism, as I suggest in the following section.

Looking forward

'Chayanov's analysis ... is incomplete (and) cannot be completed by simply proceeding along the same road', suggests Teodor Shanin, for this reason: 'Rural society and rural problems are inexplicable any longer in their own terms and must be understood in terms of labour and capital flows which are broader than agriculture ... (where peasant economy) is inserted into and subsumed under a dominant political economy, different in type' (1986, 19). I have argued that this was already widely recognised among

---

[25]Although there is a plausible sense in which scholars like Scott and Shanin have sought to create a kind of Chayanovian political sociology.

Chayanov's Marxist predecessors and contemporaries who created and developed a theoretical and historical approach in which such concerns were established from the beginning. Shanin (2009, 97, this collection) also refers to today's world of 'fewer peasants as well as of fewer "classical" industrial proletarians'. This points to an apparent paradox: that this world of fewer 'classical' industrial proletarians and peasants is one of 'more' capitalism, so to speak, and of more global capitalism, than ever before. Shanin's point here is to propose the continuing relevance of Chayanov, whose model of the family labour household he extends to the pursuit of 'livelihood' (simple reproduction) in an ever growing non-agricultural 'informal sector'. In looking forward from Lenin and Chayanov to today's world of globalising capitalism, I start from the continuing centrality of class relations and dynamics but in the conditions of a different 'historical epoch' than that which they addressed.[26] On the former, in capitalism production and reproduction are structured *universally*, but *not exclusively*, by relations of class. To paraphrase Balibar (as quoted by Therborn 2007, 88): in a capitalist world, class relations are only '*one determining* structure' but a structure that shapes '*all* social practices' (emphasis in original). In effect, what makes class unique is not that it is the only 'determining structure' but that in capitalism it is the only structure of social relations that permeates all others.[27] On the latter, the different 'historical epoch' we confront today is that of capitalist globalisation.[28]

Globalisation

First, globalisation – for all the confusion and controversy that surrounds the term and its uses (and abuses) – is central to, indeed entails, a qualitative shift in agrarian class relations without, of course, displacing or transforming all their previous forms. I come back to this below, but note here the limits of Lenin and Chayanov on the internationalisation of agriculture, its global divisions of labour, financial and commodity circuits, and so on. This was partly due to the historical circumstances of their time, when the primary issue for them in the geographical areas they focussed on (above) was international *trade* in agricultural commodities rather than, say, foreign direct investment in their production and processing (more evident then in colonial zones and parts of Latin America), or today's concerns with technical change and 'transfers of technology', or the consequences of global income distribution, hence 'effective demand', for specialised export production in different parts of the world.[29]

---

[26]This section selects from, and radically abbreviates, elements of the arguments in Bernstein (2008, in press).

[27]There remains, however, a persistent and recalcitrant theoretical issue: if commodification is a general ('world-historical') process that entails both class and other social relations, the latter can *not* be theorised through the same procedure of abstraction as the class relation of wage labour and capital (exemplified in Marx's *Capital*) even though they are ubiquitous in shaping specific and concrete forms of class relations in 'actually existing capitalisms' (and transitions to capitalism).

[28]I share the approach to globalisation that sees it as the restructuring of capital on a world scale since the 1970s, hence the latest phase of imperialism.

[29]Lenin noted that *The Development of Capitalism in Russia* deliberately excluded consideration of its international dimensions, both trade and foreign investment. Within the primarily market(ing)/trade focus of his book on cooperatives Chayanov (1991) made interesting observations about what we now call commodity chains, including the currently fashionable theme of standards and 'branding'.

Second, the majority of 'peasants'/'small farmers' (and of those in an ever expanding 'informal economy') in a globalising 'South' are a component of what I term 'classes of labour', and a component that is neither dispossessed of *all* means of reproducing itself nor in possession of *sufficient* means to reproduce itself.[30] The former is not exceptional (see note 18). The latter marks the limits of their viability as petty commodity producers. 'Classes of labour', then, comprise 'the growing numbers ... who now depend – directly *and indirectly* – on the sale of their labour power for their own daily reproduction' (Panitch and Leys 2000, ix, emphasis added). Classes of labour in the conditions of today's 'South' have to pursue their reproduction through insecure, oppressive and increasingly 'informalised' wage employment and/or a range of likewise precarious small-scale and insecure 'informal sector' ('survival') activity, including farming; in effect, various and complex *combinations* of employment and self-employment. Many of the labouring poor do this across different sites of the social division of labour: urban and rural, agricultural and non-agricultural, as well as wage employment and self-employment. This defies inherited assumptions of fixed, let alone uniform, notions (and 'identities') of 'worker', 'peasant', 'trader', 'urban', 'rural', 'employed' and 'self-employed'.

Third, class differentiation of petty producers is a *necessary tendency* of capitalist dynamics even when this is not registered in evident (or extensive) trends of class formation. For example, the ideal model or aspiration of agrarian populism of the robust 'family' (or 'middle peasant') farm is typically compromised by the analytical problem that it is naturalised, *à la* Chayanov, as a kind of norm from which other social forms (capitalist farms on one hand, 'landless peasants' on the other) deviate. However, the relatively stable 'family' farm – when it occurs – has to be problematised both analytically and concretely, that is to say, historically. This includes investigating whether and how the formation and reproduction of such farms are *the result of processes of differentiation* as 'entry' and reproduction costs rise in the course of commodification, there is competition for land and/or the labour to work it, and so on.

Fourth, a further aspect of this, that complicates the 'classic' (Leninist) three-class model of peasant differentiation, is the extent to which the operations, and relative stability, of all classes of the peasantry (increasingly) depend on activities and sources of income from outside their own farming: investment and accumulation in the case of rich peasants, 'survival' and simple reproduction at higher or lower levels of income and security for middle and poor peasant respectively. These are not aspects of peasant incorporation in capitalist social relations novel to the current moment of globalisation, but they have become increasingly central. This means (as Shanin (1986), above, indicates) that it is impossible today to even begin to conceptualise agriculture, and its classes of labour *and* capital, independently of the circuits and dynamics of capitalism more (indeed most) broadly.

The consequences of globalisation for earlier paradigms also include, first, different conditions of possibility for industrialisation (and different paths of

---

[30]I prefer the term 'classes of labour' to the inherited vocabulary of proletarianisation/proletariat (and semi-proletarianisation/semi-proletariat), as it is less encumbered with problematic assumptions and associations in both political economy (e.g. functionalist readings of Marx's concept of the reserve army of labour) and political theory and ideology (e.g. constructions of an idealised (Hegelian) collective class subject).

industrialisation), more detached from paths of agrarian change and the 'classic' preoccupation with the contributions of agriculture to industrial accumulation; and, second, patterns of 'de-agrarianisation' – or 'de-peasantisation', in Lenin's term – associated with different combinations of (a) pressures of reproduction on small-scale farming and (b) opportunities for employment outside 'own account' farming. This links with the second and fourth observations above, and is a key dimension of the wide range of variation in the conditions and prospects of classes of labour, including the salience of their agrarian components where applicable.[31]

Of course, the various elements of the fragmentation of classes of labour indicated have profound implications for the political sociology of the struggles their members engage in, actually and potentially, and it is here that 'political' agrarian populism has reappeared with such vigour to take up the cause of 'peasants against globalisation'.[32] Moreover, it has done so in ways, and for reasons, that highlight the inherited problems in the political stances of Marxist agrarian political economy, now exposed more than ever by the demise of its Stalinist or Maoist exemplars and influences, its apparent inability to imagine a socialist 'agrarian programme' to replace them, and the dangers of retreat into the comfort zone of 'reactive theoretical critique', class purism and ultra-leftism.

Politics and populism: challenges of the current moment

The dynamics, modalities and effects, not least the environmental effects, of globalisation as it affects international food regimes/systems, and the problems and prospects of 'peasants'/'small farmers', are the subject of much topical interest registered *inter alia* in the recently invented field of 'political ecology', and especially its more radical populist wing committed to supporting political movements of 'people of the land' – an emblematic signifier for the target constituencies of transnational peasant and farmers movements, defined by a political project opposed to globalisation. That opposition also informs notions of development imaginaries or alternative futures for farming and farmers.[33] In this sense, 'people of the land' is a central notion in the discourse and programmatic ambitions of Vía Campesina and its intellectual champions. Philip McMichael, for example, formulates a 'new agrarian question' which opposes 'the corporatisation of agriculture ... (that) has been globally synchronised to the detriment of farming populations everywhere' by 'revalorising rural cultural-ecology as a global good' (2006a, 473, 472). The agents of the latter comprise a 'global agrarian resistance', an 'agrarian counter-movement', that strives to preserve or reclaim 'the peasant way' (2006b, 474, 480, and *passim*). The 'new agrarian question' is world-historical in that, first, it transcends the capital–labour relation and, second, the social forces ('of global agrarian resistance') that it mobilises have the capacity, at least potentially, to generate that transformation, according to McMichael (and others).[34]

Marx and Lenin, of course, would recognise transcending the capital–labour relation as the sense of a world-historical beyond capitalism, even if they would have

[31]On determinants and patterns of 'de-agrarianisation' see, for example, the contrast between Southeast Asia (Rigg 2006) and sub-Saharan Africa (Bryceson 1999).
[32]In the title of Marc Edelman's outstanding monograph on Costa Rica (1999).
[33]For example, in the context of the global food economy and its 'crisis', compare the rigorous and provocative analysis by Weis (2007) with the sloppiness of Patel (2007).
[34]See also McMichael (2009, 139–69) in this collection.

been startled by the suggestion that this could be driven by the contradiction between capital and 'peasants', ('family') farmers or 'people of the land' mobilised in 'global agrarian resistance'. Notions of 'the peasant way' resonate lineages of agrarian populism that have always appeared and re-appeared in the long histories of modern capitalism.[35] When counterposed to what is undoubtedly, in key respects, a new phase represented by the globalisation of agriculture, advocates of 'the peasant way' argue that it does not represent nostalgia ('worlds we have lost') but that contemporary rural social movements incorporate and express specific, novel and strategic conceptions of, and aspirations to, modernity, and visions of modernity alternative to that inscribed in the neo-liberal common sense of the current epoch. This is a plausible thesis, always worth investigating in particular circumstances, but the principal weakness of 'the new agrarian question' *qua* 'the peasant way', as articulated to date, is its lack of an adequate political economy.

First, it tends to present 'farming populations everywhere' as a single social category that serves, or is necessary to, both the analytical and political purposes of 'resistance' to globalisation and neo-liberalism.[36] Indeed, 'farmers' thus constitute not only a single category but a singular one: they are deemed to experience, to challenge, and to seek to transcend the social and ecological contradictions of a globalising capitalism in a uniquely *combined* fashion. While differences within and between 'farming populations' – differences of 'North' and 'South', of market conditions, of gender relations, and sometimes even class relations – are acknowledged, this tends to be gestural in the absence of any deeper theorisation and more systematic empirical investigation of the conditions in which farming and agriculture are constituted by specific forms and dynamics of the capital–labour relation, and not least how they express, generate, reproduce and shape class differentiation. For 'the new agrarian question', whether by intent (explicitly) or by default (implicitly), class and other social differentiation is subordinate to what *all* 'farmers' ('family farmers') and their struggles have in common: 'exploitation' by capital (which they share with labour?) and a special relation with and respect for nature (which distinguishes them from non-agrarian classes of labour, or simply urbanites?). This is evident in many statements by advocates of this vision that represent contemporary variations on long-established themes of 'agrarianism'. For example, Annette Desmarais refers to 'people of the land' as a unitary global social bloc, also known as 'peasants' apparently as when she applauds 'over 5,000 farmers including European, Canadian, American, Japanese, Indian and Latin American *peasants*' marching on a GATT meeting in Geneva in 1993 (2002, 93, emphasis added).

Second, there is little specification of the 'alternative' systems of production that the 'peasant way' may generate as the basis of a future post-capitalist, ecologically friendly social order. Rural community, and an associated localism, are championed, of course: antithesis to the thesis of the global corporatisation of agribusiness and the drive to individualisation of neo-liberal ideology, but where is any plausible formulation of the social and material coordinates of a synthesis (negation of the

---

[35] Although Chayanov seems to have been largely forgotten.
[36] I distinguish globalisation in the sense of new forms of the restructuring of capital (above) and neo-liberalism as an ideological and political project. Conflating the two, which is unfortunately all too common, precludes the possibility that the former can proceed in the future without the latter, despite their close connection in the current conjuncture.

negation)?[37] That is, advocacy of 'the peasant way' largely ignores issues of feeding the world's population, which has grown so greatly almost everywhere in the modern epoch – and principally because of the revolutions in agricultural productivity (as well as medicine) achieved by the development of capitalism. In response to provocations on this matter of the demographic – or, better, population/productivity – challenge, McMichael (2006b, 415) observes that 'Longer-term questions of raising agricultural productivity to provision cities are yet to be resolved.' Why is this longer-term? And how might it be resolved?[38] While Friedmann's 'brief formulation of the population question of today' concludes that 'We may be heading for ... demographic collapse. What this means for the "global" phase of capitalism, capitalism *tout court*, and even human survival is what we need to think about now' (2006, 464). This, once more, is to (re-)state a problem (however fundamental) rather than point to its resolution.[39]

Third, celebrations of 'global agrarian resistance' and the transformational aspirations attached to it, lack any plausible formulation and analysis of how it could work as a political project (though see note 40). Interestingly, the MST (*Movimento dos Trabalhadores Rurais Sem Terra*) in Brazil is especially emblematic for both those who advocate land struggles as the cutting edge of semi-proletarian politics in the 'South' today (Moyo and Yeros, 2005) and those who aspire to transcend the capital–labour relation through 'revalorising rural cultural-ecology as a global good'. Both are frequently given to long quotes from MST documents in ways that elide that necessary distinction or distance between sympathy with the programmatic statements of the organisation and its leadership and the demands of analysis. As Wendy Wolford (2003) points out, many discussions refer to the 'imagined community' articulated in such statements as accurate empirical representations of the experiences, beliefs and practices of its socially heterogeneous membership, in effect attributing to 'the movement' a unity of vision and purpose that is unwarranted, and unhelpful. Too many accept the 'official' ideology of the MST (as of Vía Campesina) at face value from political sympathy, rather than combining sympathy with the critical inquiry necessary to adequate investigation, analysis and assessment.[40]

---

[37]Some lines of thought in considering this question are sketched by McMichael (2006b, 186–87), as usual with reference to the manifestos of Vía Campesina and its central notion of 'food sovereignty'.

[38]Elsewhere – as an instance of 'global agrarian resistance'?! – McMichael (2006c, 186) asserts that urban gardens 'provision 35 million people in the US alone' without specifying what 'provision' means here. van der Ploeg (2008) is a more serious statement of this line of argument, not least because of the author's deep knowledge of farming and of agricultural industries in different parts of the world that provides a more plausible contemporary echo of Chayanov's 'practical theory'.

[39]From their earlier work Friedmann and McMichael have sought to incorporate an ecological dimension as central to the history of international food regimes. The use of titles like 'Feeding the empire ...' (Friedmann 2004) and 'Feeding the world ...' (McMichael 2006b) makes the absence of any demographic consideration the more surprising. Interestingly, the population question is central to the ecological economics of Joan Martinez-Alier (2002) who seeks to wrest it from its Malthusian heritage, while insisting that global environmental sustainability requires fewer people (and zero economic growth?). At the same time, he is among the most evidently romantic, and unhistorical, advocates of 'the peasant way' (Bernstein 2005).

[40]As does Marc Edelman (1999), including the valuable methodological reflections of his concluding chapter on 'Peasant Movements of the Late Twentieth Century'; see also Edelman (2009, 245–65) in this collection, and a number of the contributions to Borras *et al.* (2008), including the editors' valuable introductory essay.

In short, 'people of the land' ('family' 'farmers', 'peasants', and so on) are posited as a unitary and idealised, and ostensibly world-historical, 'subject', in a strange (or maybe not so strange?) echo of qualities once attributed to 'the international proletariat'. Nonetheless, for all their shortcomings and difficulties, as briefly outlined, populist formulations of a 'new agrarian question' that seek to understand, combat and transcend globalising tendencies of the organisation of agriculture, raise fundamental issues about what is changing in the world of contemporary capitalism: the modes of operation and powers of corporate agribusiness, their social and ecological effects, and the social bases of 'resistance' and 'alternatives'. By the same token, this challenge the agrarian question inherited from 'classic' Marxism via Lenin, and especially its capacity to address a world so different from that which generated its original concerns with paths of transition from pre-capitalist (agrarian) social formations to capitalist agriculture and industry (and state socialism in the unique adaptation and impact of the 'classic' agrarian question registered by the Soviet experience).

The 'classic' agrarian question of Marxism at its most doctrinaire can reduce to a strongly deductive 'model' of the virtues of economies of scale in farming. Reconsideration by historical materialism of its historic, and uncritical, attachment to the benefits of large-scale farming is long overdue for various reasons, including the following. First, it is salutary to recover a properly materialist rather than technicist conception of scale in agriculture as an effect of specific, and variant, forms of social relations. Second, the scale and distribution – *and* uses – of capitalist landed property in particular circumstances are often shaped by speculative rather than productive purposes. Third, the productive superiority of large(r)-scale farming is often contingent on conditions of profitability underwritten by direct and hidden subsidy and forms of economic rent, and indeed ecological rents. Fourth, materialist political economy needs to take much more seriously the environmental consequences and full social costs of the technologies that give modern capitalist farming the astonishing levels of productivity it often achieves. These types of issues illustrate the challenges of, and demands on, an agrarian political economy less confined by its historic sources and preoccupations and more committed to problematising and investigating what is changing in today's (globalising) capitalism. They are *not* presented as elements of a general argument *against* large-scale farming as I am sceptical about *any* 'models' of (virtuous) farm scale constructed on deductive or *a priori* grounds.

And politics? The political economy in this paper is not deployed in any 'anti-peasant' spirit or prescriptive stance on petty commodity production. Nor do any of my observations suggest withdrawing political sympathy and support for progressive struggles because they fail to satisfy the demands of an idealised (class-purist or other) model of political action. Rather, I have suggested that part of the problem with the 'new' agrarian question sketched is how it posits a unitary and idealised, and ostensibly world-historical, 'subject': 'farmers' or 'peasants' or 'people of the land'. The point, then, is first, to recognise and, second, to be able to analyse, the contradictory sources and impulses – and typically multi-class character – of contemporary struggles over land and ways of farming that can inform a realistic and politically responsible assessment of them. This means rising to the challenges posed by a re-energised and radical agrarian populism, to engage both seriously and critically with the agrarian movements of the present time, and thereby to recover the spirit of Lenin's 'fresh and creative impulses' of the early 1920s, and of Chayanov's contributions to 'practical theory'.

References

Banaji, J. 1976a. Chayanov, Kautsky, Lenin: considerations towards a synthesis. *Economic and Political Weekly*, 11(40), 1594–607.
Banaji, J. 1976b. Summary of selected parts of Kautsky's *the agrarian question. Economy and Society*, 5(1), 1–49.
Bernstein, H. 2005. Review of Martinez-Alier (2002). *Journal of Agrarian Change*, 5(3), 429–36.
Bernstein, H. 2008. Agrarian questions from transition to globalisation. *In:* A.H. Akram-Lodhi and C. Kay, eds. *Peasant livelihoods, rural transformation and the agrarian question*. London: Routledge, pp. 239–61.
Bernstein, H. (in press). Rural livelihoods and agrarian change: bringing class back in. *In:* N. Long and Ye Jingzhong, eds. *Rural transformations and policy intervention in the twenty-first century: China in context*. Edward Elgar.
Bernstein, H., T. Brass and T.J. Byres. 1994. *The Journal of Peasant Studies: a twenty volume index, 1973–1993*. London: Frank Cass.
Bernstein, H. and T.J. Byres. 2001. From peasant studies to agrarian change. *Journal of Agrarian Change*, 1(1), 1–56.
Borras, S.M. Jr., M. Edelman and C. Kay, eds. 2008. *Transnational agrarian movements confronting globalisation*. Special issue of *Journal of Agrarian Change*, 8(2–3).
Bourgholtzer, F., ed. 1999. *Aleksandr Chayanov and Russian Berlin*. Special issue of *The Journal of Peasant Studies*, 26(4).
Brass, T., ed. 1994. *The new famers' movements in India*. Special issue of *The Journal of Peasant Studies*, 21(3–4).
Brass, T. 2005. *The Journal of Peasant Studies*: the third decade. *The Journal of Peasant Studies*, 32(1), 153–80.
Bryceson, D. 1999. African rural labour, income diversification and livelihood approaches: a long-term development perspective. *Review of African Political Economy*, (80), 171–89.
Byres, T.J. 1991. The agrarian question and differing forms of capitalist transition: an essay with reference to Asia. *In:* J. Breman and S. Mundle, eds. *Rural transformation in Asia*. Delhi: Oxford University Press, pp. 3–76.
Byres, T.J. 1996. *Capitalism from above and capitalism from below. An essay in comparative political economy*. London: Macmillan.
Byres, T.J., ed. 2004. *Redistributive land reform today*. Special issue of *Journal of Agrarian Change*, 4(1–2).
Byres, T.J. 2009. The landlord class, peasant differentiation, class struggle and the transition to capitalism: England, France and Prussia compared. *The Journal of Peasant Studies*, 36(1), 33–54.
Chandra, N.K. 1992. Bukharin's alternative to Stalin: industrialisation without forced collectivisation. *The Journal of Peasant Studies*, 20(1), 97–159.
Chayanov, A.V. 1966. *The theory of peasant economy*, eds. D. Thorner, B. Kerblay and R.E.F. Smith. Homewood, IL: Richard Irwin for the American Economic Association.
Chayanov, A.V. 1976. The journey of my brother Alexei to the land of peasant utopia. *The Journal of Peasant Studies*, 4(1), 63–116.
Chayanov, A.V. 1986. *The theory of peasant economy*, second edition, eds. D. Thorner, B. Kerblay and R.E.F. Smith. Madison, WI: University of Wisconsin Press.
Chayanov, A.V. 1991. *The theory of peasant co-operatives*. London: I. B. Tauris.
Cox, T. 1979. Awkward class or awkward classes? Class relations in the Russian peasantry before collectivisation. *The Journal of Peasant Studies*, 7(1), 70–85.
Cox, T. and G. Littlejohn, eds. 1984. *Kritsman and the agrarian Marxists*. Special issue of *The Journal of Peasant Studies*, 11(2).

74   *Henry Bernstein*

Danilov, V. 1991. Introduction: Alexander Chayanov as a theoretician of the co-operative movement. *In:* A.V. Chayanov, *The theory of peasant co-operatives*. London: I. B. Tauris, pp. xi–xxxv.

Desmarais, A.-A. 2002. The *Vía Campesina*: consolidating an international peasant and farm movement. *The Journal of Peasant Studies*, 29(2), 91–124.

Djurfeldt, G. 1981. What happened to the agrarian bourgeoisie and rural proletariat under monopoly capitalism? Some hypotheses derived from the classics of Marxism on the agrarian question. *Acta Sociologica*, 24(3).

Edelman, M. 1999. *Peasants against globalisation. Rural social movements in Costa Rica*. Stanford, CA: Stanford University Press.

Edelman, M. 2009. Synergies and tensions between rural social movements and professional researchers. *The Journal of Peasant Studies*, 36(1), 245–65.

Engels, F. 1951. *The peasant question in France and Germany. In:* K. Marx and F. Engels, *Selected works*, Vol. 2. Moscow: Foreign Languages Publishing House, pp. 381–99.

Friedmann, H. 2004. Feeding the empire: the pathologies of globalised agriculture. *In:* L. Panitch and C. Leys, eds. *The socialist register 2005*. London: Merlin Press, pp. 124–43.

Friedmann, H. 2006. Focusing on agriculture: a comment on Henry Bernstein's 'Is there an agrarian question in the 21st century?'. *Canadian Journal of Development Studies*, 27(4), 461–65.

Harrison, M. 1975. Chayanov and the economics of the Russian peasantry. *The Journal of Peasant Studies*, 2(4), 389–417.

Harrison, M. 1977a. Resource allocation and agrarian class formation: the problems of social mobility among Russian peasant households, 1880–1930. *The Journal of Peasant Studies*, 4(2), 127–61.

Harrison, M. 1977b. The peasant mode of production in the work of A.V. Chayanov. *The Journal of Peasant Studies*, 4(4), 323–36.

Harrison, M. 1979. Chayanov and the Marxists. *The Journal of Peasant Studies*, 7(1), 86–100.

Kautsky, K. 1988. *The agrarian question*. London: Zwan Publications.

Kerkvliet, B.J. Tria. 2009. Everyday politics in peasant societies (and ours). *The Journal of Peasant Studies*, 36(1), 227–43.

Kingston-Mann, E. 1980. A strategy for Marxist bourgeois revolution: Lenin and the peasantry, 1907–1916. *The Journal of Peasant Studies*, 7(2), 131–57.

Kitching, G. 1982. *Development and underdevelopment in historical perspective*. London: Methuen.

Kitching, G. 2001. *Seeking social justice through globalisation*. University Park, PA: Pennsylvania State University Press.

Lehmann, D. 1982. After Chayanov and Lenin. New paths of agrarian capitalism. *Journal of Development Economics*, 11(2), 133–61.

Lenin, V.I. 1967a. *The development of capitalism in Russia. The process of the formation of a home market for large-scale industry. Collected works*, vol. 3. Moscow: Progress Publishers.

Lenin, V.I. 1967b. Report on the substitution of a tax in kind for the surplus-grain appropriation system. *In: Collected works*, vol. 32. Moscow: Progress Publishers, pp. 214–28.

Lenin, V.I. 1967c. Summing-up speech on the tax in kind. *In: Collected works*, vol. 32. Moscow: Progress Publishers, pp. 229–38.

Lewin, M. 1968. *Russian peasants and Soviet power. A study of collectivisation*. London: George Allen & Unwin.

Lipton, M. 1977. *Why poor people stay poor. A study of urban bias in world development*. London: Temple Smith.

Mann, S.A. and J.M. Dickinson. 1978. Obstacles to the development of a capitalist agriculture. *The Journal of Peasant Studies*, 5(4), 466–81.

Martinez-Alier, J. 2002. *The environmentalism of the poor*. Cheltenham: Edward Elgar.

McMichael, P. 2006a. Reframing development: global peasant movements and the new agrarian question. *Canadian Journal of Development Studies*, 27(4), 471–83.

McMichael, P. 2006b. Feeding the world: agriculture, development and ecology. *In:* L. Panitch and C. Leys, eds. *The socialist register 2007*. London: Merlin Press, pp. 170–94.

McMichael, P. 2006c. Peasant prospects in the neoliberal age. *New Political Economy*, 11(3), 407–18.

McMichael, P. 2009. A food regime genealogy. *The Journal of Peasant Studies*, 36(1), 139–69.

Moore, B. Jr. 1966. *Social origins of dictatorship and democracy. Lord and peasant in the making of the modern world*. Boston, MA: Beacon Press.

Moyo, S. and P. Yeros. 2005. The resurgence of rural movements under neo-liberalism. *In:* S. Moyo and P. Yeros, eds. *Reclaiming the land: the resurgence of rural movements in Africa, Asia and Latin America*. London: Zed Books, pp. 8–64.

Nolan, P. 1976. Collectivisation in China: some comparisons with the U.S.S.R. *The Journal of Peasant Studies*, 3(2), 192–220.

Panitch, L. and C. Leys. 2000. Preface. *In:* L. Panitch and C. Leys, eds. *The socialist register 2001*. London: Merlin Press, pp. vii–xi.

Patel, R. 2007. *Stuffed and starved. Markets, power and the hidden battle for the world's food system*. London: Portobello Books.

Patnaik, U. 1979. Neo-populism and Marxism: the Chayanovian view of the agrarian question and its fundamental fallacy. *The Journal of Peasant Studies*, 6(4), 375–420.

Ploeg, J.D. van der. 2008. *The new peasantries. Struggles for autonomy and sustainability in an era of empire and globalisation*. London: Earthscan.

Rigg, J. 2006. Land, farming, livelihoods, and poverty: rethinking the links in the rural south. *World Development*, 34(1), 180–202.

Scott, J.C. 1976. *The moral economy of the peasant*. New Haven, CT: Yale University Press.

Scott, J.C. 1985. *Weapons of the weak. Everyday forms of peasant resistance*. New Haven, CT: Yale University Press.

Scott, J.C. and B.J. Tria Kerkvliet, eds. 1986. *Everyday forms of peasant resistance in South-East Asia*. Special issue of *The Journal of Peasant Studies*, 13(2).

Shanin, T. 1972. *The awkward class. Political sociology of peasantry in a developing society: Russia 1910–1925*. Oxford: Clarendon Press.

Shanin, T. 1986. Chayanov's message: illuminations, miscomprehensions, and the contemporary 'development theory'. *In:* A.V. Chayanov, *The theory of peasant economy*, eds. D. Thorner, B. Kerblay and R.E.F. Smith, second edition. Madison, WI: University of Wisconsin Press, pp. 1–24.

Shanin, T. 2009. Chayanov's treble death and tenuous resurrection: an essay about understanding, about roots of plausibility and about rural Russia. *The Journal of Peasant Studies*, 36(1), 83–101.

Sivakumar, S.S. 2001. The unfinished Narodnik agenda: Chayanov, Marxism, and marginalism revisited. *The Journal of Peasant Studies*, 29(1), 31–60.

Smith, R.E.F., ed. 1976. *The Russian peasant 1920 and 1984*. Special issue of *The Journal of Peasant Studies*, 4(1).

The Journal of Peasant Studies 2005. *The Journal of Peasant Studies* an eleven volume index, 1993–2004. *The Journal of Peasant Studies*, 32(1), 181–241.

Therborn, G. 2007. After dialectics. Radical social theory in a post-communist world. *New Left Review*, (NS) 43, 63–114.

Thompson, E.P. 1971. The moral economy of the English crowd in the eighteenth century. *Past & Present*, 50, 76–136.

Thorner, D. 1966. Chayanov's concept of peasant economy. *In:* A.V. Chayanov, *The theory of peasant economy*, eds. D. Thorner, B. Kerblay and R.E.F. Smith. Homewood, IL: Richard Irwin for the American Economic Association, pp. xi–xxiii.

Weis, T. 2007. *The global food economy. The battle for the future of farming*. London: Zed Books.

Wolf, E. 1969. *Peasant wars of the twentieth century*. New York, NY: Harper & Row.

Wolford, W. 2003. Producing community: the MST and land reform settlements in Brazil. *Journal of Agrarian Change*, 3(4), 500–20.

Chayanov's treble death and tenuous resurrection: an essay about understanding, about roots of plausibility and about rural Russia

Teodor Shanin

The paper presents Chayanov's 'Theory of Vertical Cooperation' as the main conceptual alternative to Stalinist collectivisation of the 1930s. It also brings back the drama of physical and intellectual destruction of the brilliant Russian school of agrarian economists 1890–1920s who paid the price for opposition to the 1930's attack on the peasant majority of the Russian population. It then proceeds to the social and political roots of ideological blindness concerning the failures of collectivisation and of its impact on the history of contemporary Russia.

A Muscovite experience: September 1987

Once in a while personal experiences matter publicly and should be put on record. In September 1987, as I arrived in Moscow, I was asked to meet Alexander Nikonov – the President of the Soviet Academy of Agricultural Sciences (the VASKhNIL). He briskly came to the point. Alexander Chayanov was officially rehabilitated a few weeks before and could I address in Russian a few colleagues about his works and his impact on Western scholarship. I agreed. Some days later, when I arrived to speak, I found an audience of 600 cramped in the Academy's main hall at the Yusupov's old palace. The President opened on the outstanding relevance of Chayanov for the country – his theory of peasant cooperation addressing the contemporary rural crisis and contributing directly to the debated strategy for transformation of agriculture under Perestroika. Chayanov's books were being urgently prepared for re-publication. On my turn, I spoke of Chayanov's life, of his studies and his novels, of peasants and scholars, of fashions and substance in analytical thought. I spoke also about contemporary agriculture and the theories of it, about collectivisation, about models of cooperation and about those who came to publish Chayanov elsewhere, while his own country had him banned. I finished speaking about the place and the price of truth in the life of societies, and about scholars as a peculiar international brotherhood of those whose chosen occupation is matters of truth. The response was as strong as my own emotions. I was speaking at a place where Chayanov was arrested to go eventually to his death. I was telling them about the man whom they now came to accept as their most talented colleague, the name of whom was spoken for generations in whispers and whose actual works were unknown to most of them. Hundreds of hands were taking copious notes. People clapped and cheered. Chayanov's son, an old man now, stood up to thank me for defending the honour of his father and I felt my tears. Questions from the floor followed thick and fast to keep me busy, an academic colloquium was rolling on. One more day and deed of the early days of hope and excitement of Perestroika.

What followed was a wave of lectures and publications concerning Chayanov. Scholars, journalists and the 'educated public' came to talk of his biography and writings, his views concerning agriculture and his novels, his achievements and his relevance. Chayanov's first biography went rapidly into print.

Yet, the emotions by now all but forgotten, and the works of a good man, by now on the open shelves, are not the end of the matter. The very facts of re-discovery, of re-learning and even of the emotions involved, the processes of official and informal re-recognitions, are questions which should be considered in turn as an issue of knowledge within social processes, and of the trade scholars pursue. There are also the relevant questions, still fresh and unanswered, of social change and social justice in contemporary agriculture as well as broader issues of the logic of social economies which are 'expolar' – i.e. defined by neither state nor capital, both in Russia and in other lands as well. That is what may make Chayanov's resurrection different from a 'media event' or an extra picture on a wall. No doubt Chayanov himself would have approved of such fleeting cheers into further questions and possible new insights.

So what is the structure of the actual event of Chayanov's extermination and resurrection? What can be learnt from it about the mental universe of the social sciences and of Russia, rural and un-rural, at its new stages of self-recognition?

One way to answer those questions is to re-tell the story of Chayanov and his mode of analysis as one of treble death and an official resurrection. Chayanov died first, physically, executed as an enemy of the people, for making clear the true reasons for the fact and the form of Stalin's collectivisation. He was gotten rid of for a second time, however, in a 1950–70s period of the obligatory half-truth which aimed to cover up the guilty secrets of a generation of Soviet administrators and its lap-dog science. He died once more in the hands of the bulk of the Western theorists and practitioners of the large-scale business of 'development' – yes, their new advisers to Russia's government too, the 'establishment' as well as its radical critics. All of those did not simply refuse Chayanov's message but twisted it through trivialisation and, by doing so, committed it to the margins of thought. Last, after his official rehabilitation, he was promptly 'iconised' rather than utilised. This is why Chayanov's resurrection has been both significant and tenuous, and why it is a part and an index of an effort to have theories of social transformation re-thought.

A Muscovite scholar: 1888–1931–1937

Considering the number of people in the Western social sciences who heard the name of Chayanov, it is remarkable how few actually read him seriously rather than browsed through his single book translated into English in the late 1980s, or else picked up a view of him 'by second hand'. Even less know something about Chayanov as a human being, except of the single fact of his murder. Yet he is well worth knowing and not as an archaeological exhibit only. His concerns, methods and solutions make him, in fact, into our contemporary par excellence.

Alexander Chayanov was a major representative of a brilliant beginning-of-a-century generation of Russian intelligentsia. A grandson of a peasant whose son went to town and made good; a graduate of the then excellent Moscow Agricultural Academy, he became already, at the age of 24, nationally known by his works on the place of flax production in peasant economy and about the demographic determinants of it. From then on, he went from strength to strength in academic achievement and reputation. But he was much more than an agrarian master-economist of a country in which 80 percent of the population made their living in agriculture. A man of extremely rich mind, trained within the best humanist traditions of Europe, by the age of 40 he combined important analytical work, field research and methodological studies concerning peasantry with five novels, a 'utopia', a play, a book of poetry, and a half-finished guide to Western painting and a history of the city of Moscow. He spoke a number of languages, travelled extensively in Europe – before and after 1917 – and was closely knit into the cultural encounters of the educated Muscovites. He was a Moscow intellectual of the day at their best, deeply committed to the cause of improvement of the livelihood of the mass of common people, of human liberties and of his country's educational standards. Chayanov's rural focus of attention was rooted accordingly in a basic moral stand. As to his scholarly endeavour, Chayanov's thought breached disciplinary frontiers between economics, sociology, history, arts, agriculture and epistemology. His particular personal strength lay in a remarkable power of disciplined imagination and ability to put it in words, an outstanding and original models-creating ability, a capacity which bridged between his scholarly and artistic achievements and made him a theorist and leader among his peers.

As from 1919, under the new Soviet regime, Chayanov headed the social sciences 'Seminarium', later an Institute, in the country's Academy of Agricultural Sciences. This made him the central figure of his academic field. He never joined the new political Establishment and stayed himself, or what was then referred to as, a 'non-party expert'. His research work blossomed – most of his major studies and all his novels were written within a decade or so. In 1930, at the age of 42, Chayanov was dismissed from the directorship of his Institute to be, a year later, arrested for high treason and for the sabotage of Russian agriculture. After serving his period of imprisonment he was sent into exile, to be re-arrested and executed in 1937 (for a long time the family was misled to believe that he was shot in 1939). His wife was made to divorce him and to take another surname. This did not save her from being arrested in turn and sent into exile, from which she was to return only after Stalin's death. Of Chayanov's two sons, one died defending Moscow, having volunteered to the army despite his ill health. The second son, Vasilii, fought in the war, came back with bravery decorations and lives now, with his children and grandchildren, near Moscow in a family house built by Alexander Chayanov himself – a remarkable symbol of continuity.

To know who a scholar really is and to define his social place and significance, it is sometimes best to establish what he was killed for. Chayanov was killed for his *theory of differential optimums and vertical cooperation*. This carried a statement of theory, a programme of transformation of early Soviet agriculture as well as an indirect but clear and well substantiated condemnation of Stalin's programme of collectivisation.

Bolsheviks' theory of progress vis-à-vis theory of vertical cooperation: a scholar is silenced

To place Chayanov's lethal confrontation with the powers that be, one must first step back to consider the environment he addressed: its social and political structure, its new Establishment and the ideological vision which helped to give unity to supporters of Stalin's version of collectivisation. The newly created Soviet Union which had just emerged from civil war was sprawling, exhausted and poverty stricken, but in the process of rapid economic recovery. This came fastest in agriculture. By the mid-1920s, rural production was mostly back to its pre-war levels. But the political structure of the country had changed. The old privileged strata disappeared. The Communist Party members were, in Lenin's own words, still but a drop in the bucket of Russia's population. The strength of this new ruling stratum lay in the power of the state/party organisation and in the shortage of alternative foci of political power. The only major social organisations which survived, indeed, flourished anew outside the direct control of the new party-state, were the family farms and the peasant communes of rural Russia. Actual farming was mostly in the hands of the peasant households, whose most significant social frame were the village communes. The roughly 400,000 communes defined the daily life of the Russian countryside in the 1920s to a degree much greater than the state plenipotentiaries within the local Soviets, Party Branches and other local authorities which formerly ruled it. But the village communes' impact and power were localised and conditional. The state power was out of their direct influence – they were 'to be led' by 'the proletariat', i.e. by what stood for it. In control of practically all of Russia's land and with their customs and demands made into state law by the Land Code of 1922, the Russian peasants came to enjoy during the New Economic Policy (NEP) period of 1922–27 much of what they fought for in the 1905–07 revolution, Duma parliaments and the Civil War, why the majority of them opposed Stolypin reforms, but they understood well enough that political power was not in their hands. The tension between the monopoly of political power at the national level in the hands of the Communist hierarchy and the actual power in the rural localities, the division of the economic resources between the state-controlled industries and the rural small-holders, expressed the major political dilemmas of the Soviet Union in the 1920s (in the ideological language of the day it was referred to as 'the question of workers and peasants alliance').

In the face of the Soviet society of the NEP, its state leaders and ideologists of the 1920s shared amongst themselves (and attempted to impregnate their 'local cadres' with) a set of images defined by them as 'orthodox' Marxism. Three of its assumed social laws underlay the rural strategies adopted. These concerned progress, size, economic growth and investment and were linked to each other within a logical sequence. First, social progress towards a better future was being defined by the necessarily increasing manufacturing and energy capacities, the industrial societies as 'simply showing to the un-industrialised ones their own future' (Marx 1979, 9). Second, large size units of production were necessarily more effective but also better in their proletarian content and political symbolism. The reconstruction of production on the way to the socialist future was treated accordingly as a matter of growth of sizes, scales of technology and of capital investment. Next, the resources for this investment were to be taken from somewhere and, as Preobrazhensky pointed out, without the colonial exploitation and the international trade monopoly which the old predators had used to that purpose, only the peasants were left to pay the price of the Soviet industrialisation. Anyway, peasantry's essential characteristics, i.e. its being petit-bourgeoisie: small-sized and non-industrial and thereby reactionary and utopian, made its disappearance both inevitable and good, that is progressive. De-peasantisation offered accordingly a reliable index of

Progress, i.e. of a truly excellent world coming into being (but not before the peasantry had performed its social function in providing the industrialisation with resources within the 'primitive accumulation' of a socialist type). The breath-taking short-sightedness of this industrialisation model, which knew nothing of complexities and fully disregarded the non-economic factors as well as the long-term results (both economic and non-economic) of such a 'grab and demolish' strategy of social transformation, was made more plausible by evidence seemingly drawn from the industrial West, by war scares and, importantly, by the characteristics of the medium and lower ranks of the Bolshevik party 'cadres' – the backbone of the Soviet party-state.

There were bitter frustrations in this badly overworked and mostly underpaid group facing difficulties of daily management they were little trained for, of limited resources and of peasant stubbornness vis-à-vis the endless demands from 'the Centre'. The general mood of the Soviet lower officialdom, the police and the local party and its youth organisation's members, later the collectivisers on the march, was one of angry belief that peasants had it too good. The same was felt by them toward what was left of Russia's educated middle classes – the nicely-living perpetuators of endless doubts and of long phrases spoken with the air of self-significance. Experts' quibbles and scholars' deliberations sounded like disloyalty, if not outright sabotage, while revolutionary courage ('like in 1919') and simple choices could carry the day. Time was short, a war possible, the difficulties multiple, communism still very far away – an image of a great leap forward commended itself easily to the human products of history of mass Bolshevism in the days of Civil War and War Communism, of cavalry charges and of a New Jerusalem which seemed just around the corner.

Those were not only matters of pure perception and moods, however. The dream of a great leap forward expressed and released the pent-up ambitions of a new generation of party activists who came into the ring too late to claim a revolutionary pedigree for their own leadership status. Nor had they the education needed to become 'an expert'. Stalin's marching army of supporters, admirers, executors and executioners, the party lower and middle cadres of the late 1920s, were, to a major degree, peasant sons usually recruited via army service and/or the Komsomol (Young Communist League). Village-bred lads promoted NCO and after military service unwilling to go back into the daily rot of their father's family farms, with few classes of schooling, a short vocabulary of acquired Marxist phrases, much energy and some common wit were typical of it. They were young, brash, not very literate and painfully aware of it. Such cast of mind valued above all loyalty and obedience, strict order and simple solutions which one could see and touch directly. Also, in the immensity of the country the spirit of revolutionary upheaval was only now reaching some of its faraway corners and firing its young people with new Messianic zeal and grand expectations. The deepest dislike of the new cadres was reserved for the peasants – stupid, slow, led by those who grasp most and, for the intelligentsia – seen as too clever by half. To enhance the country and to promote oneself was to bring to heel the former group and to replace by 'loyal comrades' (yourself included) the latter one. Each expert dismissed, or an 'old revolutionary' purged, was one promotion more to a top position otherwise unachievable. Stalinism in the 1930s was for many not a matter of surrender to fear, but a political stance felt intuitively true as much as personally profitable. Crude radicalism, totality of obedience and careerism combined to produce the

Communist Party's new cadres of the 1930s, a new social mobility and a new political hierarchy for the USSR to follow.

In the face of this political establishment, its ideas, cravings and moods, the core of Chayanov's general argument concerning the future of Soviet agriculture was presented in full in his 1927 book *Osnovnye idei i formy organizatsii sel'skokhozyaistvennoi kooperatsii* ('The basic ideas and forms of agricultural cooperation' but better expressed as *The Theory of Agricultural Cooperatives* and published as such). The general view offered there will be summarised shortly. Chayanov accepted that the Soviet peasant agriculture and rural society of the late 1920s were in need of massive restructuring to upgrade and to acculturate them. He accepted also the *formal* goals of Stalin's collectivisation programme: that is to increase productivity, to secure well-being and to enhance social justice in the Soviet countryside. But in Chayanov's view the methods suggested were wrong on all scores.

In the official vision, the increase in the size and the mechanisation of units of production was to guarantee the achievement of high productivity and of the rural well-being, while social justice and egalitarian democracy were to come through the destruction of the exploitive rural rich (the 'Kulaks'). Chayanov argued that it was not true that increase in size of production units necessarily enhances productivity in agriculture – different branches of farming would have different optimal unit-sizes while in rural production dis-economies of scale were as harmful as the undersize. Different optimal sizes are characteristic of the different activities which come together in farming. The advancing social division of labour takes mostly the form of some aspects and participants in farmers' occupation being singled out and specialised, with the possible selective increase in this unit's size and/or capital input, reflecting thereby the best use of resources – a 'vertical' segmentation adjusting to the *differential optimums*. At the same time, a universal increase in size of units may actually decrease the overall productivity. The 'large-scale only' is as bad as the 'small-scale only' where farming is concerned. Also, large-scale units created overnight would not in Russia find local leaders able to manage them and managers would have to be 'imported' into villages. They would lack local roots and specific knowledge of local conditions of farming. They would also be fully linked into and dependent on a state apparatus – bureaucratic, detached and necessarily repressive. There is no reason to assume that such new local managers would be less 'unequal' or less exploitive than the neighbours-exploiters of old. Basically, it is the self-management and particular eco-systems' efficiency that are linked in a peasant countryside with relative well-being. Peasants know it. Their opposition to policies and declarations which contradicted their daily experience concerning production, productivity, power and exploitation in an environment they know best would be as harsh as it would be destructive of the agricultural resources (directly as well as indirectly – through 'feet-dragging'), which the collectivisation ostensibly aimed to increase.

Chayanov's alternative programme to achieve agricultural transformation of the country was to advance composite cooperation from below by the smallholders (he used the expressions 'vertical cooperation' and 'cooperative collectivisation' for it). This was based on Chayanov's observations of Russia's actual cooperative movement in 1910–14 and 1922–28, as well as of the spontaneous processes of what he called the *vertical division of labour* within the market-related peasantry when aspects of its most profitable economic activities (rather than the whole process of farming) are being picked out by outsiders-entrepreneurs. The cooperative movement offered, to this view, a democratic alternative to specialisation, which takes control by outsiders-entrepreneurs, to stagnation and an exploitive and productivity-limiting state-centralisation as well.

The best solution to the problem of increase in rural productivity for Russia lay, in Chayanov's view, in a flexible *combination* of large and small units, defined by the *different* optimal sises within different branches of agricultural production, i.e. the adjustment of units to sizes best suited to production (e.g. fodder at, say, village level, multi-village units when such are justified, e.g. for forestry, family farms' production for eggs, etc.). Combined production would mean also, for example, that fodder produced most effectively on a large and mechanised cooperative farm can be used for the production of milk by the family farms, to be processed then by the cooperatively-managed local butter factory and sold in town or abroad by a region-wide marketing cooperative. To give structure to such combinations and to secure their democratic nature Chayanov supported a *multi-level cooperative movement*, a cooperative of cooperatives, organised 'from below' and facilitated but not managed by the government. A socialist government's policies and the dominant socialist perceptions within the country could have secured, to Chayanov's view, the quick advance of such rural cooperation. This farming scheme would provide an open-ended system, able to adopt new agricultural techniques, while using, rather than bulldozing-out, the existing rural social structures and spontaneous processes. It would advance productivity, enhance social equality and, at the same time, act as a school for new democratic leadership of the localities. Through an ongoing process, this system was to link to the national plans of industrialisation then considered as well as with the envisaged social reconstruction and cultural change. Corresponding with peasant experience, using peasant institutions and wide open to peasant input of initiative and of cadres it would be acceptable within rural communities and able to tap their energies for continuous reform.

Importantly, Chayanov's vision was not only prescriptive but descriptive as well. During the 1920s a rich array of cooperative farms advanced and spread in the Soviet countryside. Building on pre-revolutionary organisation, rooted in local initiative and led by a remarkable generation of devotees of cooperative development, many thousands of cooperatives for supply, selling, credit, and production incorporated by 1928 over half of the Soviet rural population. Their network proceeded to increase and 'become more dense', until they were forcibly out-rooted by the collectivisation. This massive phenomenon, as well as Chayanov's status as a scholar, may explain why a version of 'chayanovian' strategy of rural reconstruction was adopted also by Bukharin's wing of the Communist Party leadership and seemingly assumed by the initial version of the first Five Year Plan (prepared under the supervision of Russia's outstanding Marxist-but-not-Bolshevik economist, V. Groman, to be set aside by Stalin and substituted by his own version and its supra-figures, fully out of touch with what was to take place). The grim tale of the demolition of Bukharin's alternative and of the physical destruction of its supporters as well as of all of the country's leading economic advisers is known enough.

Chayanov's analysis, which challenged directly the route the collectivisation was actually to take, and its official legitimation, was based on an accumulated expertise of Russia's outstanding rural studies literature. There was no better knowledge on offer, which may explain why both analysis and the extensive evidence gathered in the 1920s were, on the whole, simply brushed aside by those in ultimate authority. Chayanov and those who thought like himself were mostly answered by the claim that their position was hypocritical – that they did not actually wish for the rural reforms to happen. Chayanov himself was viciously attacked as the rural exploiters' defender and an organiser of a clandestine and counter-revolutionary Labouring Peasants' Party (TKP), which he never was. The true reason for this smokescreen and for the fury with which 'chayanovism' was then being condemned lay in the fact that Chayanov's argument slipped the cover from a major secret – the actual grand hypocrisy of the day. Stalin's collectivisation programme did not really aim at the goals it declared: 'for the sake of progress' it actually aimed to break the backbone of the peasants' social power and to 'pump-across' ('*perekachat*' was the term used then) peasant resources into industrial construction, the army and the needs of the party-state apparatus in an exercise which, vis-à-vis its own rural producers, adopted many of the features of early colonialism at its most grasping and its least effective. Beneath the claims to ultimate science of social progress lay a pirate's idea of capital accumulation through the grabbing and the consuming of the golden egg, be what may the consequences to the existing rural economy and society. This strategy was put to use by a crafty and unscrupulous man in his rise to total power (and still in the shadow of his predecessors' call for his dismissal as being too brutal to direct their party). The decision to 'collectivise' was about that. Chayanov made this clear in the most effective of ways – through offering an alternative backed by scholarship and in no contradiction with peasants' interests and actual choices.

Chayanov's second death in the soviet ideological scene: 1960s–85

What followed the 1929–30s collectivisation policy turning was much destruction, extensive rural famine, many deaths and a great silence. Guided by the brilliant insight of the immortal Stalin, Soviet agriculture was acclaimed by all within his reach to be the best in the world. Or else . . .

When, in the mid-1950s, the new rulers of the USSR began to take stock of Stalin's actual inheritance, it took them little time to conclude that agriculture and rural society were the country's sore point and worst impediment. None of the formal aims of Stalin were actually achieved in the countryside – the rural population was poor, the agricultural production stagnant, the equality and democracy-at-root non-existent. Results of the 1941–45 war, bad as they were, did not explain those conditions of Russian agriculture, nor did the post-war reconstruction improve matters spontaneously or at speed. Malenkov, Khrushchev, Brezhnev, all of the follow-up leaders of the USSR, were to give much attention and many words to 'further improvement', i.e. to the continuing crisis of agriculture. Massive resources were by now being poured into it, and yet, by the 1970s, the USSR found itself importing food. With the exception of manual labour, agriculture did not, in fact, seem to provide even the expected 'capital accumulation' for the sake of the industrialisation of the 1930s. Later, and especially since the 1960s, agriculture was increasingly becoming a massive drag on country well-being and economic growth. Worse, the long-term negative results of Stalin's collectivisation began now to show: ecological decline, demographic crisis, selective migration gutting villages and depopulating whole rural areas (e.g. the north and centre of European Russia).

The over-urbanisation of the 1960s–80s ran contrary to the socially most effective solutions and produced many further ills usually associated with Third World countries.

As the failure of state policies to resolve the rural problems became more explicit, these policies increasingly became politically poignant. Khrushchev lost his job when his maize programme failed and made him look ridiculous, while bread lines were beginning to form in major towns. Brezhnev despoiled Siberia, exchanging its riches for imported grain. Gorbachev called for the restructuring of collectivised agriculture along lines fully vindicating Chayanov's criticism and suggestions. In 1987 he announced that the first steps of the economics of Perestroika must be a breakthrough in agriculture, it stands or fails on it. It failed and failed.

Yet, this is not just a sorry tale of truth prevailing and scholarship vindicated but with little made of it. For the question is: why did so little change? And before that stands the question of what made the silence concerning Chayanov's analysis of collectivisation last for a third of a century after Stalin's death, while the agrarian crisis deepened and became more and more difficult to resolve?

By the late 1950s, the terror of arbitrary arrests and executions was over. Khrushchev condemned Stalin and the city of Stalingrad became Volgograd. New leaders experimented at least twice with major efforts at economic reforms. The manufacturing industry and urban life were, for a time, on the mend, which made clearer still the disastrous state of agriculture and of the villages. The food supplies were increasingly inadequate, and a Muscovite joke of those days about an infant reporting both parents were out, the father having flown to the moon, the mother queuing for sugar, said it all. Yet, no pressure to reconsider at full depth the issues of the countryside, past and present, was generated by the Soviet scholarly community in the 1950s–70s even before the late-Brezhnev period made such a discussion, once again, barely possible. A number of remarkably successful self-generated local experiments with alternative methods of running agriculture were actually suppressed then, often with great brutality. In line with all of that, Chayanov was 'rehabilitated' but only in part, i.e. from his 1937 capital charge but not from the 1928–33 vilification and sentence. His works stayed accordingly out of bounds. His name was now being mentioned, but only negatively – as a representative of the petit-bourgeoisie utopianism proven wrong by the brilliant development of Soviet collective farming.

An Indian tradition defines the Brahmin's caste as 'born twice'. Scholars can be killed twice, through physical death but also through their heritage being wilfully forgotten. And it is the second type of death which would probably strike the greater of them as being worse than physical death. Why then, while evidence mounted as to his predictions and prescriptions being right, was Chayanov sentenced to oblivion by his professional peers and by the politicians of the first post-Stalin generation in the USSR? We can restate this question to say: why was there not until 1987 or so any signs of fundamental reconsideration of the agricultural system established by Stalin's collectivisation (rather than an endless debate as to how to make it work in the face of its persistent failures)?

The social context, the public mood and the social carriers of the 1950s to the early 1980s great silence in the face of a systematic rural crisis differed considerably from those who offered the social and political background to Chayanov's murder. The country's political structure was now, for a time, securely monolithic. The general economic conditions seemed to improve, if very slowly. The romantic era was over; rationalism and science were in fashion. At the top, on Lenin's tomb, standing on parade, youthful faces and half military dresses were replaced by elderly podginess, suits and epaulettes. The mood of the 'party cadres' was increasingly centred on the enhancement of personal well-being. Chauffeured cars and visits abroad spread as the new badge of authority of the middle ranks. The administrative hierarchy was increasingly manned by university graduates, professional expertise was being praised, work in the numerous research institutes well rewarded. There was also some measure of licence to argue, while those who refused to toe the line paid for it only by their promotion prospects ('like in the West'), rather than by prison or death.

Those who expected an instant explosion of rationalism and critical analysis from the better-educated and better-off new elite, made freer by government's adherence to 'socialist legality', were to be disappointed. What blocked a going-to-the-root review of Soviet agriculture and any recovery of the insights and analytical achievements of 1920s was, to begin with, the power of a political generation which reached its full maturity and majority in the 1950s–70s. The young collectivisers of the 1930s who survived the 1937 Purge and often benefited by it, were now at the top of many hierarchies of power and networks of patronage. Any view and any name which indicated that there was something badly wrong with the collectivisation was casting doubts on their political biographies and delegitimated the authority they held. This was not restricted to the political bosses only. To the scholars-in-charge the emotional investment in the explanations and justifications they had offered for decades was immense, underlining their defensive postures. For who would wish to be called a cheat by one's own students, or to face up to the fact that one's own cowardice contributed to the current crisis and possibly to the murder of the Chayanovs of one's own generation? The few colleagues who insisted on offering critical commentary could be marginalised with the help of the Party watchdog departments, the institutes' directors, journals' editors and the Glavlit censors.

The worst cases would face a total ban on publication, on working with students and on travel abroad. The young blades could be told which views earned promotions and, if necessary, be hammered into obedience. The defenders of the established truth found their countryside equivalent in those in whose interest it was for collectivised agriculture to be sustained, come what may – the little tsars of the countryside, the district Party secretaries and, in their entourage, the chairmen of collective farms and the directors of the state farms.

But there was more to it than the grip of Stalin's generation's old hands. There was also the power of intellectual inertia of endlessly repeated dead thought. While the country changed and so did its carriers of political power and organised knowledge, all of the dogmas of Stalin's generation's *Weltenschaung* concerning Progress (usually a crude rehash of Karl Kautsky's and Friedrich Engels's Darwinian historiography and positivistic epistemology) were still in place. As long as this general outlook persisted and filled all of the permitted intellectual space, the very logic and 'scientificity' of analysis dictated misconceptions where Soviet agriculture was concerned (and, of course, not only in this field).

At the centre of this *Weltenschaung* stood an assumption of hierarchy of forms of economic organisation: the State economy first as socialism's synonym, the state-controlled cooperatives next as its lesser version, the family economic units (with an invalidating term 'petit-bourgeois' attached to it), and finally, the capitalist enterprises 'exploiting wage labour for private profit'. This view fitted well the state-bureaucratic obsessions with party/state control. Even when supra-centralisation was put in some doubt (e.g. under Khrushchev), this ideological hierarchy of preferences kept sway when agriculture was concerned. Indeed, an additional tacit hierarchy facilitated it. This extra hierarchy ordered all branches of production in terms of what can be only described as aesthetic *cum* symbolic attributes, rather than of the productivity, of social characteristics or of needs – with heavy industry (the more smoke, the more wonderful) at the very top and with farming very close to the bottom. (This extended also into farming itself, where tractors were naturally at the top of the scale of prestige and pay, while actual land husbandry was close to the bottom.) These hierarchies linked directly into the dogmas professed at the beginning of collectivisation and still very much alive. Industrialisation, understood as large-scale manufacturing, was still seen as the only way to make a country rich and powerful with 'all else' bound naturally to disappear or 'to follow'. It was also made into the chief index of socialism's progress. 'Large' was beautiful and necessarily more productive than 'small'. There was still one and only road of Progress, i.e. goodness and, while some past mistakes were now being admitted, the USSR was still leading the world by showing it its future. And there was the assumption of 'Socialist Primitive Accumulation' necessarily preceding the industrialised glorious future – as a concept, a mood, and a moral judgment granting to the modernisers a peculiar charter to walk over humans for their own sake.

There was still more than the blocking force of this combination of men of the past and of past ideas, for the ability to perpetuate these ideas under new circumstances must be understood in relation to the actual conditions of the day. Defence of Stalin's collectivisation (with Stalin's own name now usually avoided) set well with a system of centralised and bureaucratised administration. In a system in which good citisenship and socialist convictions were synonymous with the execution of the orders from above, anything which smacked of decentralisation and local centres of authority was deeply suspect. The most plausible solution to the difficulties of over-centralised agriculture was not, in this context, the rethinking and resetting of its structure but a still larger dose of centralisation: the enlargement of the *kolkhoz*, its statisation, the decommissioning of the smaller villages (to be re-allocated into larger units), etc.

When in trouble, one could also ask to increase the centrally allocated industrial inputs: tractors, fertilisers, etc. 'Not to rock the boat' was basic intuition and a 'need' felt throughout a bureaucratic system. Unchallenged authority was treated as a necessary ingredient of smooth social functioning. Guilt towards those purged and fear to admit to one's own silences, opportunistic dodges, small cruelties or large crimes gave it powerful emotional roots. It was the strength of this combination of interest and emotions, resources and controls, cynical lies and plausible common sense, which made an agrarian system survive unchallenged for 30 years after Stalin, while it was moving towards its eventual state of 80 billion roubles in subsidies, 50 million tons of grain in imports, demographic decline and ecological disaster.

The 'free world': imaginations, frontiers and limits

Have we reached then the end of explanation, of reasons for Chayanov's outlook and programme being 'brushed under the carpet'? Not quite so, because some of the basic problems of agrarian policy, uncovered by the Soviet experience of collectivisation, were being repeated also outside the realm of control by the Soviet bureaucracy and scholarship. There must have been, thereby, something more than the Soviet establishment's power and dynamics at play. Or, to have it once more personalised, Chayanov died for the third time in the Western establishment's 'development theories' of 1960s–80s concerning the Third World, as well as in alternatives offered by most of their radical critics before IMF power came to marginalise debate. Nobody there was being arrested or dismissed because of Chayanov. His major book translated into English in 1966 made quite a splash, but silence settled fast over his work. Worse still, in the West, Chayanov's views were being systematically misrepresented (or, was it misread, or was it not-quite-read?) by those who did refer to him at all.

To recognise what it means we must step on in our explanations, from the sole effect of causalities of fear which dictate obedience and of self interest which silences objections into the realm of ideology, i.e. of coherent and rationally held misconception and of cognitive paralysis created by disabling words and world-views. For, in the West and its intellectual dependencies, some assumptions and policies we so often associate with Stalin's terror or the post-Stalinist Soviet bureaucracy have been, in fact, present as well. It was the Shah of Iran and his officials who set up the *Collective Boneh* in Fars and the state farms in Khuzistan, i.e. schemes practically identical to the Soviet collective farm (*kolkhoz*) and state farm (*sovkhoz*), to be blessed there by the US experts and advisers. Clearly, we are not talking then of something inherently Communist or particularly Marxist when such organisations are concerned. Later, in Africa, Tunisia's state-worshipping nation-alists set up their own *kolkhoz* quite on a par with the collectives of an Ethiopian military regime (which chose to describe itself then, not unlike its enemies who overthrew it, as 'Marxist-Leninist'). The so-called peasant cooperatives of Ecuador or of Egypt, which are usually neither peasant nor cooperative, but state institutions of input/output control imposed by officialdom on reluctant peasantry, showed much similarity also, and so did many actualities of the Ujamma in early 'populist' Tanzania. There was also marked consistency of failure to achieve the official goals each of these schemes set for itself. Once we rid ourselves of excessive attention to labels, we find in the contemporary 'developmentism' much of Soviet agrarian debates and their resolutions by Stalin and by Brezhnev.

Barely ever were the actual lessons of Soviet collectivisation and its critique by Chayanov taken on board in the 'developing societies' section of the 'Free World'. When Chayanov was cited in the 'developmentist' literature he was usually used as a synonym of the 'small is beautiful' outlook and of programmes which were actually never his, i.e. he was treated as a defender of small-holders *per se*. This caricature was very often employed as an anti-model, and a punch-bag by brash defenders of capitalist progress and/or state intervention, 'Socialist' or non-'Socialist'. In fact, Chayanov was no defender of 'small' versus 'large', he only objected to the 'large is beautiful' formula (and sounds remarkably up-to-date thereby when we look at the 1990s social and economic analysis). His was the 'combined is beautiful' strategy of development, based not on dreams but on thorough knowledge of agriculture and rural social organisations all through Europe. He was no 'populist' either insofar as party allegiances and the substantive political views of Russia's populists were concerned (e.g. their belief in the exclusive virtues of the Russian peasant commune). The label of 'neopopulism' attached to him by his foes and his murderers remains that of a Caribbean peasant saying: 'you call a man "a dog" to hang him'. Yet, his actual message could have saved the rural objects of 'agriculture's development policies' in many countries in the world over much destruction and grief.

Let us turn, once again, a persistent miscomprehension into questions and ask what made for such a broadly shared yet misleading view of Chayanov, followed by his *de facto* rejection? There have been at least four elements facilitating the continuous misreading-for-purpose-of-rejection of Chayanov's works within the academic communities of the 'First' and the 'Third' Worlds.

First, there has been the assumption of the archaic nature and thereby of the necessary disappearance of family economies. This was inbuilt into a general 'theory of progress' taken as given by most contemporary social scientists. Family economy was seen as belonging to the past on the strength of an extrapolation from the nineteenth-century industrialisation, with all else (inclusive of the late twentieth-century evidence of so-called 'informal' or 'expolary' economies as well as of the actual prevalence of small-unit farming in Europe) brushed aside. Peasant farming was being 'talked out' of reality with particular zeal by the Third World's versions of 'orthodox' Marxists through the uses of the term 'petty-bourgeoisie' (which carried the double negative of being capitalist and being backward as well). Rather than face Chayanov's analytical conclusion that family farming was neither capitalist nor necessarily inefficient, it was simpler to attach to him an invalidating label of 'smallholders-lover'.

Second, there has long been the monopoly of state-wide models of political economy, their deductive logic-from-above expanded to every unit operating in its context and used as ultimate explanation of all and sundry. Chayanov's particular point of epistemological brilliancy – his effort to build economic models 'from below' (via the diverse logic of specific enterprises and their complex combinations) was too much to comprehend for too many.

Third, any reduction of rural realities to an economicistic model clashes with the way Chayanov and his allies challenged the recognised disciplinary frontiers. History, sociology, agronomy, economics were to them one where the actual life of farmers was concerned. Chayanov called it 'social agronomy'. Once again, to many of the contemporary experts, knowing more and more about less and less, it was too much to handle.

Finally, those to whom state planning and free market, alternatively or as a combination, are the only possible forms in which social economy can function, find unacceptable Chayanov's outlook, which treated family-economy strategies in agriculture as showing a discreet operational logic. Consequently, they usually fail to see that Chayanov never assumed in actual reality a total autonomy of peasant economies with economies at large (models are, of course, a different matter). It was they who disregarded his insights, he never disregarded the substance of theirs – be it the market economy context, the new technological advances, or the place of capital and wage labour in peasant life.

The peasant agenda and a 'post-modern' epistemology

The lessons of Chayanov's treble death can be now summarised before turning to the issue of the contemporaneity of his message.

First, repressive dictatorships masquerading as a socialist paradise showed their true colour through policies and ideologies of state centralisation and the de-humanising of social sciences by their total bureaucratisation and claims of their total objectivity. Those who challenged this were treated as enemies. The most perceptive of critics paid with their lives. Scholarship and a dictatorship do not coincide well, at least in the long term. To be effective, dictatorships need not only the use of force and fear but also the demolition of alternatives from which opposition and hope can grow.

Second, repressive politicians who wish to control and exploit their people are never the sole agent of oppression. Its ideologisation by official intellectuals and the institutions of organised science has its own logic, power and momentum, its own ways of control and exploitation. These must be considered when issues of understanding and misunderstanding are involved. There are, as Kuhn asserted, conservative paradigms and set agendas of perception within the communities of scholars, but one must remember as well the determinant of corruption by privileges and the power hierarchies at roots of scholarly lies and self-deceptions.

Third, an implicit bridge of an ideological nature exists between the murderous crudities of Stalinism and much of the well-meaning advice of Western specialists concerning worlds unlike their own. The ideological substance of this bridge in analysis, emotions and trained intuitions, has been the contemporary versions of evolutionism expressed as the 'theory of progress' and/or 'modernisation'. Even when stripped of Stalin's hooliganism and Brezhnev's corruption it remains a charter for arrogant inhumanity to which peasants have been one of the major victims – 'for their own good', of course.

All of this explains the role Chayanov's work came to play in the 'peasantological' breakthrough within the Anglo-Saxon academic literature of the 1960s–70s, as an analytical approach one can describe as a Peasant Agenda came into its own. A considerable measure of conceptual continuity can be shown between Chayanov's and his friends' insights and the analytical achievements of the 1960s–70s and farther on. At the core of this continuity lay the refusal of deconceptualisation of peasantry and/or its reduction to footnotes. Attention was focused on the particularities of family farms as socio-economic entities, on the specific characteristics of rural societies and their modes of transformation. This approach has also assumed the need for some discreet analytical structures for their comprehension. The Peasant Agenda broad tradition addressed in a way new to the 1960s social conditions of the countryside, while at the same time generating as the time proceeded, important insights into some broader epistemological issues. Chayanov's view was followed and developed (and often rediscovered) as to the particular, discreet and parallel, yet related, economic modes, which do not demolish each other but combine in a continuous manner. The same can be said about the from-the-bottom-up perception of social economy, the fundamental multi-disciplinarity involved in the notion of 'social agronomy', and a particularly non-deterministic comprehension of social structures which, paradoxically, links directly into the current debate of post-modernism and of mathematical theories of chaos. General theory apart, this approach offers also important analytical input into the analysis of contemporary industrial and 'post-industrial' society such as issues of 'informal' (or 'expolary') economies, of the particular place of labour in family economies, of the decentralised yet integrated organisation of production, etc. Within the field of political ideology Chayanov put at its strongest a vision of cooperative movement as an alternative 'socialism from below', decentralised and communalised yet highly effective in its economic results.

This is why one could sum up, in the days when Chayanov was still Russia's 'un-person' but on the threshold of becoming its favourite son, not long ago and yet in what seems like the archaic past, to say: in fact there are still hundreds of millions of peasants and as many may exist in the year 2000 but, paradoxically, Chayanov's fundamental methods and insights may prove particularly enriching for worlds of fewer peasants as well as of fewer 'classical' industrial proletarians, while the subject of his actual concern, the Russian peasantry, has all but disappeared, which will make a good epitaph for a memorial of a great scholar when his countrymen remember to build him one.

All of it holds true.

The current debate about agriculture in the USSR: Chayanov's iconisation, Chayanov's use

Considering the extent of Chayanov's contemporaneity, what was the actual impact of his recovery in the USSR as from 1987? Most immediately, how did it link into the 'agrarian debate'? To answer this, one must move from the sole considerations of the rational choices of policies to that of the political and ideological struggles of the day.

The most direct way by which a Soviet citizen had been introduced daily to the current economic crisis was through the food supplies – the shortages, the price-inflation, the inadequate quality, the over-use of nitrates, etc. There had been also a widespread feeling that agriculture could do much better and that improvements in it could have been actually achieved faster than in the industry. At the 1988 Communist Party Conference – the beginning of what was billed as the new radical stage – Gorbachev declared rapid improvement of agriculture to be the first material change that must signal the general economic successes of Perestroika and to bring it to every household. This is where Perestroika was to pass or to fail a test everybody in the USSR could recognise, establishing public confidence in its authors. It failed and, in direct consequence, so did they.

There were six general lessons initially drawn by Perestroika's radical reformers of the 1980s from the sorry state of Soviet agriculture and of the past efforts to overcome it. First, the increase in the size of the units of production undertaken under Stalin and, once again, under Brezhnev, did not result in increase in productivity. Second, the steep rise in chemical inputs and mechanisation in the post-Stalin era did not secure it either, for after some improvement stagnation set it. No simple formula of 'the more you put in, the more you get out' seemed to work. Third, the extension of services offered 'from above' by the state organisations (and usually paid for by the producers and through state subsidies, e.g. the so-called Agro Industrial Complexes) did not improve matters either. It seemed rather that to become more effective agriculture had to be de-bureaucratised. Fourth, it was agreed generally that in the provision of food supplies the personal interests of a farmer must go hand-in-hand with national needs. But a single farmer is clearly no match for the local bureaucracy controlling services and supplies. Matters of production and supply are also matters of authority and control (hence the growing demands for privatisation of lands and the efforts to set up rural small-holders' associations or even a Peasant Party). Fifth, the farmers' ability to deliver the goods had been deteriorating also because of extensive ecological decline. Sixth, the drop in quality of rural populations and labour via selective migration into towns, which had been taking away the younger, the brighter and the better educated, made the countryside into the country's slum. To deliver foodstuff effectively the whole nature of rural social life had to change.

The severe problems of stagnant production, extensive waste, failure to deliver produce to the right place and in right conditions, as well as the long term deterioration of rural environment and population, have been increasingly said to be solvable only insofar as 'the human factor' would come to play a new and different role within agriculture. New technology, new skills, and a direct link through fair remuneration of the individual's and the nation's economic interests had been important, it was being said, but insufficient on their own. The rural population would need to recover its 'feeling of being a master' (or should we say, using Western terminology, becoming a subject rather than only an object of social and production processes and policies). To Perestroika radicals the recovery by actual farmers of authority and responsibility vis-à-vis the bureaucratic machine and its little local tsars, has been increasingly seen as the one way to save what needs saving, to advance what needs advancing and to invest what needs investing, and thereby to secure the long term qualitative improvement of communal welfare, while providing the country with the food supplies needed.

For a time this general view seemed to sweep all before it. Vis-à-vis the old scale of preferred forms of production: state ownership at the top, the state-directed cooperatives next, then the family farms and a capitalist economy, Gorbachev had declared the first three equal in their socialist credentials, i.e. legitimate, and only the fourth contrary to it. Re-peasantisation (*'okrest'yanivanie'*) became, after 1987, the Soviet government's official goal and media's pet. Curtailment of bureaucratic management, genuine re-cooperativisation and freedom of family farming, were declared to be the major state targets of the rural transformation of Russia.

Chayanov's return into the world of the living has to be seen (also) as part of debate over the alternative agricultural strategies. This was linked to a new historiography of the Soviet countryside which came to see Stalin's collectivisation as a major disaster to Soviet agriculture and the major reason for its many failures. It bore out fully Chayanov's criticism and gave new substance to the alternatives he offered. The assumption that, while inputs and prices matter, the socio-economic structures of farming must be also put right to enable effective and flexible response to the changing nature of agriculture and to the market demand, went well with Chayanov's major insights. Concepts of 'differential optimums', 'vertical differentiation' and the basic preference for *combinations* of large and small units within a self-governing cooperative structure, spoke directly to the needs as perceived and were able to give them substance and specificity. It also fitted well into what was substantively a 'third way' being suggested, one of a socialist alternative to the Bolshevik party-state which was neither the return of the pre–1917 past nor simply a move to become 'Western'.

It was in that spirit that Alexander Yakovlev, who for a time appeared as Gorbachev's alter ego, offered a fair index of the basic ideas which to him underpinned Perestroika of 1987–88. He spoke about the need to leave behind the 'dogma of total obedience of individual to the state', the necessity to achieve social changes 'not by all means', the duty to treat humans as the subject of change rather than its object. He accentuated the idea that the main division lies not between the supporters of the different forms of economy, but between social and economic systems of production directed towards the solution of human needs and those in which humans are used for somebody else's economic profit. He also described the Communist Party's attitude to peasants as the worst of its 'failures and crimes', and defined his own general programme as returning to general human values and as 'ethical socialism'. In line with it the term used then by Gorbachev for agricultural reform was that of 'cooperative collectivisation', i.e. the one actually introduced by Chayanov. Now, however, it was the collective and state agriculture which were to be 're-cooperativised', i.e. decentralised and made democratic, rather than the family farms of the 1920s helped to combine in new ways.

But clearly, ideas on their own, even the most sensible, do not singly make for a better world. The failure of the Soviet economic reforms in general and especially so in agriculture, and the shortage of daily supplies, was resulting in growing popular fury and in the decline of hope at 'the top' and at 'the bottom'.

Ironically, for an environment which tries hard to forget its past ideological masters, to understand what happened to the Soviet reforms of 1987–90 one must take here a leaf from Marx. The views of Chayanov as much as those of Yakovlev challenged basic interests and triggered deep fears in the holders of the Soviet 'forces of production', of power and of privilege. In matters concerning agriculture a powerful lobby of the chairmen and the directors of the collective and the state farms expressed its interest in a powerful, conservative backlash. Here was the very group of neo-controllers, neo-exploiters and new go-betweens, whose rise and character Chayanov actually predicted and made into a major reason for his objections to Stalin's collectivisation. By now in the Russian countryside they controlled the land and the equipment, housing and transportation, and through these dominated humans and goods. Gorbachevian weakening of the party-state made them even stronger and they were not going to give up their positions lightly. While Moscow talked reforms, their defensive response was sufficient to have those blocked. By the end of 1989, as 'nothing much happened' while food supplies proceeded to dwindle, the chairmen and directors' lobby moved over to an attack. They established powerful presence in the Supreme Soviet and took over the newly created 'Peasant Union'. Delivered by its leader, Starodubtsev – the junta's member-to-be in August 1990, their message was clear: if the Government wanted more food it should invest more in agriculture, leaving resources there in the hands of those who already controlled it. Also, the only acceptable way to introduce family farming was to do so under full control and as part of operations of the large production units. Or else ... (a food-delivery strike was even mentioned). The empty shelves during an excellent agricultural year 1990–91 gave it all a peculiarly phantasmagorical background. Within a short while, the government was giving way.

In the face of the food-supply crisis an important ideological shift was taking place on the other side of the political divide. Perestroika's radicals – its 'Left' by self-definition, straddled in fact two different strands of interpretation. On the one hand stood those who had seen the country's future as some type of integration of its Soviet and non-Soviet past and present, goals and deeds, while on the other hand were those who aimed to out-route the regime which failed lock, stock and barrel.

The first spoke of a change due to bring out for once the humanist potentials of socialism, while the second group took its cues from 'the West', especially the USA. Be what may its reasons – conservative spanner in the wheels, ethnic strife, Gorbachev's personal weakness, or inadequacies of their own programmes, after 1988 Perestroika's radicals were facing the decline of hope for the rapid transformation of the Soviet society. The severity of the economic crisis gave new strength to their enemies in the conservatives' lobbies. Within the radicals' ranks it gave strength to visions of a totally free market, i.e. to what was perceived as 'learning from the West', but was badly informed to the point of sounding bizarre to Western observers. This view embraced such symbols of popular well-being as Thatcher of the UK and Pinochet of Chile. Yet Perestroika's defeats gave it power of argument along the lines of 'if not us, it must be them who got it right. Look at their shops!'

The political battle, in which the alliance of conservative communists and the lobbyists for those who control bureaucratic structures and major enterprises of the old regime have been fighting what is mostly a nineteenth-century-sounding freemarketeers' position, has left precariously little space for the 'third way' political programmes and/or original solutions. During 1989–91 the USSR had been moving in this sense along lines not unlike those of the other countries of Eastern Europe.

This is the context in which, in matters of agriculture and rural society, Chayanov's contemporaneity and professional brilliancy failed to breach the ramparts of establishment's interests as well as of its critiques' simplistic perceptions. What was to be done thereby with one of Russia's brightest sons, whose martyrdom and newly discovered personal qualities was giving him the halo of a saint? The answer: Chayanov was rapidly 'iconised' Increasingly his name came to be mentioned respectfully and put aside, bowed to and fast forgotten, hung on walls but never considered in depth. Or else he was mentioned offhand as supportive of views he would not even recognise. Or else again, he was assigned to a symbolic history-play of goodies and baddies, to be now the goody that was 'for the Russian peasants' against Bolshevik baddies and 'their' Karl Marx who were 'against the Russian peasants'.

Is it to be Chayanov's fourth and final death through veneration? This will depend on his motherland's future history, in which rural population and agriculture are necessarily but a part. Chayanov was no utopian; indeed, his analytical tools and even his prescription make still better sense than anything else on offer within the contemporary commonwealth of ex-Soviet states. What makes Chayanov into utopia is the power of the old regime with its interests, privileges and fetishes, while an ideology of Soviet Thatcherism forms its only alternative. Should this binarity of choices hold fast, there will be indeed no place for Chayanov but as an icon? But the mole of history digs deep and, as shown time and time again, it is the power of original and realistic thought, as well as those who carry it, which make the world turn. This is where Chayanov's analytical impact is set to outlive his enemies and his venerators and to play on its role as a fertile, conceptual input into the world's future shapes. This is the stuff the social world is made of, for better and worse.

Reference

Marx, K. 1979. *Capital*. Harmondsworth: Penguin.

Development strategies and rural development: exploring synergies, eradicating poverty

Cristóbal Kay

This essay reviews some of the main interpretations in development studies on agriculture's contribution to economic development. It explores the relationship between agriculture and industry as well as between the rural and urban sectors in the process of development. These issues are discussed by analysing the so-called 'Soviet industrialisation debate', the 'urban bias' thesis, the development strategies pursued in East Asia and Latin America from a comparative perspective, the impact of neoliberal policies on rural–urban relations and the 'agriculture-for-development' proposal of the *World Development Report 2008*. The main argument arising from analysing these issues is that a development strategy which creates and enhances the synergies between agriculture and industry and goes beyond the rural–urban divide offers the best possibilities for generating a process of rural development able to eradicate rural poverty.

Introduction

Primitive accumulation has, historically, been central to the earliest stages of the transition to capitalism, and has involved immense suffering and social waste. Yet, while is has been a necessary condition for successful capitalist transition and its accompanying structural transformation, it has never been more than a preliminary to them. Such transition has required full-scale capitalist industrialisation and a transformed, productive agriculture; and this has entailed cumulative capitalist accumulation. These have necessitated certain kinds of class formation. All of the relevant processes have been mediated, in one way or another, by an emerging capitalist state. Critical have been the creation of accumulation-oriented capitalist classes and of a proletariat (both rural and urban) – the latter possible by the separation of peasants from their means of production. (Byres 2005, 89–90)

The world is at a turning point as for the first time in human history the urban population is today larger than the rural population. However, poverty has still overwhelmingly a rural face and the rural economy and society still perform a vital part in the development process and in people's well-being. Current concerns about global warming, deforestation, the food crisis, genetically modified organisms

Some ideas in this article were first presented in a seminar of the International Development Studies Speakers Programme, Saint Mary's University, Halifax, Nova Scotia, Canada. I am grateful for the comments received from the participants at that seminar as well as for the written comments of Ben Cousins and Saturnino M. Borras Jr. which were most helpful. I also benefited from the expert editing by Diana Kay. The usual disclaimers apply.

(GMOs), agrofuels, food sovereignty, famines, rural poverty and international migration, among others; reveal the continuing relevance of the agrarian and rural problem. While recognising the relevance of all these issues this paper does not deal specifically with most of them. Instead it is concerned with the broader issue of finding a development strategy which is able to generate a development process capable of improving people's well-being, reducing inequality and eradicating poverty. The various current concerns mentioned are best dealt within such a comprehensive framework. Hence, in this article I discuss development debates which already took place almost a century ago, as well as contemporary ones, as they provide significant insights for constructing and proposing alternative development strategies which are capable of tackling the contemporary problems mentioned. Neoliberal policies have utterly failed to resolve these urgent problems and may have made them worse. Hence there is urgency for exploring and implementing alternative development strategies.

In this essay I examine the role of agriculture in economic development as proposed by development thinkers as well as in some concrete development experiences. While some development economists have argued that agriculture is the key to development, others have argued that it is only by industrialising that development can ultimately be achieved. Those upholding the former position can be described as 'agrarianists' while the latter can be named as 'industrialisers'.[1] While the agrarianists tend to neglect industry's development and hence the role that agriculture can perform in the process of industrialisation, the industrialisers tend to neglect agriculture's development and hence the role that industry can perform in the process of agricultural development. It is argued in this article that the most successful development strategy is one in which the State is able to exploit creatively the synergies between both sectors by developing their complementarities and enhancing their dynamic linkages.

I start by examining briefly the arguments of the industrialisers and then proceed to analyse the agrarianist position.[2] I discuss in particular the agrarianist critique of the industrialisation process in Less Developed Countries (LDCs) by focusing on the 'urban bias' (UB) thesis proposed by Michael Lipton (1977). The ensuing debate generated by the UB thesis is presented next. As a way to overcome the limitations of the agriculture–industry dichotomy, I then proceed to outline and illustrate my 'synergy' thesis by contrasting and evaluating the development process of East Asia and Latin America since the end of the Second World War. This is followed by an analysis of how the transformation brought about largely by neoliberal policies since the 1980s reveals more clearly the limitations of the UB thesis for explaining rural poverty and as an agrarianist proposal for eradicating rural poverty. Finally, the

---

[1]The slogan used for the agrarianists is 'agriculture first', i.e. agriculture has to be developed first and given priority over industry, while for the industrialisers it is 'industry first'. Saith (1992, 102–7) distinguishes seven incarnations of the 'agriculture first' development discourse which includes the works of Bukharin, Chayanov, Mellor, Lipton and the World Bank, among others, all of which are referred to in this article.

[2]In an authoritative article on development strategies and the rural poor Saith (1990) refers to the industrialisers as the 'industrial trickle-down' development strategy and to the agrarianists as the 'agriculture trickle-down' position. The former believe that industrialisation will trickle down to the poor and the latter believe that it is agricultural growth that will trickle down to the poor. In his view both strategies have failed to trickle-down to the poor and hence the need for alternative development strategies. In this article I try to sketch out a possible alternative development strategy which is able to eradicate poverty.

recent 'agriculture-for-development' strategy presented in the *World Bank Development Report 2008* is discussed by revealing the lingering influence of the agrarianist UB thesis. The critical examination of this 'new agrarianist' position serves to highlight further the limitations of this approach in the contemporary period. It also illustrates the even greater relevance of the 'synergy' approach for the design and implementation of a development strategy which is able to achieve development with the eradication of poverty.

'Industrialisers' and the primacy of industry in development

In the early post-war period development economists in devising a development path for the LDCs tended to draw upon the various historic examples of development of the now developed countries (DCs) such as Great Britain, Japan and the Soviet Union. As most LDCs at the time were largely agricultural countries development economists were particularly interested in examining the role that agriculture could play in the process of industrialisation. However, there is controversy among economic historians as to the timing, magnitude and impact of the resources flows between agriculture and industry in the early stages of economic development. For example, even in the case of Great Britain, which experienced the first industrial revolution in the world, there is no consensus on this matter due to conflicting bodies of evidence and their interpretation.[3] While some historians, such as Kerridge (1967), argue that Britain's agricultural revolution happened prior to the industrial revolution, others, like Deane (1965), argue that the agricultural revolution was contemporaneous to the industrial revolution and is best regarded as forming part of the same process of economic development (Overton 1996). In short, 'the performance of agriculture in Britain has become central to most of the key debates on the Industrial Revolution' (Clark 1993, 266). These historical controversies reveal how important it is to discuss the intersectoral relations between agriculture and industry in the process of economic development (Mellor 1973) as the remainder of this article aims to substantiate regarding LDCs.

Industrialisation was seen as the road to modernity and development for countries emerging from colonialism and described at the time as 'backward'.[4] The lessons which development economists took from the successful development experience of the DCs were that the transfer of a large agricultural surplus[5] was a precondition for initiating a process of industrialisation in LDCs (Ghatak and Ingersent 1984). It was generally taken for granted that development strategy should give primacy to industrialisation. This euphoria for industrialisation in the

---

[3]The industrial revolution in Great Britain is generally dated from the last two decades of the eighteenth century to the first three decades of the nineteenth century.

[4]Almost all Latin American countries had already achieved independence by the first half of the nineteenth century and many were already well on the path of industrialisation during the decolonisation in Africa and Asia.

[5]An agricultural surplus can be defined and measured in various ways (Nicholls 1963, Saith 1985, 17–28, Karshenas, 1990, 1994). A common and simple meaning of agricultural surplus refers to the total value of agricultural production minus what the agricultural sector retains for its own consumption and reproduction. The gross agricultural surplus thus refers to that part of agricultural output that is not retained by the sector itself and which is transferred to other economic sectors through a variety of means. The net agricultural surplus is equal to the above less what the agricultural sector purchases from other sectors, such as industrial consumer and investment goods as well as services.

immediate post-war years, especially in the newly independent countries of Asia and Africa, focused attention on how agriculture could contribute to industry and left largely unexamined the question of what if any contribution industry could make to agriculture (Kuznets 1964). Textbooks on economic development have tended to view agriculture as a subsidiary sector whose task is to underpin an industrialisation process in LDCs.[6] So for example, in their classic article, Johnston and Mellor (1961) argued that agriculture's function in economic development was to supply food, raw materials, capital, labour and foreign exchange for industry as well as creating a home market for domestically produced industrial products. Some development economists argued that industrialisation in itself would stimulate agriculture by offering employment and higher wages to those who migrate from the rural areas to the urban industrial areas and a market for agricultural commodities.

One of the pioneers of development economics, Kurt Mandelbaum who later changed his surname to Martin, argued in 1945 for the industrialisation of 'backward areas' by transferring the surplus and less productive labour of the rural sector to the more productive industrial sector (Mandelbaum 1945).[7] Sir Arthur Lewis (1954, 139) later developed this idea in his classic dual economy model which distinguished between a 'traditional' sector and a 'modern' sector. Lewis argued that a key characteristic of LDCs was their 'unlimited supply of labour'. The transfer of this surplus labour from the traditional sector to the modern sector could be done at almost no cost to the traditional sector as they contributed little or nothing to agricultural production. In the modern sector this new labour would achieve a much higher productivity due its technological superiority but their wages would remain close to the subsistence income they received in the traditional sector. This transferred labour by raising output well above their wages in the modern sector would thereby enhance profits, capital accumulation and economic growth. Lewis's model has subsequently been used to analyse the relationships between agriculture (identified as the traditional sector) and industry (identified as the modern sector) by many development economists such as Ranis and Fei (1964). However, Lewis himself did not make these overlapping identifications as his model aspired to be more general. Accordingly, it left open the possibility of a modern sector within agriculture and a traditional sector within the urban sector, as in the so-called informal economy.

Viewing agriculture solely within the prism of surplus extraction for industry could seriously endanger the general process of economic development (Nicholls 1964). While an insufficient surplus transfer from agriculture to industry might hold back industrialisation (Mundle 1985), too great an extraction might lead to agriculture stagnating and thereby run the danger of killing the goose which lays the golden eggs and stifling economic growth. On the one hand, agriculture would be left

---

[6]See, for example, the readers on agricultural economics by Fox and Johnson (1970), Eicher and Witt (1964), Wharton Jr. (1969) and Eicher and Staatz (1984).

[7]Another pioneer of modern development economics Paul Rosenstein-Rodan (1943), also an advocate of the industrialisation of 'underdeveloped countries', already before Mandelbaum and Lewis wrote about an 'excess agrarian population' (disguised unemployment) which could be seen as a source of development by transferring it to the emergent industry where their labour would be more productive. In his later writings on LDCs, Martin (1982) advocated agrarian reform for tackling the problems of the 'labour-surplus condition', rural poverty and inequality as well as stressing the importance of agriculture in providing a marketable surplus to the growing non-farm population thereby underpinning the process of industrialiation.

with too few resources to invest and hence be unable to provide an adequate supply of food and raw materials to the nonagricultural sector. On the other hand, rural incomes may not grow enough or even decline, thereby restricting the home market for industry. The relevance of this dilemma and a way to deal with it is suggested by Martin (2002, 6, 7) who argues that 'financial resource outflows from agriculture and agricultural productivity gains can go together ... provided that the productivity gains in agriculture do not themselves necessitate large-scale capital investments *within* agriculture.' Unfortunately his warning has not always been heeded and its implications are forcefully illustrated by a debate which took place in the former Soviet Union to which I turn next.

The Soviet industrialisation debate

During the 1920s a major debate on development strategy took place in the Soviet Union. Known in the literature as the 'Soviet industrialisation debate' (Erlich 1960), it involved two opposing conceptions of the relationship between agriculture and industry in the process of development in the context of a socialist transition. The Soviet experience and in particular the debate between Bukharin and Preobrazhensky (both Marxist intellectuals and prominent members of the communist regime) reveal the central importance of this issue. The revolutionary regime implemented a New Economic Policy (NEP) during the years 1921 to 1929 so as to restart the economy which had been ravaged by the First World War, the revolution and the subsequent civil war (Bernstein 2009, 55–81, this collection). The main concern of NEP was 'to satisfy the middle peasantry as much as possible without damaging the proletariat's interests' (Mandelbaum 1979, 539, quoting from a speech by Lenin at the Tenth Congress of the Russian Communist Party held in 1921). The ideas of Nikolai Bukharin much influenced the shaping of this policy which aimed to restore food production by allowing a free market and higher prices for agricultural commodities, so as to entice peasants to invest, expand production and to sell much of the increased output to the market rather than keeping it for self-consumption. Hence the industrialisation process had also to satisfy the needs of the peasantry as otherwise they would be reluctant to release their marketable surplus. Bukharin argued that industrialisation could only proceed at a pace at which the agricultural sector was able to produce, and peasants were willing to provide, a marketable surplus.[8] For,

> the more solvent the peasants are, the faster our industry develops. The greater the accumulation in our peasant economy – in other words: the sooner the peasants overcome their poverty, the richer they become ... the more they are able to purchase the commodities made by urban industries – the more rapidly the accumulation in our industry takes place. (Bukharin as quoted by Mandelbaum 1979, 540)

---

[8]Shanin (2009, 83–101), in his fascinating essay in this collection, argues that 'a version of "chayanovian" strategy of rural reconstruction' much influenced the followers of Bukharin within the leadership of the Communist Party (2009, 89). No wonder that several decades later and in the context of LDCs, during the debate between Marxists and followers of Chayanov, which took place mainly during the 1970s and manifested itself with particular force in the debate between '*campesinistas*' ('peasantists') and '*descampesinistas*' ('de-peasantists') or '*proletaristas*' ('proletarianists') in Latin America, a 'Chayanovian Marxist' current emerged which combined ideas from both currents of thought (Kay 2000). For a subtle and empirically-based analysis which combines ideas from Chayanov and Marxism, see Deere and de Janvry (1979) and Deere (1990). For a critique of 'Chayanovian Marxism', see Lehmann (1986).

Evgeny Preobrazhensky (1965) argued that such a policy would favour the rich peasants or *kulaks* whose aim was to become capitalist farmers and who would thus oppose any socialist transition. Peasant farming was also seen as impeding rapid progress in agriculture. The revolution was becoming hostage to the rich peasants who were not expanding production quickly enough for the requirements of a rapid industrialisation process. Preobrazhensky argued that as under capitalism, the transition to socialism required its own 'primitive socialist accumulation' (Harrison 1985).[9] This was to be achieved by manipulating the terms of trade against agriculture and in favour of industry. He proposed to entice peasants to join collectives, which would be supported with mechanisation and the provision of fertilisers. Collective farming was thus able to achieve economies of scale enabling faster growth in agricultural output which could then be transferred to industry through unequal exchange.

This dispute within the regime was brought to a brutal conclusion with Stalin's forced, rapid and massive collectivisation of the peasantry.[10] Contrary to what was assumed at the time and for several decades after, collectivisation did not lead to an increase in the net transfer of an agricultural surplus from agriculture to industry (Ellman 1975). As collectivisation failed to raise agricultural production the increased supply of agricultural commodities to the urban areas was at the expense of consumption in the countryside. Industrialisation was achieved but at an extremely high social cost. Stalin's forced collectivisation was in a way self-defeating as it led to a fall in agricultural output. Although it did succeed in extracting a higher proportion of agriculture's output to be delivered to the cities this was obtained only through compulsory delivery quotas imposed on the collectives. To compensate for the massive slaughter of draft animals used for ploughing and other economic activities, as an expression of peasant opposition to forced collectivisation, industry had to deliver an increasing quantity of tractors. Standards of living in the countryside deteriorated dramatically leading to famine and flight from the land. This massive migration to the cities provided the cheap labour force that fuelled the country's rapid industrialisation. Only by lowering the living standards of both rural and urban workers was it possible to increase the investment resource for industry (Ellman 1978). It is an open question whether Bukharin's peasant way, albeit transitory, or a voluntary collectivisation process could have achieved similar rates of economic growth without demanding such high sacrifices from the people and having such self-destructive consequences (Chandra 1992).

This debate had a major influence in development thinking, especially in those countries attempting a socialist path of development, as it was one of the first major debates on development strategies and planning. More generally, several of the issues raised by the Soviet industrialisation debate – such as the relative merits of

---

[9]There exists a large body of work which studies the relationship between capital accumulation, including of a 'primary', 'primitive' or 'original' kind, and the peasantry in the Marxist and associated literature, see, for example, Saith (1985, 1995), FitzGerald (1985) and Wuyts (1994).

[10]First Bukharin but later also Preobrazhensky became victims of Stalin's purges. Chayanov also was a victim of Stalinism. For a remarkable reflection on Chayanov's shifting influence in rural studies and policies, see Shanin (2009, 83–101, this collection). For a lucid discussion of the 'Lenin–Chayanov debate' and its contemporary relevance, see Bernstein (2009, 55–81, this collection). For the related and more general debate between the 'neopopulist' Chayanov and the 'agrarian Marxists', see Cox (1979), Harrison (1979) and Patnaik (1979). There is a large literature critical of the neopopulist agrarians, see Bernstein and Brass (1996/1997), Brass (1997, 2000) and Cowen and Shenton (1998a, 1998b), among others.

100   *Cristóbal Kay*

peasant farming and collective farming, of small-scale agriculture and large-scale agriculture as well as the timing, sequence and type of industrialisation – are central to more recent debates on the relations between agriculture and industry and the sequencing of the industrialisation process. These issues will come up at various points throughout the remainder of this essay.

'Agrarianists' and the primacy of agriculture in development

In the early post Second World War period, the views of the industrialiser group predominated. In the initial phase of import-substituting-industrialisation (ISI), the so-called 'easy phase', industry grew quite fast as it benefited from protectionism and a series of supportive government measures such as credit and public investments in the required infrastructure. But after one or two decades of rapid industrialisation problems came to the fore. Among these were the saturation of the domestic market which meant that firms could not make full use of their installed capacity, problems arising from the fact that industry had diversified too quickly, the inability to exploit economies of scale which meant that firms could not compete in the international market, and shortages of foreign exchange which restricted the importation of raw materials, spare parts and capital goods thereby strangling industrial expansion (Little, Scitovsky and Scott 1970).

As the shortcomings of the ISI became increasingly apparent in the 1960s and as agricultural production began to falter, the voices of those who emphasised the dangers of neglecting agriculture in the development process became more audible. According to the agrarianists, development strategy in LDCs should have prioritised agriculture given that the majority of the population was rural, labour productivity was low and rural poverty levels were high. Adherents of this argument, as well as neoclassical economists, pointed out that LDCs enjoyed comparative advantages in agriculture and other primary commodities and advocated that they should continue to specialise in the export of these commodities and import the necessary industrial products from the DCs. Some authors within this strand did not deny that some LDCs could industrialise at some time in the future but for most countries ISI had been premature. Furthermore, if industrialisation was warranted it should not rely on protectionist measures as this would lead to inefficiencies and rent seeking. It is within this context that the UB thesis of one of the most prominent agrarianists was formulated.[11]

---

[11]A distinction may be made between 'neoclassical agrarianists' such as T.W. Schultz (1964) and 'neopopulist agrarianists', such as A.V. Chayanov (1966, orig. 1925), Teodor Shanin (1973, 1974, 1988) and Michael Lipton (1968a, 1977). Both prioritise agricultural development but whereas neopopulists believe that small-scale peasant household farming is superior to large-scale commercial farming (the 'inverse relationship'), neo-liberals allow for the possibility of economies of scale and efficient large-scale farming. Instead of viewing peasants as traditional and backward Schultz saw them as 'poor but efficient'. Neopopulists are in favor of State support for smallholders while neoclassicals and neoliberals prefer a minimalist State and argue that market forces should be given free reign so as to encourage competitiveness and eliminate inefficient producers who might include smallholders. Byres (2003a) distinguishes four neoclassical views in development economics, all of which he finds wanting as they fail to perceive the contradictions of capitalism and its exploitative nature. Furthermore, neoclassicals, like neoliberals, do not endorse a developmentalist State which he sees as crucial for achieving the required structural transformation for development (Byres 2006).

The urban bias thesis

In this section I will discuss the UB thesis put forward by the eminent British development economist Michael Lipton as it is one of the most forceful and comprehensive critiques of the industrialisers. In particular he was critical of government policies in LDCs which he argued favoured industry at the expense of agriculture. His UB thesis gained much popularity but also generated much controversy, especially during the late 1970s and the 1980s. However, its lingering influence can be detected in current development debates as I will show in the latter part of this paper. It is for these reasons that the UB thesis merits more detailed discussion. Lipton (1968b) made an early critique of the emphasis placed by development economists on industrialisation. Writing on India, Lipton accused government planners of UB. In his own words 'farm policy is made by the towns, and to some extent for the towns' (Lipton 1968b, 141). Such policies prioritising industry were explained by the fact that power resided in the urban sector and by the widespread acceptance of an ideology that viewed industry as epitomising modernity and peasants as traditional and conservative.

In Lipton's view UB occurs through the State which, controlled by the urban class, under-allocates resources to, and extracts surpluses through a variety of means from, the rural class. 'UB involves, first, an *allocation*, to persons or organisations located in towns, of shares of resources so large as to be inefficient and inequitable, or second, a *disposition* among the powerful [urban classes] to allocate resources in such a way' (Lipton 2005, 724, emphasis in original). He stated that since about 1950 governments in LDCs had allocated public resources such as expenditure on health, education and infrastructure in favour of urban areas. In addition the urban sector was favoured through what he called 'price twists' and exchange rate, taxation, subsidy and credit policies. Price twists were the consequence of State measures, which caused outputs from rural areas to be under-priced, and inputs into rural areas to be over-priced when compared to a market norm. In short, governments through deliberate action turned the terms of trade against agriculture in favour of industry.

Lipton objected to UB on grounds of efficiency and equity. It is this inefficient allocation and inequitable distribution of resources that perpetuates low rates of growth and poverty in LDCs. Lipton claimed that the priority given to industry since Indian Independence in 1947 had had a damaging effect on the growth potential of the economy as a whole. The under-allocation of resources to agriculture not only held back growth, as additional investment in agriculture would achieve a higher rate of return than in industry, it also hampered industrialisation as foreign exchange was diverted towards importing food rather than to essential raw materials for industry.

Furthermore urban-based industrialisation policies have an adverse impact on the development of rural areas and particularly on peasants. He argued that investment in peasant farming would yield higher returns than investment in large farms (whose owners, as will be seen, he considered to be part of the urban class). Large farms create less employment and output per hectare as compared to small farms.[12] Additional government expenditure in education and health in the rural

---

[12]This proposition that small-scale farms are more efficient than large-scale farms is referred to as 'the inverse relationship' (between farm sise and efficiency) in the specialist literature and is much supported by the neopopulists, see Byres (2004).

sector would also yield a higher rate of return as well as have a greater impact on poverty reduction as compared to its allocated in the urban sector.

One of the effects of UB was to increase inequality between rural and urban areas, but also within the rural sector itself. In his words, 'business and trade union leaders both want low food prices. They can "buy" them from the big landlords in return for loopholes in land reforms and in laws limiting interest rates; for low agricultural taxes; and for subsidised inputs for the big farmers' (Lipton 2005, 141–2). While bigger and so-called progressive farmers might have something to gain, 'inequality within the village, and between village and town, has been worsened, and growth has suffered' (Lipton 2005, 142). The reduction or elimination of UB would improve equity, as income distribution was more equal in rural areas as compared to urban areas.

Lipton's UB thesis generated much controversy inspiring many studies on the topic, particularly in Asia and Africa where urban bias was held to be most acute.[13] Its emphasis on the exploitation of the peasantry and rural poverty may also have touched an emotional chord. The concept of UB began to circulate widely in development circles and beyond and was closely associated with his work. His UB thesis gained momentum at the same time as neoclassical economists launched an attack on ISI and may well have inspired a subsequent series of World Bank studies on price policies in developing countries revealing their distorted nature and negative impact on agriculture and economic growth (Krueger 1991, Schiff and Valdés 2004).

The UB thesis has been challenged on empirical grounds as well as on theoretical grounds. With regard to the empirical validity of Lipton's claims, UB is beset by methodological and measurement problems. Lipton's evidence and his interpretation of it have been strongly contested by Byres (1974), Mitra (1977), Varshney (1993) and Corbridge (1982), among others. For Byres (1979, 232) 'Urban bias is a myth – a purely ideological construct – which has no empirical support; hardly any of its "effect" can be sustained by the evidence, while those that can are not explained by urban bias.' Byres turned the tables on Lipton stating that India's policy was rural biased and not urban biased.

Urban class and rural class

Many critiques focus on Lipton's class analysis and his consequent views on State and policy formulation. Lipton's views on class are indeed perplexing. Most controversially Lipton assigns groups such as landlords and rich farmers – commonly thought of as rural – to the urban class on the grounds that they are 'bought off' by the State through such measures as special dispensations on taxation, favourable prices, subsidised inputs and property rights and others, such as urban workers – commonly viewed as urban – to the rural class. As Byres (1979, 236) put it, 'with deft sleight of hand, large proportion of the urban population are spirited away. It seems the 'urban jobless' ... are, in reality part of the single rural class.' Such realignment of the rural rich to the urban sector and of the urban poor to the

---

[13]The publication of Lipton's (1977) massive book of over 450 pages gave new impetus to the rural–urban debate. His UB thesis led to the publication of two special issues of the prestigious *Journal of Development Studies*, the first published in 1984 (Vol. 20, No. 3) and the second in 1993 (Vol. 29, No. 4), dozens of book reviews and countless references in journal articles and books.

does not backfire and harm agricultural development itself. Arguing for 'industry first' omits to see how extraction of a surplus might lead to agricultural stagnation but arguing for 'agriculture first' omits to see how a transfer of an agricultural surplus to industry can contribute to agricultural development. Hence, a judicious development strategy has to find the right relationship between agriculture and industry in the development process. This is a relationship which will vary according to the particular phase of the development process and as structural conditions and international circumstances change. By focusing one-sidedly on either agriculture or industry analysts fail to examine the complex and dynamic interactions between them. Therefore they are unable to grasp the importance of exploring the multiple intersectoral resource flow possibilities and their varying impact on particular trajectories of economic development. Hence the efficiency of resource use within sectors as well as the allocation of resources between sectors has to be discussed when trying to understand the different development performances and potentials (Karshenas 1996/1997).

While the more extreme agrarianists stress that only agriculture matters and do not appreciate or even deny the contribution that industry can make to agricultural development, the more extreme industrialisers make the opposite mistake. But some analysts consider the nature of the inter-sectoral relationship between agriculture and industry as being of prime importance for explaining differences in the development performance between countries (Johnston and Kilby 1975, Bhaduri and Skarstein 1997). Lewis (1958, 433) had earlier highlighted the importance of stressing neither agriculture nor industry but both as he argued that '[i]ndustrialisation is dependent upon agricultural improvements; it is not profitable to produce a growing volume of manufactures unless agricultural production is growing simultaneously. This is also why industrial and agrarian revolutions always go together, and why economies in which agriculture is stagnant do not show industrial development.' It is the analysis of the various dynamic links which can be forged between agriculture and industry and their particular sequence which are the key for uncovering a country's development potential and hence for designing appropriate development strategies for its realisation.

Although the debate on whether agricultural development is a prior requisite for industrialisation or whether both can be concurrent processes is still unresolved, few specialists question that the performance of the agricultural sector has a major bearing on a country's industrialisation. To achieve a successful industrialisation a country will have to resolve the problems associated with the generation, transfer and use of an agricultural surplus. This is particularly important in the initial stages of industrial development. Once an industrial sector is established it can generate the necessary surplus for investment from within the sector itself so that the need to extract an agricultural surplus becomes less urgent. At later stages of economic development the flow is often in the opposite direction, i.e. an industrial surplus helping to finance agriculture.

There are several ways in which an agricultural surplus can be transferred to other economic sectors: voluntarily or compulsorily, in a visible 'on the table' manner or an invisible 'under the table' manner. These different mechanisms for transferring an agricultural surplus are made not only to illustrate the variety of resource transfers which exist but also because some mechanisms are more appropriate or more efficient in achieving certain developmental goals as compared to others. Thus besides analysing the best ways of generating and increasing the agricultural surplus it is necessary to discuss the most suitable mechanisms for its transfer to the sector with the best growth and distributional potential as identified by the development strategy. Among the questions which policy makers need to consider are how to ensure that sufficient incentives are provided to farmers so that agriculture produces a required surplus, how to ensure that the extraction of the surplus does not lead to agricultural stagnation, and how to ensure that the surplus is not used to finance an inefficient industrialisation process. Linkages have to be developed between agriculture and industry in such a way as to bring about a virtuous cycle of economic growth and to reinforce the complementarities between these sectors.

Hence, the discussion over development strategy should focus on how to achieve and maximise intersectoral synergies as well as on how best to ensure an equitable distribution of the fruits of progress between the rich and the poor whether urban or rural. A related issue is to identify which class or coalition of classes is best able to design and implement such an equitable well-being promoting development strategy. Finally the question arises of how best to promote the necessary State capacity to carry such a national development project forward. I will illustrate the significance of these three issues by comparing the development experience of Latin America with that of the East Asian newly industrialising countries (NICs), specifically South Korea and Taiwan.[23] Such a comparative analysis within a political economy framework can help explain the uneven economic performance of the two regions. The contrast is indeed remarkable. While in the aftermath of the Second World War income per capita in Latin America was several times higher and the incidence of poverty substantially lower than in the East Asian NICs, within three or four decades the situation had been dramatically reversed.

Contrasting East Asia and Latin America[24]

I will argue in this part that the different development strategies pursued, the different timing and extent of agrarian reform and the different sequence of the industrialisation process followed by the two regions had profound effects on their development performance and ability to drastically reduce poverty.

State capacity and development strategies

To achieve a sustainable process of economic development is not just a matter of transferring resources from agriculture to industry. It requires a development strategy that generates a dynamic interaction between the two sectors. In South Korea and Taiwan, the State played a pivotal role in the process of surplus creation, extraction and transfer from agriculture to industry. It created both the conditions for productivity growth in agriculture as well as securing the transfer of much of this growth to the industrial sector via such mechanisms as taxation and manipulation of the terms of trade in favour of industry. Not only did the State play a crucial role in the process of industrialisation, it also had an absolute grip

---

[23]The World Bank (1993) refers to the East Asian NICs as a 'miracle' while others characterise them as a myth as their achievement is due to 'blood, sweat and tears', i.e. the exploitation of their rural and urban labourers (Krugman 1994).
[24]This part draws from a previous essay of mine, see Kay (2002).

over the agricultural sector, especially as the landlord class had lost their land and political power in an earlier agrarian reform. Although peasant farming was extended even further after land reform, the State exercised control over the peasantry through a variety of economic, political and institutional mechanisms (Apthorpe 1979). The State changed class relations and established the economic and political conditions favourable to rapid industrialisation. As landlords no longer had political power the South Korean and Taiwanese governments could afford to ignore the demands of agriculturalists. Urban labour did not fare much better under conditions of political unfreedom that effectively repressed any form of industrial protest, although their economic conditions were better than those of the peasantry.

By contrast in Latin America even in the period of ISI, when governments were most favourably inclined towards industrialisation, the State had to make economic concessions to landlords providing them with generous subsidies and other economic benefits. Thus the Latin American State was unable to extract proportionally such a high surplus from agriculture as compared to South Korea and Taiwan. Furthermore, populist regimes in Latin America, while mainly favouring industrialists, were unable to dictate industrial policy as in South Korea and Taiwan. They thus gave in to demands for increasing protectionism required due to industry's rising lack of competitiveness. Also the populist regimes could not ignore the demands of the expanding industrial working class, which gained certain rights as well as access to some of the benefits of the welfare State. The increasing inefficiency of the industrial sector and its declining dynamism meant that the situation became increasingly untenable. The crisis of ISI and the populist State paved the way for neoliberal economic policy in Latin America but by then Latin America had already fallen economically well behind the East Asian 'miracle' countries. But neoliberalism has also failed to deliver in Latin America as the gap with South Korea and Taiwan continues to widen.

What is remarkable about the South Korean and Taiwanese case is that the State managed not only to squeeze agriculture but that it did so while at the same time ensuring agriculture's sustained growth and thus the production of a large economic surplus.[25] This allowed industry's spectacular expansion, which in its initial stages was financed through the peasant squeeze. As argued earlier when discussing the UB thesis, relations between agriculture and industry are often viewed as conflicting and oppositional, with a gain in one sector being won at the expense of the other. However, in a dynamic context where synergies are created between the sectors, both can gain as the experience of South Korea and Taiwan testifies. This was generally not the case in Latin America as the squeeze was less effective and often self-defeating. During the ISI period landlords were able to limit the transfer of an agricultural surplus at least as far as their own interests were concerned while ensuring that the main squeeze was born by the peasantry and rural workers which given their poverty left very little to squeeze. Squeezing peasants and capitalist farmers was often counter-productive as the loss of economic incentives resulted in agricultural stagnation. Thus too high a squeeze

---

[25]For a succinct and masterful analysis of the capitalist agrarian transition of South Korea and Taiwan within a comparative materialist political economy perspective, see Byres (2003b). For more general reflections on the analytical potentials of the comparative method within a political economy framework and which influenced my own work, see (Byres 1995).

might deny agriculture the resources to create a surplus and thus be counter-productive (Teranishi 1997).

Agriculture–industry synergies

The South Korean and Taiwanese policy makers were aware that to avoid this dilemma it was necessary to ensure sustained increases in efficiency in agriculture as well as in industry. They thus had a dynamic view of the interaction between agriculture and industry in which the institutional set up and technological innovation were central. Governments thus ensured that the conditions were conducive to the adoption of new technologies and stimulated shifts in production patterns to higher value crops over the whole of the farming community (Oshima 1987). As for industrialisation they tried to ensure that any resources transferred to industry were invested in industries that had great potential for growth and for success in export markets. In contrast to Latin America where protectionism was generally exercised across the board, in South Korea and Taiwan it was highly discriminatory.

These Asian governments also encouraged the creation of industries that would allow improvements in agriculture such as the chemical fertiliser, farm machinery and equipment industries. Furthermore, agricultural-supporting industries received a higher allocation of foreign aid funds than other types of industry (Cheng 1990). Much industrialisation in Taiwan was also rural based thereby being more attuned to the needs of the agricultural sector. Once a successful industry is established the need for extracting a surplus from agriculture diminishes and the flow of resources might even revert. This has been the case in post-war Japan and in recent decades also in South Korea and Taiwan as comparative advantages shifted from agriculture to industry (Bautista and Valdés 1993).

By contrast, Latin American policy makers generally failed to create such synergies. They were unable or unwilling to drastically reform the land tenure system and modernise agriculture. They also failed to resist the pressures from industrialists for higher rates of protectionism and lacked a sufficiently vigorous export-oriented industrialisation (EOI) strategy, if any at all (Ranis and Orrock 1985). By failing to break through into the industrial export market Latin America's economic growth continued to be hampered by foreign exchange constraints which limited the import of capital goods and thus raised the country's investment rate (Jenkins 1991). The key obstacle to Latin America's industrialisation was not so much the lack of capital as the lack of foreign exchange. This neglect of agricultural exports together with the failure to shift earlier to an EOI strategy are some of the key reasons why Latin America fell behind the East Asian NICs.

The fact that policy makers in South Korea and Taiwan decided early on to become competitive in international markets had the great advantage that it created an industrial structure that took advantage of their cheap labour supply. This was a major factor in their comparative advantage relative to the industrial countries where labour was expensive and at the time in short supply. The transformations in South Korea's and Taiwan's agriculture enabled surplus labour to be released to the industrial sector thereby keeping wages low, while at the same time ensuring that agricultural production continued to grow so as to provide an adequate supply of food to industrial workers. This satisfactory supply of food meant that food continued to be relatively cheap and thus an upward pressure on industrial wages was avoided. This in turn allowed industrialists to reap high profits, remain competitive and use these profits to finance industrial investment and thus sustain a high rate of industrial growth. Furthermore, the high rate of labour absorption of South Korea's and Taiwan's industrial sectors meant that at a certain point the labour surplus was reduced or even eliminated and thus wages began to rise. Thus, after some time, growth did trickle down thereby further improving equity (Kuznets 1988).

Agrarian reform

The foundations for a more equitable income distribution were laid by the agrarian reform. Income inequalities in Taiwan, and to a lesser extent in South Korea, are probably among the lowest in the world and this has not only had positive effects on social and political stability but also provided a solid foundation for their industrialisation. This relatively equitable income distribution widened the size of the domestic market for industrial commodities, which is particularly important in the initial stages of an industrialisation process, as well as stimulating rural industry (Saith 1992). According to White (1987, 64, 65) 'perhaps the single most important element in the East Asian success has been the implementation of rather comprehensive agrarian reforms' which also had 'powerful growth-releasing and poverty-reducing effects'.

By contrast, the more limited extent of agrarian reform in Latin America, coupled with the fact that it was implemented several decades after industrialisation had started, denied the region this potential widening of the internal market and also created a distorted and inefficient industrial structure which produced commodities largely catering for high-income groups and required capital-intensive and foreign-exchange intensive technologies. This meant that a large proportion of the surplus rural population, which migrated to the urban centres, was unable to find industrial employment. In South Korea and Taiwan by contrast the industrial structures were geared to the production of mass consumer goods, where greater possibilities for using labour-intensive technologies exist.

Increases in agricultural productivity in South Korea and Taiwan were achieved with only limited capital requirements, such as the greater use of fertilisers and improved seeds. Changes in agricultural productivity in Latin America, by contrast, were more demanding of scarce capital resources and often also required more foreign exchange. Governments favoured the large-scale commercial farm sector, which invested in technological innovations of a mechanical kind and required the importation of tractors, combine harvesters and other machinery. By contrast in South Korea and Taiwan technological change in agriculture was widely diffused among peasant farmers as a consequence of the redistributionist agrarian reform and the active promotion of improved technologies by the State. Rural expenditure was disbursed in a far more egalitarian manner and the State made far more substantial investments in rural infrastructure, such as irrigation and roads, compared to Latin America (Aoki *et al.* 1997).

Sequence of industrialisation

Latin America fell behind the East Asian NICs not only because it neglected agriculture but also because it failed to shift in time from an ISI to an EOI development strategy. After the exhaustion of the easy or primary phase of ISI based on the consumer-goods industry during the 1960s, some Latin American countries managed to raise their savings rate due to the higher capital accumulation requirements for financing the investment in the intermediate-goods and above all in the capital-goods industrial sector. A similar process happened in South Korea and Taiwan with the difference that both countries were able to continue with, as well as deepen, this shift to a more capital-intensive, labour-skill-intensive, foreign-exchange-intensive and large-scale industrialisation process while Latin America was unable to do so due to foreign exchange and market constraints.

By already moving into exports during the consumer-goods industrial stage the East Asian countries were able to earn the additional foreign exchange necessary to finance the import of the intermediate-goods and capital-goods required for the next stage in the industrialisation process. They also gained valuable experience in international markets and by being exposed to a greater extent than the Latin American economies to world competition had a powerful incentive to become more efficient and hence competitive. This early shift to an EOI strategy meant they were able to access a much wider market thereby being able to reap the benefits of economies of scale which are particularly important in the manufacturing of products such as cars, ships, steel, chemicals, and electronics, most of which South Korea and Taiwan started to produce. The comprehensive and inclusionary educational system of South Korea and Taiwan also ensured the necessary supply of skilled labour required for some of these industries whose wages were still relatively low compared to the developed countries as well as to Latin America.

In summary, three key factors explain the difference in performance between the Asian NICs and Latin America. The first is South Korea's and Taiwan's superior State capacity and policy performance. The second is Latin America's failure to create an agrarian structure more conducive to growth with equity. The third is South Korea's and Taiwan's greater ability to design an appropriate industrial policy as well as developing the synergies between agriculture and industry. While Latin America got off to an early start with industrialisation it was unable to overcome quickly enough the limitations of ISI and shift to a more export-oriented and competitive industrial structure. All of the three factors that I have identified are closely interconnected. South Korea and Taiwan managed to develop the positive linkages between them while in Latin America these factors were often in conflict (Saad-Filho 2005). While the East Asian NICs succeeded in creating a virtuous and mutually-reinforcing, upwardly-moving spiral between these factors, resulting in high living standards for the majority, the Latin American countries failed to do so, thereby perpetuating poverty.

Contrary to the neoliberal interpretations of the East Asian 'miracle', I am arguing that South Korea's and Taiwan's superior performance was not due to 'getting prices right', avoiding protectionism, promoting an unrestricted free market and a minimalist State. Quite the contrary, their success is due to a strong interventionist and developmentalist State, flexible protectionism, 'getting prices wrong' and governing the market (Amsden 1989, Wade 1990, Chang 2002).[26]

---

[26]By 'getting relative prices "wrong"' the State deliberately manipulated prices in favour of a particular sector or group so as to achieve certain goals of its development strategy, which evolved and thus relative prices could be 'distorted' in favour of another sector or group (Amsden 1994).

Crossing frontiers, fording the divide, creating linkages

The lesson I want to convey from this comparative analysis is that the form of dualistic thinking as expressed in the UB thesis is an increasingly unhelpful way of thinking about development. I have already discussed the usefulness of thinking in terms of synergies between agriculture and industry and between the rural and urban sectors. In this section I will discuss how transformations in rural livelihoods over the last two or three decades have strengthened the explanatory power of analytical frameworks which explore linkages, interactions and synergies and further eroded the UB thesis.

As argued earlier, the UB thesis rested crucially on the existence of distinct and major differences and inequalities between the rural and urban sectors.[27] For Lipton, the rural–urban divide is profound and persistent in LDCs and its blurring is rare and exceptional (Moore 1984). Yet recent transformations within rural and urban sectors and their growing interaction have made analyses which rely on a strict separation of these spaces problematic. Rural–urban borders have become more permeable making it plausible to speak of the ruralisation or rurification of the urban and the urbanisation of the rural in LDCs.[28] It is becoming more common, especially in times of food crisis, for agricultural activities to take place in urban areas and the term 'urban agriculture' is used to indicate this. Peri-urban areas and intermediate cities are springing up which act as transmission belts between the larger cities and the rural hinterlands (Lynch 2005, 2008). In sum, rural and urban spaces are being reconfigured.

In the era of neoliberal globalisation an escalating interaction and fluidity between the rural and urban sectors in terms of capital, commodities and labour can be observed (Hart and Sitas 2004). The increasing dependence on inputs purchased from industry, the continuing industrialisation of agriculture through agro-processing plants, the spread of rural industries, the expanding integration of agricultural producers into global commodity chains, the growing intrusion of agro-food corporations and supermarkets into the countryside are tying the urban and rural sectors more closely together than ever (Goodman and Watts 1994, Reardon and Berdegué 2002, Friedmann 2005). Hence it becomes more difficult to draw a line between where one ends and the other begins.

Furthermore, rural households have increasingly constructed their livelihoods across different sites, crossing the rural–urban divide and engaging in agricultural and non-agricultural activities (Bernstein 2009). Straddling the rural–urban divide is a survival strategy for the poorer peasantry ('distress migration') or part of an accumulation strategy for the richer peasantry. Rural household incomes are increasingly made up from rural non-farm activities arising from outside agriculture (wage or salary employment such as working in agroprocessing plants and construction; self-employment such as marketing, rural tourism and other business activities; urban-to-rural and international remittances and pension payments to

---

[27]In the case of DCs it could be argued that structural transformations resulting from the development process, as well as State policies seeking to spread the benefits of development more widely and evenly, have reduced rural–urban inequalities in the DC. Lipton (1993, 240) accepts that rural rather than urban bias now predominates in DCs.

[28]To indicate this ruralisation of urban areas in LDCs Bryan Roberts (1995) uses the term 'cities of peasants'. Lipton in some ways alludes to this point by saying that the urban poor can have rural allegiance.

retirees or other urban-to-rural transfers) and off-farm activities which generally arise from wage employment on other farms (Ellis 2000).[29] Hence an increasing source of employment and income for rural people is derived from non-agricultural and urban sources. Multi-locational and multi-spatial households that cut across the rural–urban divide by combining farm and non-farm activities and rural and urban residence are increasingly frequent. This diversification of rural livelihoods has been characterised as a process of 'deagrarianisation' (Bryceson 2000) or as signifying the emergence of a 'new rurality' (Kay 2008). Moreover, urban residents, largely living in shanty-towns on the edges of cities, often engage in seasonal rural work, especially during the harvest period, through labour contractors who act as intermediaries.

In addition, rural labour and urban labour straddle the rural–urban divide through migration, often of a circular kind (Standing 1981). Roberts and Long (1979) already three decades ago employed the concept of 'confederation of households' to highlight the interaction between rural and urban livelihoods through kinship ties largely of indigenous people. Members of peasant communities migrate to urban areas establishing a foothold there and they act as a transmission belt for subsequent migrants from that community. The exchange of goods and services, which flows in both directions, cements the ties of solidarity and cooperation between family and community members.

Rural persons are not only constructing livelihoods across the urban–rural divide within the national territory but also across national boundaries, migrating to richer countries within the region: North America, Europe and the oil-rich states.[30] Remittances have become an important source of income for rural households and some are increasingly dependent on transnational links for their livelihoods, largely through international migration of members of the household. In some countries, remittances form one of the main sources of foreign exchange, surpassing the value of agricultural exports (Akram-Lodhi and Kay 2009a). This important international dimension is missing in the UB analysis.[31]

The growing rural–urban flows erode the rigid distinction between urban and rural development. Populations and activities once described either as rural or urban are now more closely intertwined both across space and across sectors than is usually thought and distinctions are often arbitrary. These new and intensified forms of urban–rural entanglement lead some analysts to think beyond the urban–rural divide (Tacoli 2003, 2006). Some authors are developing a promising territorial approach to facilitate an analysis of the linkages between the rural and the urban and for

---

[29]For a pioneering work on the rural non-farm economy and which provides a rich analytical framework for its study, see Saith (1992).

[30]This issue of migration, especially international migration which has become of increasing global significance, needs to be explored further. Issues such as the characteristics of the migrants (age, gender, ethnicity, previous income and employment conditions), their conditions of employment in the receiving country (income, stability, discrimination, access to social welfare, etc.), the impact of remittances on the household and sending country, the role of public policy of the sending and receiving country are all worthy of investigation. In addition, the impact of remittances on poverty, inequality and social differentiation, and whether they are used for capitalising peasant farmers and other small scale enterprises are all relevant questions.

[31]In reply to some critics Lipton (1984, 147–8) mentions that urban–rural remittances are an insignificant proportion of rural incomes in most LDCs. While this may have been the case at the time of writing, this is no longer so today given the increasing links between rural and urban areas. It is also surprising that Lipton makes no reference to international remittances.

designing development programmes which take advantage of these interactions and enhance potential synergies (Schejtman and Berdegué 2004, Losch 2004, Tapella and Rodríguez 2008, Fernandes 2008, Ruis and Delgado 2008).

Regardless of my criticisms of the UB thesis and what I consider to be its increasing irrelevance, there are those who argue the opposite. Bezemer and Headey (2008, 1343) declare that Lipton's 'strong conclusion is still valid today: urban biases are the largest institutional impediment to growth and poverty reduction in the world's poorest countries. Yet thirty years on from Lipton's original conclusion, the importance of urban bias is still insufficiently recognised in development theory and practice today.' Despite Bezemer's and Headey's lament, the UB thesis is gaining new life and influence in the contemporary context as will be discussed next.[32]

The challenge of rural poverty and the World Bank

A key and persistent challenge for rural development is the eradication of rural poverty. Given the powerful influence of the World Bank in shaping development ideas and influencing development policies in LDCs it is pertinent to examine the *World Development Report 2008: Agriculture for Development*, shortened to WDR 2008 in what follows. It is a timely publication even though it refers only marginally to the global food crisis.[33] This is the first time since 1982 that the yearly World Development Report focuses on agriculture and, despite my critical remarks, is a most welcome study. It was long overdue and has the merit of focusing discussion on a much neglected sector in research and policy and on highlighting the plight of the rural poor. The report presents a wealth of statistical material and covers a wide range of issues. Nevertheless, it has major shortcomings largely deriving from its neo-institutionalist analytical framework. Some critics point to its many contradictions, its failure of imagination, its ideological and rhetorical content and its remarkable continuity with previous World Bank development reports.[34] As stated by Akram-Lodhi (2008, 889): 'Despite its aspirations, the WDR 2008 is not a paradigm-shifting reimagining of the policy and practice of rural development.'

By selecting this theme the World Bank acknowledges that the neoliberal policies of the last decades have failed to give a new impetus to agriculture and above all to reduce rural poverty. Its key message that 'agriculture continues to be a fundamental instrument for sustainable development and poverty reduction' in the twenty-first century (World Bank 2007, 1) is something of an overstatement as only one of the three proposed pathways out of rural poverty is based on farming, while the other two rely on rural nonagricultural activities and on outmigration. It notes, following the rural livelihoods approach (Scoones 2009, this collection), that rural smallholders

---

[32]Bezemer and Headey (2008) make no reference to WDR 2008 as their final revised paper was accepted for publication on 17 July 2007 just before the report was published.

[33]The report does note, however, that 'global models predict the possibility of rising food prices' (World Bank 2007, 69). It has to be acknowledged that the food crisis largely manifested itself after the report went to press. Commodity prices rose sharply between 2007 and mid 2008, thereafter falling again as the financial and economic crisis unfolded. For the background to the global food crisis, see Weis (2007) and Patel (2007a). For useful analyses of the world food crisis, see Bello (2008), Magdoff (2008), von Braun (2008) and Wiggins and Levy (2008), among others.

[34]There have been a great many reviews of the WDR 2008, see Patel (2007b), Havnevik *et al.* (2007), Oxfam International (2007), Veltmeyer (2008), Akram-Lodhi (2008) and Riszo (2009), among others.

increasingly combine these three activities in their livelihood strategies.[35] But to what an extent can these often mutually reinforcing multiple pathways become, in the report's view, routes out of poverty?

In my view, the policy proposals presented by the report are unlikely to benefit the majority of the rural poor, especially the poorest of the poor. The report's pathways out of poverty have already been followed by the rural poor. Rather than lifting them out of poverty, they have only staved off a further deterioration of their livelihoods. As seen, an increasingly common livelihood strategy of the rural poor has been to migrate to rich countries to engage in a variety of mainly wage labour activities. The remittances have become an important source of survival for the members of the rural household in the country of origin. Much of this international outmigration is illegal, risky and disruptive of family and local community life. Yet governments in LDCs have done little to stem this drain of human resources by providing better employment opportunities for the rural poor in the home country or by negotiating for better conditions for migrants in the receiving countries. The report has little, if anything, to say on these important matters for the migrants' well-being.

In the shadow of the UB thesis

Although unacknowledged, the WDR 2008 is imbued with many aspects of Lipton's UB thesis, revealing its lingering seduction. The report refers in a Liptonian manner to 'macroeconomic, price, and trade policies [that] unduly discriminate against agriculture', to the 'urban bias in the allocation of public investment' (World Bank 2007, 38), to the 'reduced but continuing policy biases against agriculture', to the 'underinvestment and poor investment of public resources in agriculture' (World Bank 2007, 226), and to situations were 'smallholder interests tend to be poorly represented, and policy is biased toward urban interests and those of the landed elite' (World Bank 2007, 43). While the report shares many aspects of Lipton's UB thesis, particularly his attack on the surplus extraction from agriculture, it differs across three major issues which are relevant for my analysis.

First, the report is not neopopulist, or only partly so, as it is not beholden to the inverse relationship between farm sise and farm productivity.[36] This is a major departure for the World Bank as in the past its analyses and policies have assumed the existence of such an inverse relationship. This shift in thinking arises largely from new developments in technology, processing and marketing. 'As agriculture becomes more technology driven and access to consumers is mediated by agroprocessors and supermarkets, economies of scale will pose major challenges for the future

---

[35]The rural livelihoods approach was largely developed by scholars in the UK during the 1990s. It stresses the agency of actors such as the ability of peasants and rural workers to construct their own livelihood strategies by drawing on a variety of resources at their disposal, including 'social capital' (Chambers and Conway 1992, Bebbington 1999, Ellis 2000). While structural constraints are acknowledged they are not always taken into account in the analysis. This approach has been used for the analysis of rural poverty and policy proposals for its reduction by international and governmental development agencies as well as by NGOs.

[36]The WDR 2008 cannot fully shake off the influence of neopopulist perspectives as it states smallholder family farming has historically been superior to any other form of farm organisation (World Bank 2007, 91).

competitiveness of smallholders' (World Bank 2007, 92). It thus accepts that small farms may be increasingly unable to compete with larger and more capitalised farms. Instead of blaming UB for the crisis of peasant farming and for the process of depeasantisation, it acknowledges that this may be due to the superior efficiency of larger farms. The report makes a distinction between subsistence-oriented farmers and market-oriented small farmers. While the former may have increasingly to rely on wage labour for their subsistence and eventually leave farming altogether, the report does argue that a number of public policy measures (which it sets out) can be taken to enhance the competitiveness of the market-oriented smallholders under the new circumstances.[37] In a somewhat contradictory fashion the report still holds out some hope for subsistence farmers so long as they are able to become market-oriented farmers by seizing the new opportunities created by the global food system and policy biases against smallholders are reversed. In Latin America such policies were pursued by some governments under the slogan of 'reconversion' which aimed to find a farming pathway out of poverty by shifting from traditional subsistence production patterns to non-traditional crops geared mainly to the profitable export market. Few of these policies ever succeeded (Kay 2006).

Second, the report acknowledges the heterogeneity of the peasantry and the growing importance of rural nonagricultural and off-farm activities as well as migration and remittances for their livelihood strategies.[38] It thus addresses rural–urban linkages, but only a few of them, and the migration issue. Smallholders increasingly move beyond the farm by diversifying their employment and income opportunities. Whether smallholders diversify their activities as a way out of distress or to grasp new opportunities is the pertinent question. Several studies have indicated that the increasing engagement of the peasantry in diverse activities is due to the crisis of peasant farming which is unable to compete with corporate agriculture in the era of neoliberal globalisation (Akram-Lodhi and Kay 2009b). An increasing number of peasants are no longer able to make a living from farming, forcing household members to seek wage employment often under very precarious and exploitative conditions. While the report is aware of some of these tendencies it nevertheless argues that public policies and institutional reforms enabling markets to get prices right and removing imperfections in credit, insurance and land markets, should make it possible for peasant farming to move out of poverty and in some instances even to flourish. The WDR 2008 is keen to promote rural non-farm activities but as Saith (1992, 114) has already concluded some time ago 'the rural non-farm sector must not be viewed as a panacea for the fundamental problems of rural development and poverty alleviation. The problem of development is a

---

[37]This distinction between subsistence farmers and market-oriented peasant farmers echoes a distinction made by neoliberal economists in Latin America in the 1980s between viable and unviable peasant farmers. They argued that government resources should only be directed to the viable peasant farmers as the best option for unviable peasant farmers was to sell their land and abandon farming. Later the more politically correct distinction between peasant farmers without and with productive potential was adopted (Kay 1997). In the 1970s there was a lively debate in Latin America, and especially Mexico between Marxists and neopopulists influenced by Chayanov over the future of the peasantry: for some this seems to be depeasantisation and proletarianisation and hence demise of the peasantry, for the others capitalisation and accumulation and hence survival of the peasantry (Kay 2000).
[38]The WDR 2008 presents a typology of family farmers but this heterogeneity in livelihood strategies is not analysed within a social class differentiation process.

problem of the whole: it cannot be solved by tinkering with a single little part. The tail cannot wag the dog.'

Third, the report does address some of the international dimensions which have an impact on rural development and poverty in LDCs. The report recognises the negative consequences of agricultural protection and subsidies in DCs for the agricultural sector of LDCs. However, it puts too much faith in the potential positive impacts derived from trade liberalisation, especially concerning the rural poor. It is likely that most benefits would be captured by capitalist farmers and not trickle-down to the rural poor (McMichael 2009a, 2009b).

Limitations of the agriculture-for-development policy agenda

The report proposes a series of policy measures which aim to improve the functioning of markets at local, national and global level. It also addresses some institutional and governance issues geared towards enabling markets and bringing about 'level playing fields'. Owing to the institutionalist framework adopted issues of class, class conflict and class domination of the State are beyond its scope. Given the highly unequal distribution of assets and political power in LDCs it is most unlikely that a so-called 'level playing field' can ever be attained. Even if such a situation could be reached the rural poor lack the means to avail themselves fully of the new opportunities unless a developmentalist State brings about the required structural transformations (Chang 2003). The various policy measures proposed by the WDR 2008 do not add up to such a transformative development strategy which I have argued is required for the eradication of poverty. This can be illustrated in two of their main policy proposals.

First, regarding land policies the report continues to advocate the decades' old neoliberal World Bank policy of enhancing tenure security, land titling and securing property rights. Yet this policy has not stemmed the continuing encroachment by the powerful on lands previously controlled by peasants and indigenous peoples. As for increasing access to land for the rural poor, the report proposes enabling land rental markets, strengthening land sales markets, and market-led land reform through the voluntary willing-seller and willing-buyer mechanism (Borras 2003). All these measures have as yet had only a limited impact on land redistribution and if not accompanied by other State supportive measures, such as the provision of credit and technical assistance, make it difficult for the beneficiaries to succeed (Akram-Lodhi *et al.* 2007, Borras *et al.* 2008a).

Second, the report proposes a series of measures for transforming smallholding peasants into budding commercially-oriented entrepreneurs able to compete in global markets by linking them more effectively into the agroindustrial commodity chains in which supermarkets are gaining increasing influence. The expectation or assumption is that this would transform the efficiency and profitability of family farms by shifting production to new agricultural commodities using new technologies and improved inputs. Thus, the report takes a benign view of agribusiness and supermarkets which control the transnational commodity chains. As cogently put by Akram-Lodhi (2008, 1160): 'It offers a vision, in short, that will consolidate the corporate food regime and the establishment of agrarian capitalism across the worlds of global agriculture.' A similar assessment is reached by Amanor (2009, 261) who writes in relation to the governance framework promoted by the WDR 2008 that it is 'a hegemonic discourse in favour of neoliberal policy prescriptions, which serves to further integrate farm producers into the oligopolistic governance structures of international agribusiness.' This is a worrying development as the increasingly powerful global reach of corporate private governance limits public governance in LDCs which is more amenable to the concerns and influence of its citizens. The increasing reach and dominance of the global corporate food regime may benefit a minority of peasant farmers but will ultimately further the process of peasant differentiation leading to the partial or full proletarianisation of the majority.[39] To what extent transnational peasants and rural labour movements can effectively challenge the corporate global food regime and create an alternative which is more inclusionary and egalitarian remains an open question (Borras *et al.* 2008b, McMichael 2008).

As discussed earlier in this essay, agriculture by itself is unable to lift the rural poor out of poverty and hence a broader vision of rural development is required which goes beyond agriculture and beyond the rural sector. The WDR 2008 and the rural livelihoods approach partially acknowledge this reality. However, neither the rural livelihoods approach nor the report propose development strategies which focus on the synergies that can be created between industry and agriculture and which, according to the historical experiences analysed earlier in this essay, have been shown to be more comprehensive and sustainable paths out of poverty. Although some rural–urban linkages are explored, these are largely confined to labour markets and migration. There is also a brief mention that urban-based industries, especially in densely populated countries, can stimulate the rural nonfarm sector (World Bank 2007, 238). But the crucial importance of the attainment of synergies in industry–agriculture relations and the key significance of the developmental State in creating a development process which is able to achieve high rates of growth, equity and then eradicate poverty, as discussed earlier in this paper, remain unexplored.

The 'agriculture-for-development' agenda of the WDR 2008 can be seen as an updated version of the 'agriculture first' position upheld by those development economists who prioritise agriculture. The central question remains what agriculture can do for development. The question of what industry can do for agriculture is largely forgotten. The report deploys historical experiences, among other arguments, to justify this position. It argues that 'agricultural growth was the precursor to the industrial revolutions' in England and Japan and that 'more recently, rapid agricultural growth in China, India, and Vietnam was the precursor to the rise of industry' (World Bank 2007, 7). Furthermore, '[h]igher agricultural productivity generating an agricultural surplus, taxed to finance industrial development, and enabling lower food prices underpinned early development in Western Europe, the United States and Japan, and later in Taiwan, China, and the Republic of Korea' (World Bank 2007, 35).

Two comments need to be made about the report's historical interpretation. First, regarding England, as discussed earlier debate among economic historians has shifted from an agriculture first position to a position in which agriculture and industry stimulated concurrently each others' growth and transformation (Hudson 1992). This revisionist interpretation is gaining ground and being extended to other historical experiences. Second, the report fails to mention the key role played by the developmentalist State in these transformations, especially in the cases of Japan, Taiwan, China, South Korea and Vietnam. Far from creating 'level playing fields'

---

[39]For an analysis of the processes of peasant differentiation in Africa, Asia and Latin America, see Bryceson *et al.* (2000) and Akram-Lodhi and Kay (2009b).

and 'getting prices right' governments in these countries manipulated the terms of trade between agriculture and industry as well as interfering in other ways with the free operation of the market. As discussed earlier in relation to South Korea and Taiwan the State played a key role in the industrialisation process and in ensuring that industry supports the technological transformation of agriculture. In summary, these countries did not exactly follow the report's 'agriculture-for-development' strategy based on enabling free markets.

I do agree with the report that agriculture can and needs to become a dynamic sector for achieving development and poverty reduction. I also agree that a premature and unduly high extraction of an agricultural surplus can lead to agriculture's stagnation (World Bank 2007, 35). However, to achieve such a dynamic agriculture it is necessary to develop industry and its linkages with agriculture in a manner already discussed, but the WDR 2008 does not investigate this issue (Woodhouse 2009).

In summary, while the WDR 2008 contains a cornucopia of most useful information, by failing to embed their analysis in the structural processes and dynamics of the world capitalist system it does not uncover the essential causes of poverty and is unable to propose a development strategy which may be successful in eradicating rural poverty.[40]

Conclusions

My emphasis throughout this essay has been on the importance of designing and implementing a development strategy that exploits the dynamic synergies between agriculture and industry. Over the longer term thriving industrial and service sectors are required so as to sustain the dynamism of the whole economy and achieve the eradication of rural and urban poverty. Agriculture has inherent limitations as an engine for growth over the longer term while industry has a greater potential to generate technological innovations, capture dynamic economies of scale and generate external economies which can further support agricultural development and sustain the continuing development of the country.

This comparative political economy analysis has revealed how it is possible for development strategies to achieve synergies between agriculture and industry. Proposals for prioritising agriculture over industry or vice-versa will not be able to achieve the productivity and growth enhancing outcomes that are desirable and possible through a more comprehensive understanding of the relations between agriculture and industry in the development process. I am aware that I have not explicitly explored the role of the service sector which has now often become the largest sector in the economy. Services play an increasingly vital part in the generation of innovations, in facilitating the adoption of new technologies and management practices as well as in the diffusion of knowledge and information which will raise productivity in agriculture and industry. Services can also provide a bridge between agriculture and industry thereby facilitating the development of synergies between them. I hope that future research will analyse the role of services in the development of these synergies.

---

[40]It is perhaps not surprising that the WDR 2008 does not make reference to the large body of work on unorthodox and critical writings on the issues it discusses (Veltmeyer 2008). For a survey of several approaches to the study of rural poverty and its eradication, see Kay (2006).

I have endeavoured through a discussion of a particular debate and development experience in the Soviet Union, a comparative analysis of the development experience of some East Asian NICs with Latin America, and a review of the UB thesis and the WDR 2008, to show the limitations of agrarianist and industrialist approaches to development. Such perspectives fail to explore the synergies between agriculture and industry. To achieve and sustain synergies through a development strategy, the State can use the various policy instruments at its disposal to favour one group of producers over another within a sector as well as to favour one sector over another. This will depend on which group and what sector can generate the most productivity enhancing synergies for the whole economy with the aim of reducing poverty and inequality. But such priorities will shift according to the results achieved and changing market conditions as well as social and political circumstances. The neoliberal approach does not overcome the limitations of an agrarianist or industrialist perspective as the free market by itself will generally not generate or maximise the synergies between the sectors. Hence the importance of enabling the State to design and implement such synergetic development strategies which run counter to the free market proposal of the neoliberals.

I have also attempted to show that there are great gains to be achieved for development studies and rural development by analyses that explore the linkages that can be developed between agriculture and industry. Such analyses have to be context specific as each country has differing economic, social and political characteristics as well as differing State capacities for designing and implementing the growth and equity enhancing development strategies in a dynamic setting and within the limitations and opportunities offered by the constantly changing international environment.[41] Furthermore, such a 'synergy' development strategy also offers a useful framework for analysing the current concerns mentioned at the start of this article, such as biotechnology, food sovereignty and climate change, and developing proposals for tackling them.

Finally, I wish to conclude by endorsing the following reflection by Ben White on the importance of the comparative approach as a research, learning and teaching methodology and strategy.

> The *comparative approach*, requiring detailed analysis of the contrasting experiences of rural development in actual societies, with recognition of the particular historical, social and political contexts at national and local level in which agrarian changes take place, in which strategies and policies have been formed and introduced and have succeeded or failed. In this way we may hope to confront and come to terms with the diversity that exists in the real world – whatever uniform tendencies some abstract theories might suggest – and to learn from it, to see the ways in which general 'tendencies' interact with specific conditions to produce particular outcomes, and to understand in this way that 'success stories' may offer valuable lessons, but not directly transferable models for other societies to follow or for external agencies to impose. (White 1987, 69–70, emphasis in original)

---

[41]In view of the current world financial and economic crisis which may herald the end of the neoliberal globalisation era, this might be the time to write the history of the rural transformations and struggles of the rural people during this era. This history needs to be written following the example of such classic works as Barrington Moore Jr.'s *Social Origins of Dictatorship and Democracy: Lords and Peasants in the Making of the Modern World* (1969), Eric R. Wolf's *Peasant Wars of the Twentieth Century* (1969) and Terence J. Byres' *Capitalism from Above and Capitalism from Below* (1996), among others. In these exceptional works the authors have studied previous periods of epochal transformations in the rural economy, society and polity within a comparative, political economy and interdisciplinary perspective.

References

Akram-Lodhi, A.H. 2008. (Re)imagining agrarian relations? The *World Bank Development Report 2008: Agriculture for Development. Development and Change*, 39(6), 1145–61.

Akram-Lodhi, A.H., S.M. Borras and C. Kay, eds. 2007. *Land, poverty and livelihoods in an era of globalisation: perspectives from developing & transition countries*. London: Routledge.

Akram-Lodhi, A.H. and C. Kay. 2009a. Neoliberal globalisation, the traits of rural accumulation and rural politics: the agrarian question in the twenty-first century. *In:* A.H. Akram-Lodhi and C. Kay, eds. *Peasants and globalisation: political economy, rural transformation and the agrarian question*. London: Routledge, pp. 314–38.

Akram-Lodhi, A.H. and C. Kay, eds. 2009b. *Peasants and globalisation: political economy, rural transformation and the agrarian question*. London: Routledge.

Amanor, K.S. 2009. Global food chains, African smallholders and World Bank governance. *Journal of Agrarian Change*, 9(2), 247–62.

Amsden, A.H. 1989. *Asia's next giant: South Korea and late industrialisation*. New York, NY: Oxford University Press.

Amsden, A.H. 1994. Why isn't the whole world experimenting with the East Asian model to develop? Review of *The East Asian miracle. World Development*, 22(4), 627–33.

Aoki, M., K. Murdoch and M. Okuno-Fujiwara. 1997. Beyond the *East Asian miracle*: introducing the market-enhancing view. *In:* M. Aoki, K. Murdoch and M. Okuno-Fujiwara, eds. *The role of government in East Asian economic development: comparative institutional analysis*. Oxford: Clarendon Press.

Apthorpe, R. 1979. The burden of land reform in Taiwan: an Asian model land reform re-analysed. *World Development*, 7(4–5), 519–30.

Barraclough, S. 2001. The role of the state and other actors in land reform. *In:* K. Ghimire, ed. *Land reform and peasant livelihoods*. London: ITDG, pp. 26–64.

Bautista, R.M. and A. Valdés, eds. 1993. *The bias against agriculture: trade and macroeconomic policies in developing countries*. San Francisco, CA: ICS Press.

Bebbington, A. 1999. Capitals and capabilities: a framework for analyzing peasant viability, rural livelihoods and poverty. *World Development*, 27(12), 2021–44.

Bello, W. 2008. *How to manufacture a global food crisis. How 'free trade' is destroying Third World agriculture – and who's fighting back*. Amsterdam: TNI. Available from: http://www.tni.org/detail_page.phtml?act_id=18285 [Accessed 19 May 2008].

Bernstein, H. 2009. V.I. Lenin and A.V. Chayanov: looking back, looking forward. *The Journal of Peasant Studies*, 36(1), 55–81.

Bernstein, H. and T. Brass. 1996/1997. Questioning the agrarians: the work of T.J. Byres. *The Journal of Peasant Studies*, 24(1–2), 1–21.

Bezemer, D. and D. Headey. 2008. Agriculture, development, and urban bias. *World Development*, 36(8), 1342–64.

Bhaduri, A. and R. Skarstein, eds. 1997. *Economic development and agricultural productivity*. Cheltenham: Edward Elgar.

Borras, S.M. Jr. 2003. Questioning the market-led agrarian reform; experiences from Brazil, Colombia and South Africa. *Journal of Agrarian Change*, 3(3), 367–94.

Borras, S.M. Jr. 2004. La Vía Campesina: an evolving transnational social movement. *TNI briefing series no. 2004/6*. Amsterdam: Transnational Institute.

Borras, S.M. Jr., C. Kay and E. Lahiff, eds. 2008a. *Market-led agrarian reform: critical perspectives on neoliberal land policies and the rural poor*. London: Routledge.

Borras, S.M. Jr., M. Edelman and C. Kay, eds. 2008b. *Transnational agrarian movements confronting globalisation*. Oxford: Wiley-Blackwell.

Brass, T. 1997. The agrarian myth, the 'new' populism and the 'new right'. *The Journal of Peasant Studies*, 24(4), 201–45.

Brass, T. 2000. *Peasants, populism and postmodernism: the return of the agrarian myth*. London: Frank Cass.

Bryceson, D. 2000. Peasant theories and smallholder policies: past and present. *In:* D. Bryceson, C. Kay and J. Mooij, eds. *Disappearing peasantries? Rural labour in Africa, Asia and Latin America*. London: Practical Action Publishing, pp. 1–36.

Bryceson, D., C. Kay and J. Mooij, eds. 2000. *Disappearing peasantries? Rural labour in Africa, Asia and Latin America*. London: Practical Action Publishing.

Byres, T.J. 1974. Land reform, industrialisation and the marketed surplus in India: an essay on the power of rural bias. *In:* D. Lehmann, ed. *Agrarian reform and agrarian reformism: studies of Peru, Chile, China and India.* London: Faber and Faber, pp. 221–61.

Byres, T.J. 1979. Of neo-populist pipe dreams: Daedalus in the Third World and the myth of urban bias. *The Journal of Peasant Studies,* 6(2), 210–40.

Byres, T.J. 1995. Political economy, the agrarian question and the comparative method. *The Journal of Peasant Studies,* 22(4), 561–80.

Byres, T.J. 1996. *Capitalism from above and capitalism from below: essays in comparative political economy.* London: Palgrave Macmillan.

Byres, T.J. 2003a. Agriculture and development: the dominant orthodoxy and an alternative view. *In:* H.-J. Chang, ed. *Rethinking development economics.* London: Anthem Press, pp. 235–53.

Byres, T.J. 2003b. Paths of capitalist agrarian transition in the past and in the contemporary world. *In:* V.K. Ramachandran and M. Swaminathan, eds. *Agrarian studies: essays on agrarian relations in less-developed countries.* London: Zed Books, pp. 54–83.

Byres, T.J, 2004. Neo-classical neo-populism 25 years on: *déjà vu* and *déjà passé.* Towards a critique. *Journal of Agrarian Change,* 4(1–2), 17–44.

Byres, T.J. 2005. Neoliberalism and primitive accumulation in less developed countries. *In:* A. Saad-Filho and D. Johnston, eds. *Neoliberalism: a critical reader.* London: Pluto Press, pp. 83–90.

Byres, T.J. 2006. Agriculture and development: towards a critique of the 'new neoclassical development economics' and of 'neoclassical neo-populism'. *In:* J.K.S. and B. Fine, eds. *The new development economics: after the Washington consensus.* London: Zed Books, pp. 222–48.

Byres, T.J. 2009. The landlord class, peasant differentiation, class struggle and the transition to capitalism: England, France and Prussia compared. *The Journal of Peasant Studies,* 36(1), 33–54.

CEPAL 2007. *Panorama social de América Latina 2007.* Santiago: Comisión Económica para América Latina y el Caribe (CEPAL), Naciones Unidas.

Chambers, R. and G.R. Conway. 1992. Sustainable rural livelihoods: practical concepts for the 21st century. *IDS discussion paper no. 296.* Brighton: Institute of Development Studies (IDS).

Chandra, N.K. 1992. Bukharin's alternative to Stalin: industrialisation without forced collectivisation. *The Journal of Peasant Studies,* 20(1), 97–159.

Chang, H.-J. 2002. *Kicking away the ladder: development strategy in historical perspective.* London: Anthem Press.

Chang, H.-J. 2003. *Globalisation, economic development and the role of the state.* London: Zed Books.

Chayanov, A.V. 1966. The theory of peasant economy. *In:* D. Thorner, B. Kerblay and R.E.F. Smith, eds. Homewood, IL: Richard D. Irwin, Inc.

Cheng, T. 1990. Political regimes and development strategies: South Korea and Taiwan. *In:* G. Gereffi and D.L. Wyman, eds. *Manufacturing miracles: paths of industrialisation in Latin America and East Asia.* Princeton, NJ: Princeton University Press.

Clark, G. 1993. Agriculture and the industrial revolution: 1700–1850. *In:* J. Mokyr, ed. *The British industrial revolution: an economic perspective.* Boulder, CO: Westview Press, pp. 227–66.

Corbridge, S. 1982. Urban bias, rural bias, and industrialisation: an appraisal of the work of Michael Lipton and Terry Byres. *In:* J. Harris, ed. *Rural development: theories of peasant economy and agrarian change.* London: Hutchinson, pp. 94–116.

Corbridge, S. and G.A. Jones. 2005. The continuing debate about urban bias: the thesis, its critics, its influence, and implications for poverty reduction. *Department research papers in environmental and spatial analysis,* No. 99. London: Department of Geography and Environment, London School of Economics and Political Science (LSE), University of London.

Cowen, M.P. and R.W. Shenton. 1998a. Agrarian doctrines of development: part I. *The Journal of Peasant Studies,* 25(2), 49–76.

Cowen, M.P. and R.W. Shenton. 1998b. Agrarian doctrines of development: part II. *The Journal of Peasant Studies,* 25(3), 31–62.

Cox, T. 1979. Awkward class or awkward classes? Class relations in the Russian peasantry before collectivisation. *The Journal of Peasant Studies*, 7(1), 70–85.

Davis, M. 2006. *Planet of slums*. London: Verso.

Deane, P. 1965. *The first industrial revolution*. Cambridge: Cambridge University Press.

Deere, C.D. 1990. *Household and class relations: peasants and landlords in northern Peru*. Berkeley, CA: University of California Press.

Deere, C.D. and A. de Janvry. 1979. A conceptual framework for the empirical analysis of peasants. *American Journal of Agricultural Economics*, 61(4), 602–11.

Desmarais, A.A. 2007. *La Vía Campesina: globalisation and the power of peasants*. Black Point, Nova Scotia: Fernwood Publishing.

Eastwood, R. and M. Lipton. 2000. Rural–urban dimensions of inequality change. *WIDER Working Paper No. 200*. Helsinki: World Institute for Development Economics Research.

Eicher, C.K. and J.M. Staatz, eds. 1984. *Agricultural development in the Third World*. Baltimore, MD: Johns Hopkins University Press.

Eicher, C.K. and L.W. Witt, eds. 1964. *Agriculture in economic development*. New York, NY: McGraw Hill.

Ellis, F. 1984. Relative agricultural prices and the urban bias model: a comparative analysis of Tanzania and Fiji. *Journal of Development Studies*, 20(3), 28–51.

Ellis, F. 2000. *Rural livelihoods and diversity in developing countries*. Oxford: Oxford University Press.

Ellman, M. 1975. Did the agricultural surplus provide the resources for the increase in investment in the USSR during the first five year plan?. *Economic Journal*, 85(339), 844–63.

Ellman, M. 1978. On a mistake of Preobrazhensky and Stalin. *Journal of Development Studies*, 14(3), 353–58.

Erlich, A. 1960. *The Soviet industrialisation debate 1924–28*. Cambridge, MA: Harvard University Press.

Fernandes, B.M. 2008. Território, teoria y política, paper presented at the International Seminar *Las Configuraciones de los Territorios Rurales en el Siglo XXI*, Pontificia Universidad Javeriana, Bogotá, 24–28 March.

FitzGerald, E.V.K. 1985. Agrarian reform as a model of accumulation: the case of Nicaragua since 1979. *In:* A. Saith, ed. *The agrarian question in socialist transitions*. London: Frank Cass, pp. 208–26.

Fox, K.A. and D.G. Johnson, eds. 1970. *Readings in the economics of agriculture*. London: George Allen & Unwin.

Friedmann, H. 2005. From colonialism to green capitalism: social movements and emergence of food regimes. *In:* F.H. Buttel and P. McMichael, eds. *New directions in the sociology of global development*. Oxford: Elsevier, pp. 227–64.

Ghatak, S. and K. Ingersent. 1984. *Agriculture and economic development*. Brighton: Wheatsheaf Books.

Goodman, D. and M. Watts. 1994. Reconfiguring the rural or fording the divide? Capitalist restructuring and the global agro-food system. *The Journal of Peasant Studies*, 22(1), 1–49.

Griffin, K. 1977. Book review of *Why poor people stay poor: urban bias in world development* by M. Lipton. *Journal of Development Studies*, 14(1), 108–9.

Harrison, M. 1979. Chayanov and the Marxists. *The Journal of Peasant Studies*, 7(1), 86–100.

Harrison, M. 1985. Primary accumulation in the Soviet transition. *In:* A. Saith, ed. *The agrarian question in socialist transitions*. London: Frank Cass, pp. 81–103.

Hart, G. and A. Sitas. 2004. Beyond the urban–rural divide: linking land, labour and livelihoods. *Transformations*, (56), 31–8.

Havnevik, K.D., *et al.* 2007. *African agriculture and the World Bank: development or impoverishment?*. Uppsala: Nordic Africa Institute.

Hudson, P. 1992. Agriculture and the industrial revolution. *In:* P. Hudson, ed. *The industrial revolution*. London: Edward Arnold, pp. 64–97.

Jenkins, R.O. 1991. The political economy of industrialisation: a comparison of Latin American and East Asian newly industrialising countries. *Development and Change*, 22(2), 197–231.

Johnston, B.F. and P. Kilby. 1975. *Agriculture and structural transformation: economic strategies in late-developing countries*. New York, NY: Oxford University Press.

Johnston, B.F. and J.W. Mellor. 1961. The role of agriculture in economic development. *American Economic Review*, 51(4), 566–93.

Jones, G.A. and S. Corbridge. 2008. Urban bias. *In:* V. Desai and R. Potter, eds. *The companion to development studies*, second edition. London: Hodder Arnold, pp. 243–47.

Karshenas, M. 1990. Oil income, industrialisation bias, and the agricultural squeeze hypothesis. *The Journal of Peasant Studies*, 17(2), 245–72.

Karshenas, M. 1994. Concepts and measurements of agricultural surplus. *The Journal of Peasant Studies*, 21(2), 235–61.

Karshenas, M. 1996/1997. Dynamic economies and the critique of urban bias. *The Journal of Peasant Studies*, 24(1–2), 60–102.

Kay, C. 1977. Book review of *Agrarian reform and agrarian reformism* edited by D. Lehmann. *The Journal of Peasant Studies*, 4(2), 241–4.

Kay, C. 1980. Relaciones de dominación y dependencia entre terratenientes y campesinos en Chile. *Revista Mexicana de Sociología*, 42(2), 751–97.

Kay, C. 1989. *Latin American theories of development and underdevelopment*. London: Routledge.

Kay, C. 1997. Globalisation, peasant agriculture and reconversion. *Bulletin of Latin American Research*, 16(1), 11–24.

Kay, C. 2000. Latin America's agrarian transformation: peasantisation and proletarianisation. *In:* D. Bryceson, C. Kay and J. Mooij, eds. *Disappearing peasantries? Rural labour in Africa, Asia and Latin America*. London: Practical Action Publishing, pp. 123–38.

Kay, C. 2002. Why East Asia overtook Latin America: agrarian reform, industrialisation and development. *Third World Quarterly*, 23(6), 1073–102.

Kay, C. 2006. Rural poverty and development strategies in Latin America. *Journal of Agrarian Change*, 6(4), 455–508.

Kay, C. 2008. Reflections on Latin American rural studies in the neoliberal globalisation period: a new rurality?. *Development and Change*, 39(6), 915–43.

Kerridge, E. 1967. *The agricultural revolution*. London: Allen and Unwin.

Kitching, G. 1982. *Development and underdevelopment in historical perspective*. London: Methuen.

Krueger, A.O. 1991. *The political economy of agricultural pricing policy. Volume 5, a synthesis of the political economy in developing countries*. Baltimore, MD: The Johns Hopkins University Press for the World Bank.

Krugman, P. 1994. The myth of Asia's miracle. *Foreign Affairs*, 73(6), 62–78.

Kuznets, P.W. 1988. An East Asian model of economic development: Japan, Taiwan and South Korea. *Economic Development and Cultural Change*, 36(3), S11–S43.

Kuznets, S. 1964. Economic growth and the contribution of agriculture; notes on measurement. *In:* C.K. Eicher and L.W. Witt, eds. *Agriculture in economic development*. New York, NY: McGraw Hill, pp. 102–19.

Lehmann, D. 1986. Two paths of agrarian capitalism, or a critique of Chayanovian Marxism. *Comparative Study of Society and History*, 28(4), 601–27.

Lewis, W.A. 1954. Economic development with unlimited supplies of labour. *The Manchester School of Economic and Social Studies*, 22(2), 139–91.

Lewis, W.A. 1958. Economic development with unlimited supplies of labour. *In:* A.N. Agarwala and S.P. Singh, eds. *The economics of underdevelopment*. New York, NY: Oxford University Press, pp. 400–49.

Lipton, M. 1968a. The theory of the optimising peasant. *Journal of Development Studies*, 4(3), 327–51.

Lipton, M. 1968b. Strategy for agriculture: urban bias and rural planning. *In:* P. Streeten and M. Lipton, eds. *The crisis of Indian planning*. London: Oxford University Press, pp. 83–147.

Lipton, M. 1977. *Why poor people stay poor: urban bias in world development*. London: Temple Smith.

Lipton, M. 1984. Urban bias revisited. *Journal of Development Studies*, 20(3), 139–66.

Lipton, M. 1993. Urban bias: of consequences, classes and causality. *Journal of Development Studies*, 29(4), 229–58.

Lipton, M. 2005. Urban bias. *In:* T. Forsyth, ed. *Encyclopedia of international development*. London: Routledge, pp. 724–26.

Little, I., T. Scitovsky and M. Scott. 1970. *Industry and trade in developing countries*. Oxford: Oxford University Press.

Losch, B. 2004. Debating the multifuctionality of agriculture: from trade negotiations to development policies by the south. *Journal of Agrarian Change*, 4(3), 336–60.

Lynch, K. 2005. *Rural–urban interaction in the developing world*. London: Routledge.

Lynch, K. 2008. Rural–urban interaction. *In:* V. Desai and R. Potter, eds. *The companion to development studies*, second edition. London: Hodder Arnold, pp. 268–72.

Magdoff, F. 2008. The world food crisis: sources and solutions. *Monthly Review*, 60(1), 1–15.

Mandelbaum, K. 1945. *The industrialisation of backward areas*. Oxford: Basil Blackwell.

Mandelbaum, K. 1979, orig. 1928. Introduction to the correspondence of Marx and Engels and Danielson (Nikolai-on). *Development and Change*, 10(4), 515–44.

Martin, K. 1982. Agrarian reforms and intersectoral relations: a summary. *ISS Working Papers Series*, No. 1. The Hague: Institute of Social Studies.

Martin, K. 2002. Agrarian reforms and intersectoral relations. *In:* V. FitzGerald, ed. *Social institutions and economic development: a tribute to Kurt Martin*. Dordrecht: Kluwer Academic Publishers, pp. 1–8.

McMichael, P. 2008. Peasants make their own history, but not just as they please…. *In:* S.M. Borras Jr., M. Edelman and C. Kay, eds. *Transnational agrarian movements confronting globalisation*. Oxford: Wiley-Blackwell, pp. 61–89.

McMichael, P. 2009a. Food sovereignty, social reproduction and the agrarian question. *In:* A.H. Akram-Lodhi and C. Kay, eds. *Peasants and globalisation: political economy, rural transformation and the agrarian question*. London: Routledge, pp. 288–312.

McMichael, P. 2009b. Banking on agriculture: a review of the *World development report 2008*. *Journal of Agrarian Change*, 9(2), 235–46.

Mellor, J.W. 1973. Accelerated growth in agricultural production and the intersectoral transfer of resources. *Economic Development and Cultural Change*, 22(1), 1–16.

Mitra, A. 1977. The terms of trade, class conflict and classical political economy. *The Journal of Peasant Studies*, 4(2), 181–94.

Moore, B. Jr. 1969. *Social origins of dictatorship and democracy: lord and peasants in the making of the modern world*. Harmondsworth: Penguin Books.

Moore, M. 1984. Political economy and the rural–urban divide, 1767–1981. *Journal of Development Studies*, 20(3), 1–27.

Moore, M. 1993. Economic structure and the politics of sectoral bias: East Asian and other cases. *Journal of Development Studies*, 29(4), 79–128.

Mundle, S. 1985. The agrarian barrier to industrial growth. *Journal of Development Studies*, 22(1), 49–80.

Nicholls, W.H. 1963. An 'agricultural surplus' as a factor in economic development. *Journal of Political Economy*, 71(1), 1–29.

Nicholls, W.H. 1964. The place of agriculture in economic development. *In:* C.K. Eicher and L.W. Witt, eds. *Agriculture in economic development*. New York, NY: McGraw-Hill, pp. 11–44.

Ocampo, J.A. and M.A. Parra. 2007. The continuing relevance of the terms of trade and industrialisation debates. *In:* E. Pérez Caldentey and M. Vernengo, eds. *Ideas, policies and economic development in the Americas*. London: Routledge.

Oshima, H. 1987. *Economic growth in monsoon Asia: a comparative survey*. Tokyo: University of Tokyo Press.

Overton, M. 1996. *Agricultural revolution in England: the transformation of the agrarian economy 1500–1850*. Cambridge: Cambridge University Press.

Oxfam International 2007. *What agenda now for agriculture? A response to the World Bank development report 2008*. Oxford: Oxfam Briefing Note.

Patel, R. 2007a. *Stuffed & starved: from farm to fork, the hidden battle for the world food system*. London: Portobello Books.

Patel, R. 2007b. *The World Bank and agriculture: a critical review of the World Bank's*. World development report 2008, Discussion Paper, Johannesburg: Action Aid.

Patnaik, U. 1979. Neo-populism and Marxism: the Chayanovian view of the agrarian question and its fundamental fallacy. *The Journal of Peasant Studies*, 6(4), 375–420.

Prebisch, R. 1971. Economic development or monetary stability: the false dilemma. *In:* I. Livingstone, ed. *Economic policy for development*. Harmondsworth: Penguin Books, pp. 345–84.

Preobrazhensky, E.A. 1965. *The new economics*. Oxford: Clarendon Press.
Rakodi, C. 2008. Prosperity or poverty? Wealth, inequality and deprivation in urban areas. *In:* V. Desai and R.B. Potter, eds. *The companion to development studies*, second edition. London: Hodder Education, pp. 252–7.
Ranis, G. and J.C.H. Fei. 1964. *Development of the labour surplus economy: theory and policy*. Homewood, IL: Richard D. Irwin.
Ranis, G. and L. Orrock. 1985. Latin America and East Asian NICs: development strategies compared. *In:* E. Durán, ed. *Latin America and the world recession*. Cambridge: Cambridge University Press.
Reardon, T. and J.A. Berdegué. 2002. The rapid rise of supermarkets in Latin America. *Development Policy Review*, 20(4), 371–88.
Rizzo, M. 2009. The struggle for alternatives: NGOs responses to the 2008 WDR. *Journal of Agrarian Change*, 9(2), 277–90.
Roberts, B. 1995. *The making of citizens: cities of peasants revisited*. London: Arnold Publishers.
Roberts, B. and N. Long. 1979. *Peasant cooperation and capitalist expansion in central Peru*. Austin, TX: University of Texas Press.
Rosenstein-Rodan, P.N. 1943. Problems of industrialisation of Eastern and Southeastern Europe. *Economic Journal*, 53(4), 202–11.
Ruiz, N. and J. Delgado. 2008. Territorio y nuevas ruralidades: un recorrido teórico sobre las transformaciones de la relación Campo-Ciudad. *Revista Eure*, 34(102), 77–95.
Saad-Filho, A. 2005. The rise and decline of Latin American structuralism and dependency theory. *In:* J.K.S. and E.S. Reinert, eds. *The origins of development economics: how schools of economic thought have addressed development*. London: Zed Books, pp. 128–45.
Saith, A. 1985. 'Primitive accumulation', agrarian reform and socialist transitions: an argument. *In:* A. Saith, ed. *The agrarian question in socialist transitions*. London: Frank Cass, pp. 1–48.
Saith, A. 1990. Development strategies and the rural poor. *The Journal of Peasant Studies*, 17(2), 171–244.
Saith, A. 1992. *The rural non-farm economy: processes and policies*. Geneva: ILO.
Saith, A. 1995. From collectives to markets: restructured agriculture–industry linkages in rural China. *The Journal of Peasant Studies*, 22(2), 201–41.
Schejtman, A. and J.A. Berdegué. 2004. *Desarrollo territorial rural*. Santiago: RIMISP.
Schiff, M. and A. Valdés. 1998. The plundering of agriculture in developing countries. *In:* C.K. Eicher and J.M. Staatz, eds. *International agricultural development*, third edition. Baltimore, MD: The Johns Hopkins University Press, pp. 226–33.
Schultz, T.W. 1964. *Transforming traditional agriculture*. New Haven, CT: Yale University Press.
Scoones, I. 2009. Livelihoods perspectives and rural development. *The Journal of Peasant Studies*, 36(1), 171–96.
Shanin, T. 1973. The nature and logic of the peasant economy. I: a generalisation. *The Journal of Peasant Studies*, 1(1), 63–80.
Shanin, T. 1974. The nature and logic of the peasant economy. II: diversity and change; III: policy and intervention. *The Journal of Peasant Studies*, 1(2), 186–206.
Shanin, T., ed. 1988. *Peasants and peasant societies*, second edition. London: Penguin Books.
Shanin, T. 2009. Chayanov's treble death and tenuous resurrection; an essay about understanding, about roots of plausibility and about rural Russia. *The Journal of Peasant Studies*, 36(1), 83–101.
Standing, G. 1981. Migration and modes of exploitation: social origins of immobility and mobility. *The Journal of Peasant Studies*, 8(2), 173–211.
Tacoli, C. 2003. The links between urban and rural development. *Environment and Urbanisation*, 15(3), 3–12.
Tacoli, C., ed. 2006. *The earthscan reader in rural–urban linkages*. London: Earthscan.
Tapella, E. and P. Rodríguez. 2008. *Transformaciones globales y territorios*. Buenos Aires: Editorial La Colmena.
Teranishi, J. 1997. Sectoral resource transfer, conflict, and macrostability in economic development: a comparative analysis. *In:* M. Aoki, H.K. Kim and M. Okuno-Fujiwara, eds. *The role of government in East Asian economic development: comparative institutional analysis*. Oxford: Clarendon Press.

Varshney, A., ed. 1993. *Beyond urban bias*. London: Frank Cass & Routledge.

Veltmeyer, H. 2004. *Civil society and social movements: the dynamics of intersectoral alliances and urban–rural linkages in Latin America*. Civil society and social movements programme paper no. 10. Geneva: UNRISD.

Veltmeyer, H. 2008. The World Bank on agriculture and development: a failure of imagination or the power of ideology? Paper presented at the seminar on *Rural Latin America: contemporary issues and debates*, 13 June. Amsterdam: CEDLA.

von Braun, J. 2008. High food prices: the what, who, and how of proposed policy actions. *Policy brief May 2008*. Washington, DC: IFPRI.

Wade, R.H. 1990. *Governing the market: economic theory and the role of government in East Asian industrialisation*. Princeton, NJ: Princeton University Press.

Weis, T. 2007. *The global food economy: the battle for the future of farming*. London: Zed Books.

Wharton, C.R. Jr., ed. 1969. *Subsistence agriculture and economic development*. Chicago, IL: Aldine Publishing Co.

White, B. 1987. Rural development: rhetoric and reality. *Journal für Entwicklungspolitik*, 1(1), 54–72.

Wiggins, S. and S. Levy. 2008. Rising food prices: a global crisis. *Briefing paper, no. 37*. London: ODI.

Wolf, E.R. 1969. *Peasant wars of the twentieth century*. New York, NY: Harper and Row.

Woodhouse, P. 2009. Technology, environment and the productivity problem in African agriculture: comment on *World development report 2008*. *Journal of Agrarian Change*, 9(2), 263–76.

World Bank 1993. *The East Asian miracle: economic growth and public policy*. New York, NY: Oxford University Press for the World Bank.

World Bank 2001. *World development report 2000/2001: attacking world poverty*. New York, NY: Oxford University Press for the World Bank.

World Bank 2007. *World development report 2008: agriculture for development*. New York, NY: Oxford University Press for the World Bank.

Wuyts, M. 1994. Accumulation, industrialisation and the peasantry: a reinterpretation of the Tanzanian experience. *The Journal of Peasant Studies*, 21(2), 159–93.

Zeitlin, M. and R.E. Ratcliff. 1988. *Landlords and capitalists: the dominant class of Chile*. Princeton, NJ: Princeton University Press.

A food regime genealogy

Philip McMichael

Food regime analysis emerged to explain the strategic role of agriculture and food in the construction of the world capitalist economy. It identifies stable periods of capital accumulation associated with particular configurations of geopolitical power, conditioned by forms of agricultural production and consumption relations within and across national spaces. Contradictory relations within food regimes produce crisis, transformation, and transition to successor regimes. This 'genealogy' traces the development of food regime analysis in relation to historical and intellectual trends over the past two decades, arguing that food regime analysis underlines agriculture's foundational role in political economy/ecology.

Introduction

In the past year or so, in conjunction with an increasingly evident energy shortage, and global warming, a palpable food crisis has commanded public attention across the world. Rising food prices, and a cascade of food rioting, has signaled the end of the era of cheap food. It has also focused awareness of our agricultural foundations, especially as they have become increasingly dependent on fossil fuels. Recent attention to 'food miles' in public discourse has raised questions about our ability to continue to transport food, or its components, across the world. Just beneath the surface lurks a larger question, namely when public discourse will acknowledge that global agriculture is responsible for between a quarter and a third of greenhouse gas (GHG) emissions.[1] Under these circumstances, a coherent political-economy and political-ecology of food is of utmost importance – not simply to understand the dimensions of the food crisis, but also to situate the world food system and its crisis within a broader historical understanding of geo-political and ecological conditions. 'Food regime' analysis provides this possibility.

It is over two decades since the first formulation of the concept of 'food regime', by Harriet Friedmann (1987). This notion stemmed from previous research on the

The author is grateful to Haroon Akram-Lodhi for his helpful comments on the first draft of this paper.
[1]The *industrial* food system requires expenditure of 10–15 energy calories to produce one calorie of food, contributing 22 percent of greenhouse gas (GHG) emissions (McMichael *et al.* 2007). Agriculture's GHG emissions remain unaddressed in Al Gore's *An Inconvenient Truth* (2006) and Jeffrey Sachs' *Commonwealth* (2008).

post-World War II international food order, in which Friedmann (1982) charted the rise and demise of the US food aid program, as a geo-political weapon in the Cold War. Following this, a more systematic formulation by Friedmann and McMichael (1989) appeared in the European journal, *Sociologia Ruralis*. Since then, the food regime concept paper has been reprinted and translated, debated, and informed research and teaching in sociology, geography, political science and anthropology.[2] The 'food regime' concept historicised the global food system: problematising linear representations of agricultural modernisation, underlining the pivotal role of food in global political-economy, and conceptualising key historical contradictions in particular food regimes that produce crisis, transformation and transition. In this sense, food regime analysis brings a structured perspective to the understanding of agriculture and food's role in capital accumulation across time and space. In specifying patterns of circulation of food in the world economy it underlines the agrofood dimension of geo-politics, but makes no claim to comprehensive treatment of different agricultures across the world. Its examination of the politics of food within stable and transitional periods of capital accumulation is therefore quite focused, but nevertheless strategic. It complements a range of accounts of global political economy that focus, conventionally, on industrial and technological power relations as vehicles of development and/or supremacy. It is also complemented by commodity chain analyses,[3] dependency analyses,[4] and fair trade studies[5] that focus on particular food relationships in international trade. And, finally, there are studies of agriculture and food that focus on case studies,[6] questions of hunger,[7] technology,[8] cultural economy,[9] social movements,[10] and agribusiness[11] that inform dimensions of food regime analysis, once positioned historically within geo-political relations.[12] The difference made by food regime analysis is that it prioritises the ways in which forms of capital accumulation in agriculture constitute global power arrangements, as expressed through patterns of circulation of food.

This essay reviews the construction of the food regime perspective, and its development in relation to historical transformations and intellectual trends. I argue that debates over the composition and significance of a food regime have been productive in expanding its analytical reach (e.g., to include social movement, ecological and nutritional science relationships). I also note that there is an unresolved debate about whether a food regime is currently in place, suggesting that

---

[2]Frederick Buttel (2001, 171, 173) wrote: 'Beginning in the late 1980s the sociology and political economy of agriculture began to take a dramatic turn... only five years after the seminal piece – Friedmann and McMichael's 1989 *Sociologia Ruralis* paper on food regimes – was published, the sociology of agriculture had undergone a dramatic transformation... [the] article on food regimes was arguably the seminal piece of scholarship... and 'regime-type' work has proven to be one of the most durable perspectives in agrarian studies since the late 1980s, in large part because it is synthetic and nuanced.'
[3]e.g., Barndt (2008), Pritchard and Burch (2003), Fold (2002), and Fold and Pritchard (2005).
[4]e.g., Thomson (1987).
[5]e.g., Raynolds, Murray and Wilkinson (2007), Jaffee (2007).
[6]e.g., Hollis and Tullist (1986), Goodman and Watts (1997).
[7]e.g., George (1977), Lappé, Collins and Rosset (1998), Davis (2001).
[8]e.g., Kloppenburg (1988), Perkins (1997).
[9]e.g., Mintz (1986), Rifkin (1992), Dixon (2002).
[10]e.g., Borras, Edelman and Kay (2008).
[11]e.g., Morgan (1980), Burbach and Flynn (1980), Bonnano *et al.* (1994), Kneen (2002).
[12]Cf. McMichael (1994, 1995).

there are different ways in which to frame that question, which can help us to understand contemporary structuring forces in the global food system. Finally, I claim that food regime analysis is key to understanding a foundational divide between environmentally catastrophic agro-industrialisation and alternative, agro-ecological practices that is coming to a head now as we face a historic threshold governed by peak oil, peak soil, climate change, and malnutrition of the 'stuffed and starved' kind across the world (Patel 2007). This divide is, arguably, endemic to capitalism, and its food regime at large – generating a rising skepticism regarding the ecological and health impact of industrial food (Lang and Heasman 2004), and a gathering of food sovereignty movements across the world (Desmarais 2007) to reverse the modernist narrative of smallholder obsolescence etched into the development paradigm and current development industry visions of 'feeding the world' (McMichael 2008a).

Food regime formulations

Initial food regime analysis set parameters for historical analysis of opposing spatial relations within a political economy of an emerging international food system. Thus, the *first food regime* (1870–1930s) combined colonial tropical imports to Europe with basic grains and livestock imports from settler colonies, provisioning emerging European industrial classes, and underwriting the British 'workshop of the world'. Complementing mono-cultural agricultures imposed in colonies of occupation (compromising their food systems and ecological resources), nineteenth-century Britain outsourced its staple food production to colonies of settlement (over-exploiting virgin soil frontiers in the New World). Here, the establishment of national agricultural sectors within the emerging settler states (notably USA, Canada, and Australia), *modeled* twentieth-century 'development' as an articulated dynamic between national agricultural and industrial sectors.

The *second food regime* (1950s–70s) re-routed flows of (surplus) food from the United States to its informal empire of postcolonial states on strategic perimeters of the Cold War. Food aid subsidised wages, encouraging selective Third World industrialisation, and securing loyalty against communism and to imperial markets. 'Development states' internalised the model of national agro-industrialisation, adopting Green Revolution technologies, and instituting land reform to dampen peasant unrest and extend market relations into the countryside. Meanwhile, agribusiness elaborated transnational linkages between national farm sectors, which were subdivided into a series of specialised agricultures linked by global supply chains (e.g., the transnational animal protein complex linking grain/carbohydrate, soy/protein, and lot-feeding). In other words, as the 'development project'[13] universalised the 'national' model of economic development as a key to completion of the state system, following decolonisation, at the same time a 'new international division of labour' in agriculture began to form around transnational commodity complexes (Raynolds *et al.* 1993).

---

[13]The 'development project' refers to a politically-orchestrated initiative following the Second World War, incorporating postcolonial states into an imperial field of power to legitimise and expand capitalist markets as the vehicle of 'national' economic growth and modernity (McMichael 1996).

A *third, possibly emergent,*[14] *regime* (late-1980s–) has deepened this process, incorporating new regions into animal protein chains (e.g., China and Brazil), consolidating differentiated supply chains including a 'supermarket revolution' (Reardon *et al.* 2003) for privileged consumers of fresh fruits and vegetables, and fish, and generating populations of displaced slum-dwellers as small farmers leave the land.[15] Part of this conjuncture includes an emerging global food/fuel agricultural complex, now in tension with various forms of localism.[16] As 'food miles' add to rising food costs, and mass production runs standardise and process foods, movements such as Food Sovereignty, Slow Food, Community Supported Agriculture, and small-scale organic producers expand their social base on the grounds of democracy, ecology and quality. Whether inspired by alternative social visions, or political (and ecological) exigencies of a food system dependent on fossil fuels (Roberts 2008), such counter-movements contribute to the exhaustion of WTO-style agricultural liberalisation.

Arguably, each period, *and* the transitions between them, have reframed the politics of development and the scope and significance of agricultural and food technologies, including future implications (concerning environmental sustainability, food access and security, energy relations, control of technology, population displacement, nutrition and public health). In this sense, the food regime concept offers a unique comparative-historical lens on the political and ecological relations of modern capitalism writ large. We will return to this below. In the meantime, it is important to trace the lineages of the food regime concept – not only to understand its theoretical underpinnings, but also to emphasise that it is still in formation, especially now with the conjunction of energy, food, and climate crises.

Setting the terms

Friedmann's basic notion of the food regime is of a 'rule-governed structure of production and consumption of food on a world scale' (1993a, 30–1). To illustrate, consider her complex representation of the mid-twentieth century food regime during the development era:

> The postwar food regime was governed by implicit rules, which nonetheless regulated property and power within and between nations. The food regime, therefore, was partly about international relations of food, and partly about the world food economy. Regulation of the food regime both underpinned and reflected changing balances of power among states, organised national lobbies, classes – farmers, workers, peasants – and capital. The implicit rules evolved through practical experiences and negotiations among states, ministries, corporations, farm lobbies, consumer lobbies and others, in response to immediate problems of production, distribution and trade. Out of this web of practices emerged a stable pattern of production and power that lasted for two and a half decades.

---

[14] As discussed below, there is a debate regarding whether this is actually a full-fledged, or incipient food regime, or simply a hangover from the previous regime.

[15] Global slums now contain one billion people (Davis 2006).

[16] The localist project refers to the local/regional certification movement led by the Slow Food Foundation for Biodiversity (Friedmann and McNair 2008), mirrored in the move by supermarkets, from Wegmans to Wal-Mart, to replicate the lead taken in the UK by Tesco and Sainsbury in appropriating the localist project as a new form of profitability via 'quality' (Burros 2008, Frith 2006).

The rules defining the food regime gave priority to national regulation, and authorised both import controls and export subsidies necessary to manage national farm programmes. These national programmes, particularly at the outset of New Deal commodity programmes, generated chronic *surpluses*. As these played out, they structured a specific set of international relations in which power – to restructure international trade and production in one state's favour – was wielded in the unusual form of subsidised exports of surplus commodities. In this way agriculture, which was always central to the world economy, was an exceptional international sector. (Friedmann 1993a, 31)

Thus the mid-twentieth century resolution of the agrarian crisis of the 1930s, manifested in the 'dust-bowl' experience for mid-western and southern farmers in the US, involved a substantial reorganisation of the US farm sector with substantial implications for the world. The new emphasis on commodity programs rather than American rural development *per se*, laid a foundation for the surplus export regime in following decades (Winders 2009) as well as converting agricultures elsewhere to the agro-export model. The consequences have generated the recent *World Development Report* (World Bank 2008), with its new-found (nevertheless misplaced) concern for 'agriculture for development' (McMichael 2009b). Friedmann has more recently renamed the postwar regime the 'mercantile-industrial food regime' to emphasise its foundations in agro-industrialisation and its state-protectionist origins. The latter included export subsidies as a 'defining feature', transforming the US into a 'dominant exporter' and in turn transforming 'Japan and the colonies and new nations of the Third World from self-sufficient to importing countries', and Europe into a 'self-sufficient and eventually major export region' (2005, 240).

However, the operative phrase is 'implicit rules', a subtle method of establishing that a food regime, for Friedmann, involves a period of 'relatively stable sets of relationships', with 'unstable periods in between shaped by political contests over a new way forward' (Friedmann 2005, 228). By this Friedmann means that what works, under the specific historical circumstances, is not the direct expression of interest, so much as the distillation of political struggles among contending social groups. Thus she argues that

Each of the past two food regimes was the combined outcome of social movements intersecting with state strategies and strategies of profit-seeking corporations... Of course, the new regime rarely had all the results they had envisioned. Like states and corporations, social movements are rarely careful about what they ask for. (Friedmann 2005, 234)

And so, for the 'first' 'colonial-diasporic food regime' Friedmann argues that it

was framed within a general rhetoric of free trade and the actual workings of the gold standard. The world wheat market that arose in the decades after 1870 was not really anyone's goal. However, vast international shipments of wheat made possible what actors really wanted to do.... Wheat was the substance that gave railways income from freight, expanding states a way to hold territory against the dispossessed, and diasporic Europeans a way to make an income (Friedmann 2005, 231–2),

through the new social form of (settler) family farming. She argues that the second food regime was 'even more implicit', since the circulation of agricultural commodities was 'framed specifically not as trade' but as aid, transferred through a mechanism involving 'counterpart payments' by recipient states into local banks to be used at the discretion of local American advisors (Friedmann 1982). Framing surplus transfers (that eroded local food systems) as 'aid' naturalised what were a set of implicit power relationships – Friedmann (2005, 232) comments: 'When the regime works really well, the consequences of actions are predictable, and it appears to work without rules.'

Conceptual evolution

From the above statements it is clear that the concept of the 'food regime' has evolved. One might say that while the initial conception was primarily structural, it has been refined over time with historical prompting – both from intellectual debates and from the transformation of the global food economy itself. The initial Friedmann and McMichael article (1989) blended an insight from regulation theory with one from the world system perspective. Regulation theory suggested the notion of a stable set of relationships through which the food regime articulated with periods of capital accumulation. The emphasis on reducing labour costs in late-nineteenth century European manufacturing with cheap food from the colonies and settler states coincided with regulation theory's concept of 'extensive accumulation'. Counterposed to this was the industrialisation of agriculture and the construction of processed, 'durable foods' in the mid-twentieth century food regime, as part of the 'intensive accumulation' associated with the Fordist period of consumer capitalism, where consumption relations were incorporated into accumulation itself, rather than simply cheapening its wage bill. Periodising history this way, in a rather stylised conceptualisation,[17] provided the opportunity for an unusually sharp critique by Goodman and Watts (1994), who preferred to see agriculture as governed by distinct processes from manufacturing, and so, as Araghi observed, dismissed 'the concept of a global food regime *tout court* in favour of agrarian particularism' (2003, 51).[18] As Araghi noted, the critique threw the baby out with the bathwater by discounting a nonetheless significant world-historical periodisation anchored in the political history of capital (Araghi 2003, 50).

In fact, the original 1989 formulation of successive food regimes was constructed around a juxtaposition of successive moments of British, and US, hegemony in governing the capitalist world economy. The food circuits in each regime supported the dominant state's exercise of power in expanding and sustaining fields of market and ideological dominance. In the mid- to late-nineteenth century regime the British state and capital were at the center of two sets of food flows – tropical foods from the non-European colonies, and temperate foods (meat and grains) from the settler-colonial states (US, Canada, Australia, Argentina, Uruguay, South Africa). These food circuits (alongside raw material imports) underpinned the 'workshop of the world' model, assisted by 'free trade imperialism', which (for a time) allowed British

---

[17]Note that Le Heron (1993) deployed a regulationist perspective on agricultural restructuring in the second half of the twentieth century, although with considerably more concrete details on the interplay between institutional settings and forms of regulation of agriculture, at all scales, within and beyond the GATT regime. Moran *et al.* (1996) argue for a 'real regulation', based in farmer organised policy mobilisations across regions and states.

[18]Pistorius and van Wyck used the 'food regime' to frame their substantive book, *The Exploitation of Plant Genetic Information*, on the politics of 'crop development', noting 'Friedmann and McMichael (1989)... explicitly refer to the Regulation theory, but they do not explain *how* their conception of food regime fits into that theory' (1999, 22).

capital and commerce access to European economies and empires. The City of London and the sterling/gold standard lubricated capital's first world market, channeling investment to the various imperial frontiers of extraction (McMichael 1984, 21–7). Within these global circuits the first 'price-governed market' in food emerged, anchored in the American family farming frontier, which produced low-cost wheat relative to that produced on European capitalist farms (Friedmann 1978). Cheap colonial foodstuffs, extracted with catastrophic consequence for non-European cultures (Davis 2001), provisioned British and European capital, which was able to sustain accumulation through imposing patterns of 'underconsumption' on its early generation labour forces (Mintz 1986, Halperin 2005, 91, Araghi 2003).

Subsequently, the US-centered food regime deployed food surpluses to build development states in the Third World. As the model of an internally-articulated national economy, based in the dynamic exchanges between the farm sector and the manufacturing sector, the postwar 'development project' projected a national development model into the postcolonial world. At the same time, food aid provisioned emerging urban labour forces (and their political regimes) with cheap food, stimulating industrialisation, and promoting food dependency in the longer run. Agricultural commodity prices were stabilised during this period by government-managed trade in surplus foods (Tubiana 1989). Meanwhile 'counterpart funds' encouraged agribusiness expansion in the Third World, developing livestock industries supplied with American grains, followed by the introduction of 'Green Revolution' technologies to expand staple food supplies and de-politicise the countryside.

While each food regime deepened global circuits of food, based in distinct agricultural models (from plantations to family farms to specialised forms of agribusiness), securing global hegemony for Britain and the US successively, they also embodied contradictory historical relations. In keeping with the structural tenor of the 1989 article, the authors identified these as follows: in the British-centered food regime, an outsourcing of food production via the conversion of tropical colonies into exporters of sugar, tea, coffee, bananas, palm oil, peanuts and so on, assumed a new dimension in the settler colonies, which expanded on the basis of temperate foods (grains and meat). Whereas the former relationship reproduced a colonial division of labour, the latter (sovereign relationship) resulted in an internal division of labour as a home market for manufacturing capital grew on the agricultural base (McMichael 1984). The settler states, then, prefigured the twentieth-century ideal-typical national model, with articulated economic sectors (manufacturing and agriculture), informing the US model of a farm/factory dynamic as the ideal vehicle and outcome of development (cf. Rostow 1960).[19] In other words, the food regime concept offered an interpretation not only of the agrarian basis of world hegemonies, but also a historicised understanding of the evolution of models of development that expressed and legitimised those power relations.

In the US-centered food regime, the national model of development, based in modernising farm sectors (behind mercantilist/tariff walls, institutionalised in the 1947 GATT agreement), where agriculture and industry would articulate in a virtuous cycle of technologically-based growth, stood in contradiction to the construction of transnational commodity chains linking specialised agricultural

---

[19]It was ideal-typical not only because US farmers were also exporters of food, but also because it assumed that all states could, and would, follow this path (McMichael 2008, 42).

sectors in different world sites. From the Marshall Plan through the Third World Green Revolution, the US state encouraged international agribusiness with export credits and counterpart funds designed to universalise the American farming and dietary models.[20] Not only were Korean housewives taught to make sandwiches with imported American wheat, courtesy of the counterpart deposits of *won* in Korean banks, but also American grains-fed livestock (cattle, poultry and hogs) in what Friedmann (1994) has termed the international 'animal protein complex'. In short, during the moment of US hegemony, while the ideology was of national development (certainly supported by selective expansion of the Green Revolution), the reality was an internationalisation of agribusiness chains of inputs, technologies and foodstuffs, eroding the coherence of national farm sectors. The authors characterised this as a significant tension between *replication* and *integration* of farm sectors, arguing that the organising principle of the world economy was shifting from state to capital. This would explain why, for example, the US-centered food regime eventually converted national farm lobbies into corporate lobbies, as small farmers have been increasingly marginalised by agribusiness (Friedmann 2005, 244).

In short, leaving aside the regulation theory strand, food regime analysis clearly historicised, and therefore politicised, our understanding of the strategic role of agrofood relationships in the world economy. Central to this were two formative developments, the universalisation of a nation-state system (as decolonisation occurred within the terms of the 'American century'), and associated processes of agro-industrialisation. A food regime perspective, however, is not intended to offer a comprehensive understanding of food cultures and relationships across the world:

> note that this concept has a comparative macro-status, and in no way assumes that all food production and consumption conforms to this pattern. Certainly other forms of production and consumption of food may be marginalised or emboldened by the arrogances of the corporate food system, but there is a substantial arena of food production and consumption beyond that of the food regime. (McMichael 2000, 421fn)

In historicising and politicising food, the food regime perspective opened doors to further development of the concept.

One such door is the social movement door, through which both Friedmann and McMichael have passed. Friedmann's above references to social movements leads her to 'refocus historical analysis on *transitions*' between regimes, where social movements act as 'engines of regime crisis and formation' (2005, 229). By renaming the regimes, Friedmann signals the key role of workers, and farmers, respectively in shaping the 'colonial-diasporic' and the 'mercantile-industrial' food regimes. In the former, working class unrest and migration contribute, as settlers constituted the new frontier of family farmers, who 'could exist only through international trade, and would suffer most from a collapse of the regime' (2005, 236). The unraveling of this regime in the early twentieth century produced a 'new type and significance of

---

[20]Friedmann offers a sophisticated discussion of how the 'export of the US model' was 'the outcome of specific practices in the postwar food regime. At the same time, these practices also reflected historical experiences, so that the effects were quite distinct in Europe, the emergent third world, and as we shall see later, in Japan' (1993a, 35). Le Heron (1993) offers a comprehensive account of the political decisions in the making of a globalised agriculture following the collapse of the 'second food regime', with particular Antipodean detail.

farm politics', symbolised in the 'mercantilist' epithet of the second food regime. This was built on the basis of agricultural support and protectionist programs fueling agro-industrialisation behind tariff walls, breached only by a public 'food aid' program. In laying the foundations for a successor 'corporate-environmental' regime, Friedmann identifies contradictions in an unfolding 'green capitalism', where 'a new round of accumulation appears to be emerging in the agrofood sector, based on selective appropriation of demands by environmental movements, and including issues pressed by fair trade, consumer health, and animal welfare activists' (2005, 229).

Similarly, but with a different purpose, McMichael has focused on the transnational mobilisation of peasants (articulating with movements such as Food Sovereignty,[21] Slow Food, Fair Trade), in opposition to what he has termed a 'food from nowhere'[22] regime (2002), or a contemporary 'corporate food regime' (2005). This conception pivots on the original notion of a food regime embodying a historical conjuncture comprising contradictory principles. Just as the dynamics of the previous regimes centered on tensions between opposing geo-political principles – colonial/national relations in the first, national/transnational relations in the second, so the corporate food regime embodies a central contradiction between a 'world agriculture' (food from nowhere) and a place-based form of agro-ecology (food from somewhere).[23] In addition, this formulation focuses attention on the politics of dispossession of the world's small farmers, fisher-folk and pastoralists, including a counter-mobilisation in the name of 'food sovereignty' against the modernist narrative that views peasants as residual (McMichael 2006, 2008a). While Friedmann focuses on the social movements of worker/migrant farmers in the first food regime, as her transitional link to the second regime (absent colonial producers of tropical – and Indian wheat,[24] for example), McMichael focuses on social movements from the global South as the key hinge in a current food regime dynamic.

On a broader scale, McMichael's perspective implies that the simplification of industrial agriculture that began with colonial monocultures, and has been universalised through successive food regime episodes, has now reached a fundamental crisis point. It is expressed in the emergence of a transnational movement of smallholders intent on asserting the critical importance of biodiverse and sustainable agriculture for human survival (Desmarais 2007), in addition to the question of stemming the 'planet of slums' phenomenon (Davis 2006, Araghi 1995, 2008), the question of human rights to culturally and nutritionally adequate food, and the reformulation of states, *and development*, around democratised food systems

---

[21]'Food sovereignty' has been defined by the transnational peasant movement, Vía Campesina, as 'the right of peoples to define their own agriculture and food policies, to protect and regulate domestic agricultural production and trade in order to achieve sustainable development objectives, to determine the extent to which they want to be self reliant, and to restrict the dumping of products in their markets. Food sovereignty does not negate trade, but rather it promotes the formulation of trade policies and practices that serve the rights of peoples to safe, healthy and ecologically sustainable production' (2001).

[22]The 'food from nowhere' concept comes from Bové and Dufour (2001) and concerns differentiating craft from industrial agriculture.

[23]Campbell (2009) has modified the meaning of food from somewhere to include global food audit relations.

[24]Davis points out that at the turn of the twentieth century, 20 percent of England's bread was produced from wheat extracted through unscrupulous marketing of Indian grain reserves (2001, 299).

(Patel 2006). At this point in the story, the focus on peasant mobilisation is an acknowledgement that the human and ecological wake created by the 'globalisation' of the corporate food regime is the central contradiction of the twenty-first century global food system. Here, 'the trajectory of the corporate food regime is constituted through resistances: both protective (e.g., environmentalism) and proactive, where "food sovereignty" posits an alternative global moral economy' (McMichael 2005, 286).

Another food regime?

At this point it is important to address the vexed question as to whether and to what extent we can identify a 'third food regime'. From Friedmann's perspective, we have not yet seen the full-scale (hegemonic) establishment of a food regime, with 'implicit rules' (framed by social forces) imprinted in the production and consumption of traded foods (which currently divide between industrial and affluent/fresh foods). In contrast, McMichael views the recent neo-liberal world order as resting on a 'corporate food regime', containing atavisms of the previous regime, and organised around a politically constructed division of agricultural labour between Northern staple grains traded for Southern high-value products (meats, fruits and vegetables). The free trade rhetoric associated with the global rule (through states) of the World Trade Organisation suggests that this ordering represents the blossoming of a free trade regime, and yet the implicit rules (regarding agro-exporting) preserve farm subsidies for the Northern powers alone, while Southern states have been forced to reduce agricultural protections and import staple, and export high-value, foods.

Despite several formulations of a third food regime, perhaps the most important issue is that the disagreements are not entirely definitional or empirical – rather there is a broader question concerning the status of the concept itself. As a historical concept, it could serve to simply identify the agrofood foundations of historical periods, cycles, or even secular trends of capitalism. In this respect, the 'food regime' becomes not a structural formation in itself so much as an attribute or an optic on one or more historical conjunctures. And so: 'The point is not to hypostatise "food regimes." They constitute a lens on broader relations in the political history of capital. They express, simultaneously, forms of geo-political ordering, and, related, forms of accumulation, and they are vectors of power' (McMichael 2005, 272). Thus, the 'food regime' can be considered to be simply an analytical device to pose specific questions about the structuring processes in the global political-economy, and/or global food relations, at any particular moment. Here the 'food regime' is not so much an episodic structure, or set of rules, but becomes a method of analysis. Arguably, the various representations of a 'third food regime' are simultaneously expressions of deployment of the food regime as an analytical device.

As intimated in the 1989 article, the possibility of a decentralised and ecologically-grounded food regime was already an important imaginary (expressed in a range of practices in Southern peasant sites, Northern community agricultures, central Italy and even in Francis Moore Lappé's prescient *Diet for a Small Planet* of the early 1970s). Four years later, Friedmann (1993a, 51–2) articulated this possibility as follows:

> Emergent tendencies have unfolded quickly since the Uruguay Round began in 1986. These prefigure alternative rules and relations. One is the project of corporate freedom

contained in the new GATT rules. The other is less formed: a potential project or projects emerging from the politics of environment, diet, livelihood, and democratic control over economic life. Farmers (who are heterogeneous) must somehow ally themselves in the main contest over future regulation: will it be mainly private and corporate, or public and democratic? What international rules would promote each alternative? The answers depend on the ways that emerging agrofood policies are linked either to accumulation imperatives or to demands raised by popular social movements.

In the meantime, the contours of a third, or new, food regime were sighted from the Antipodean perspective (McMichael 1992, Le Heron 1993, Pritchard 1996, 1998). McMichael, focusing on 'the reconstitution of the state system to support a unified global market', surmised that the passage of a GATT regime

> may complete the institutionalisation of mechanisms and norms of global regulation by rendering all states (some being more equal than others) subject to retaliation for unfair trade practices...[where] national sovereignty would be subordinated to an abstract principle of membership in the state system that sanctions corporate rights of free trade and investment access. (1992, 356, 354)

Pritchard focused on regulatory developments in sub-sectors like dairy and wheat in specific regions. Le Heron's (1993) *Globalised Agriculture* formalised the new focus on (early) 'value chains' identified in research by Friedland (1994), Llambi (1994) and Raynolds (1994), on the rise of 'non-traditional exports' of fruits and vegetables from the global South, and shaping subsequent research on various commodities such as shrimp, poultry, canned seafood, canned pineapple and fresh fruit from Thailand (Goss and Burch 2001), green beans, baby carrots and corn, and snowpeas from Kenya (Dolan and Humphrey 2000), corporate tomatoes from Mexico (Barndt 2008),[25] and Pritchard and Burch's global analysis of different sources and forms of tomato production (2003). In the early days of 'globalisation', Le Heron (1993, 191–2), extrapolating from the New Zealand experience,[26] wrote:

> Some researchers see in the 1980s–90s the erosion of the second food regime which grew around the grain-livestock and durable foods complexes supplying rising demand in the developed nations. They attribute this decline to the duplication of productive capacity in developing countries and declining governmental powers over the principal agents of restructuring, the TNCs. The recent appearance of a particularly dynamic component of the world food system, the global fresh fruit and vegetable industry, is perhaps the harbinger of a third food regime. In this investment nexus, transnational organisations have been prominent from the beginning...Globalised agriculture can be considered a

---

[25]Dolan (2004) and Barndt (2008) represent research strategies concerned with tracing out the spatialised, racialised and gendered sub-contracting relations surrounding horticultural production systems.

[26]Pritchard has observed: 'New Zealand's agricultural sector was thoroughly liberalised from the mid-1980s, making it a kind of test-case for what an incipient global agricultural system might look like. Reflecting on these issues in his book *Globalised Agriculture: Political Choice*, Le Heron (1993) argued that the advancement of Uruguay Round objectives would lead inexorably to the de-nationalisation of food systems, as circuits of capital in agrofood production, exchange and reproduction become beholden to footloose global corporations and financial institutions' (forthcoming). Moran *et al.* (1996) use New Zealand and French rural political mobilisations to specify national regulatory variation, which they argue differentiates agro-commodity chains within food regimes at large.

by-product of the international crisis in agriculture, which accompanied and was linked to the more general crisis in the 1970–90s in world capitalism.

The point of this perspective was that in the early 1990s a discernible transnational corporate 'global sourcing' of foods was most obvious in the technologies of seed modification, cooling and preserving, and transport of fruits and vegetables as non-seasonal, or year-round, access for affluent consumers became available through the management of archipelagos of plantations across the global South – giving rise to what Friedmann (1991) called the New Agricultural Countries (NACs), as counterparts to the Newly Industrialising Countries (NICs). In this new division of world agricultural labour (cf. Raynolds *et al.* 1993) it was noted at the time that:

> transnational corporations typically subcontract with Third World peasants to produce specialty horticultural crops and off-season fruits and vegetables. They also process foods (such as fruit juices, canned fruits, frozen vegetables, boxed beef, and chicken pieces), often in export processing zones, for expanding consumer markets in Europe, North America and Pacific-Asia. (McMichael 1996, 105)

Enabling this global process was the phenomenon of the 'Second Green Revolution' (DeWalt 1985) – distinguished from its antecedent by its shifts: from public to private initiative, from staple grains to affluent foods (animal protein, fruits and vegetables, chemical feedstocks), and from domestic to global markets. The international division of agricultural labour was essentially an extension of developments within the US-centered food regime, involving a rising share (73–82 percent) of the OECD in the volume of cereal exports (1970–96), and the global South importing 60 percent of world cereal volumes. At the same time the NACs expanded their share of a tripling of the world market for fruits and vegetables, and the OECD countries became the world's major suppliers of plant varieties. But the significant shift was *political* (Pistorius and van Wyk 1999, 110–11). As Pistorius and van Wyk (1999, 51) put it, re-defining the food regime as an agro-food order:

> The advent of the Third Agro-Food Order has revealed a tendency for the state as the pivot of crop development to be replaced by private industry. Since the 1980s, the growth of public investment in agricultural R&D has declined, private industry has obtained a greater say in the allocation of public agricultural R&D funds, while private investment in agricultural research has risen rapidly. This development has been accompanied by a thorough restructuring of the organisation of the plant breeding sector, which has given rise to the formation of industrial crop development conglomerated, based in OECD countries. Given the accumulation of unrivalled financial and technological capacity within these industrial conglomerates, they seem to become the central actors and dynamic force of crop development in the Third Agro-Food order.

The privatisation of agricultural research was a key marker of the 'globalisation project'[27] – a politically-instituted process of economic liberalisation privileging corporate entities and rights in the food system, with respect to crop development

---

[27]The 'globalisation project' represents 'an emerging vision of the world and its resources as a globally organised and managed free trade/free enterprise economy pursued by a largely unaccountable political and economic elite' (McMichael 1996, 300).

and the management of 'food security' – as a service performed not by nation-states, but by transnational corporations through the world market. It suggested another formulation, the 'corporate food regime': which 'carries legacies of the previous food regimes, nevertheless expressing a new moment in the political history of capital', reversing the political gains of the welfare and development eras, by 'facilitating an unprecedented conversion of agriculture across the world to supply a relatively affluent global consumer class. The vehicle of this corporate-driven process is the WTO's Agreement on Agriculture, which... institutionalises a distinctive form of economic liberalism geared to deepening market relations via the privatisation of states' (McMichael 2005, 273). A key legacy of the previous food regime was managed overproduction, via subsidy structures hidden from plain view, formally legitimising a principle of 'free trade' across the board, as members would lower barriers to agricultural trade and submit to the minimum five percent import rule – facilitating 'orderly' disposal of Northern surpluses by transnational agribusiness. In this sense, the food regime is 'not so much a political-economic order, as such, rather it is a vehicle of a contradictory conjuncture, governed by the "double movement" of accumulation/legitimation' (McMichael 2005, 274). This particular formulation, therefore, focused on the political reconstruction of a regime institutionalising elements of the previous regime (transnationalism and managed Northern surpluses) in different, multilateral arrangements represented as state-authored market liberalisation.

To establish the central contradiction of this new formulation, McMichael (2005, 274) argued that 'the *corporate* food regime embodies the tensions between a trajectory of "world agriculture" and cultural survival, expressed in the politics of food sovereignty'. The point of this rather stark contrast was to situate the food regime within the broader historical conjuncture, whose distinctiveness is 'how dispossession is accomplished. Briefly, where the "development project" socialised security, the "globalisation project" privatises security. These phases both represent political solutions to material needs' (McMichael 2005, 275). Here, the notion of a 'corporate food regime' serves as a double reminder. First, that the concept of the food regime enables historicisation and politicisation of particular global political-economic conjunctures. And second, that for all the focus on 'internal' details of food regimes, such as key commodity chains, the rise of retailing and the politics of consumption, regulation of quality and standards, differentiation of diets, organics, niche marketing, and so on – all legitimate dimensions of the contemporary food system – the historic meaning and impact of the food regime, as lens, is more substantive. There is little doubt, for instance, that the trajectory of this so-called 'corporate food regime' is such that it poses a fundamental threat to the survival of a substantial proportion of the inhabitants of the planet (especially those who do not participate in the global marketplace), and to the ecology of the planet.

Since the initial Friedmann and McMichael article of 1989, these authors have diverged in focus, laying groundwork for distinct (but not necessarily contradictory) understandings of what 'food regime' might mean. Whereas McMichael values the notion of a 'corporate food regime' characterising the neo-liberal world order, Friedmann is more cautious about identifying such a phenomenon at present, choosing instead to speak of an emergent 'corporate-environmental' food regime (2005). Consistent with her institutional interest – both formal (rules and regulations) and informal (beliefs, norms, implicit rules), Friedmann argues that this would-be regime,

> like past food regimes, is a specific constellation of governments, corporations, collective organisations, and individuals that allow for renewed accumulation of capital based on shared definition of social purpose by key actors, while marginalising others... Unlike the postwar regime, which standardised diets, it is likely to consolidate and deepen inequalities between rich and poor eaters. (2005, 228)

Paying attention to the reconfiguration of old institutions, such as food aid, farm subsidies and marketing boards, and consumer food subsidies, Friedmann (2005, 249) surmises:

> A new regime seems to be emerging not from attempts to restore elements of the past, but from a range of cross-cutting alliances and issues linking food and agriculture to new issues. These include quality, safety, biological and cultural diversity, intellectual property, animal welfare, environmental pollution, energy use, and gender and racial inequalities. The most important of these fall under the broad category of environment.

Friedmann finds the 'lineaments' of a corporate-environmental food regime in converging environmental politics and the reorganisation of retailing food supply chains, subdivided by class diets, and represented, for example, in the US by Whole Foods ('a stunning appropriation of a 1960s counter-cultural term') and WalMart, catering respectively to 'transnational classes of rich and poor consumers' (Friedmann 2005, 252). As trade disputes among governments in the WTO have stalled, in part over national standards, giving rise to 'green protectionism' (Campbell and Coombes 1999), private capital has taken the lead in setting (and often raising) food standards in context of negotiating certification and brand image profitability in relation to 'social movements of consumers, environmentalists, and others' (Friedmann 2005, 253). Complementing public regulation of the food system, private standard setting differentiates consumers between those served by 'standard edible commodities' (WalMart) and those consuming products from 'quality audited food chains' (Whole Foods). As Friedmann puts it: 'the distinction between fresh, relatively unprocessed, and low-chemical input products on one side, and highly engineered edible commodities composed of denatured and recombined ingredients on the other, describes two complementary systems within a single emerging food regime' (2005, 258). Demurring on outcomes, Friedmann notes that 'while the rise of "quality" agrofood systems may herald a new "green capitalism", it may serve only privileged consumers within a food regime rife with new contradictions' (2005, 257). It is in the vortex of 'green capitalism' that Friedmann finds the key tension: 'states, firms, social movements, and citizens are entering a new political era characterised by a struggle over the relative weight of private, public, and self-organised institutions' (2005, 259), including how to redefine 'public', democratise agrifood systems, and incorporate Lang and Heasman's concept of 'ecological public health' (2004).

In contrast to Friedmann's scenario, or as a 'play within a play', McMichael's concept of the 'corporate food regime' has had a specific purpose, namely to focus attention on how instituting the full-scale dispossession of an alternative agriculture is licensed by the so-called 'globalisation project'. The latter, essentially a recalibration of 'development' at the global, rather than the national, scale, accentuates the narrative of peasant extinction in the modern world. Whether focusing on dispossession as the fundamental contradiction is normative or not, the issue is ultimately epistemological.[28] That is, regarding smallholding as irrational is itself irrational,[29] when we consider that industrial agriculture is undermining conditions of human survival, through its intensive dependence on fossil fuels, its accounting for about a third of GHG, its degradation of soil (intensifying dependence on petro-fertiliser), its destruction of biodiversity, and ultimately its depletion of cultural and ecological knowledges about living and working with natural cycles by wiping out smallholder diversified farming, shown to be more productive and more environmental than specialised industrial farming (Pretty *et al.* 2006, Weis 2007, 164–8, Altieri 2008, and cf. IAASTD 2008).

The 'corporate food regime' defines a set of rules institutionalising corporate power in the world food system. While the WTO is the key institution, there are associated trade agreements like the North American Free Trade Agreement (NAFTA) that replicate the asymmetry of the WTO protocols, which preserve Northern agricultural subsidies behind a façade of economic liberalisation, directed at states in the global South. As Friedmann suggests, 'farm lobbies' in the global North became agribusiness lobbies as an outcome of the postwar food regime, and managed to institute mechanisms in their respective states, and through the Uruguay Round negotiations, which were ostensibly about stemming the escalation of farm subsidies and managing the crisis of overproduction, and dumping of surpluses in the world market during the 1970s and 1980s, in the wake of the postwar regime. Prior to the formalisation of the WTO Agreement on Agriculture, the EU began switching from its original Common Agricultural Policy (CAP) farm price support policies to US-style government direct payments to producers, thereby decoupling production costs from prices, which, allowed the formation of a single '*world price*'. Thus,

> For traders, low commodity prices enable commodity dumping in the world market (assisted by export subsidies, especially European), forcing local prices down at the expense of small farmers.... Despite the rhetoric of free trade, the Northern agenda is realised through a corporate-mercantilist comparative advantage in a highly unequal world market. (McMichael 2005, 278–9)

Alongside of the displacing effects of this system of disposal of cereal surpluses from the North, has been the proliferation of agro-exporting from the global South, much of it mandated through Structural Adjustment Policies devised by the IMF and World Bank, in the name of 'feeding the world'. The consequence has been to

---

[28]That is, peasant mobilisation suggests this narrative is not being conceded. This is a political intervention, with claims transcending modernist/Marxist teleology. Bernstein (2008), for example, misconstrues an epistemic for a normative position, by associating the descriptor of La Vía Campesina's politics, 'the peasant way', with a nostalgic or populist discourse. This understanding de-historicises peasant politics by imposing a capital lens on a movement that consciously challenges orthodox, and fetishised, understandings of food as a product/input, rather than a socio-ecological relationship in which smallholders (and all of us) have a political stake (McMichael 2008b).

[29]Marx understood this well enough, when he observed: 'The moral of the tale ... is that the capitalist system runs counter to a rational agriculture, or that a rational agriculture is incompatible with the capitalist system (even if the latter promotes technical development in agriculture) and needs either small farmers working for themselves or the control of the associated producers' (quoted in Foster 2000, 165).

displace land for domestic food production, integrating small farmers into tenuous contract relations, or simply regrouping the dispossessed on agro-industrial estates, or in mushrooming slums. Conservative (FAO) estimates suggest upwards of 30 million peasants lost their land in the decade after the WTO was established (Madeley 2000). The functional consequence of such a regime of course has been to swell the reserve labour force for manufacturing and service industries at the same time, underpinning the militarised proliferating export processing zones across the global South:

> The paradox of this food regime is that at the same time as it represents global integration as the condition for food security, it immiserates populations, including its own labour force. The perverse consequence of global market integration is the export of deprivation, as 'free' markets exclude and/or starve populations dispossessed through their implementation. In turn, dispossessed populations function as reserve labour, lowering wages and offering the possibility of labour casualisation throughout the corporate empire. (McMichael 2005, 285)

Refocusing the food regime

The articulation of the products and consequences of the corporate, or neo-liberal, food regime with patterns of accumulation in manufacturing and service industries refocuses the question of the food system's compositional, and contextual, relations. Araghi (2003) addresses this head-on, arguing the food regime is a 'political regime of global value relations'. Here food is intrinsic to capital's global value relations, insofar as it is central to the reproduction of wage labour, and, indeed, other forms of labour coming under capital's sway. The focus remains on the relations of capital, rather than food itself, which is, rather, a means to the end of provisioning and managing the labour relation. In his view, what Friedmann and McMichael called the 'second food regime', was actually an interregnum in the history of capital, and was more appropriately understood as an 'aid-based food order of an exceptionally reformist period of world capitalism' (Araghi 2003, 51). In other words, global value relations – the organising principle for the British-centered regime, and arguably, for the late-twentieth century (neo-liberal) regime – were compromised in the postwar Keynesian/Fordist compact of 'embedded liberalism'.

The question arises, then, what is the appropriate institutional complex in which to situate a 'food regime'? While First World states implemented the UN 'social contract', empire was reformulated through neo-colonial relations, with the food aid program directing food exports towards securing the loyalty of Third World states on the Cold War perimeter with cheap food to lower manufacturing costs and create new food dependency. Unlike the British state, the US reconstructed the capitalist world order 'not through formal empire, but rather through the reconstitution of states as integral elements of an informal American empire' (Panitch and Gindin 2004, 17). The counterpart (of an overproducing farm sector) was rising wages in the First World serving to intensify accumulation through the consumption of animal protein and 'durable foods' (Friedmann 1991). One might argue with Araghi that in each food regime, a particular geo-political configuration organised a set of production and circulation relations of food that maintained capitalism's empire. That is, the materiality and/or expression of value relations are subject to specific socio-political configurations of the wage relation in the politics of the state system.

Nonetheless, Araghi's intervention poses a significant challenge to the stylised patterning of the first two food regimes. The value relations perspective reminds us that under capitalism, food is an exchange-value, first, and a use-value second. In other words, food regime analysis of forms of capitalist modernity is both more systematic and historical to the extent that it resists the temptation to take food as its point of departure. The exemplar of this is perhaps *Sweetness and Power. The Place of Sugar in Modern History* (1986), by Sidney Mintz, whose title signals the power relations wherein sugar, or more precisely, sucrose, has been, historically, an item of status emulation, a caloric fuel for the European proletariat (whose slave counterpart experienced similar 'underconsumption'), and an essential ingredient in modern 'fast food' consumerism. Araghi uses the 'food regime of capital' phrasing to emphasise food as not just a commodity, but as a historic commodity-relation, re-centering agriculture in the analysis of capitalist transformations. Thus Araghi (2003, 51, emphasis added) argues:

> This revised understanding of food regimes, not as components of some abstract (and nation-state based) concepts of economic regulation, but much rather as the political fact of world historical value relations, is crucial to a more nuanced and differentiated understanding of what we may call 'embedded imperialism'. The latter approach . . . – posits imperialism . . . in connection with world historical value relations and food regimes of capital. This perspective brings agriculture and food to the centre of analysis not as a result of a postmodern retreat into locality, anti-urbanism and neo-populist nostalgia for rurality, but precisely because *global agriculture and food are inseparable from the reproduction of labour power.*

For Araghi, this inseparability is the governing principle – meaning that the structures of capital accumulation across time and space involve distinctive forms of food production and consumption that cannot be understood without situating them in the broader, patterned circuits of capital. Central to this approach is the juxtaposition of relationships of 'overconsumption' and 'underconsumption' of food, corresponding to class relations on a world scale.

The 'food regime of capital approach' not only politicises the world food order, but also it focuses on the fungibility of food, as a commodity. We have seen this with the concept of 'crop development' and its relationship to corporate strategies of 'substitutionism', whereby tropical products (sugar, palm oil) are displaced by agro-industrial byproducts (high-fructose corn syrup, margarine) and cereals are rendered functionally equivalent, such as in feedstuffs and biotechnology feedstocks (Goodman *et al.* 1987). We are now seeing it massively in the current socially and ecologically inappropriate 'agrofuels project'. Here, value relations result in the conversion of agriculture to meet short-sighted alternative energy/emissions targets at a time in which the market episteme governs problem framing and solving. The agrofuels project represents the ultimate fetishisation of agriculture, converting a source of human life into an energy input at a time of rising prices (McMichael 2009a). A 'food regime of capital' perspective incorporates this development into our understanding of politicisation of agriculture within the neo-liberal regime at large.

In shifting the focus from institutional to value relations, Araghi refocuses an enduring dimension of the original food regime analysis, namely its attention to the political history of capital – as expressed or realised through the structuring of global food relations. One can read into this perspective that in the history of capital, there are cyclical moments and expressions within capital's food regime relations. Converging on this, perhaps,[30] McMichael and Friedmann (2007, 295–6) recently rephrased the food regime in a chapter on supermarkets, observing:

> just as retailers are expedient in appropriating discourses, and sometimes practices of alternative food systems, so corporate agriculture is premised on the appropriation of food cultures around the world... When considering such comprehensive transformations, the *historical* concept of food regimes points to two useful distinctions: between periods of stability and transition, and between the destruction of old relations and the emergence of new ones. While our early accounts of food regimes focused on periodising world history from the perspective of relatively stable arrangements of agricultural production, trade, and transformation into food commodities, our recent efforts turn to periods of transition. To put it simply, a food regime analysis of contemporary conditions would highlight the profound conditions of instability of the agrofood system.

In short, whether and to what extent another food regime can be identified depends on the terms of reference, which in turn depends on the terms of comparison (with previous such constructs) and one's methodological point of departure. Perhaps more promising, in the context of these developments (both historical and intellectual), one reason why it has become difficult to specify the 'food regime' as any one construct is the appearance of new dimensions in food regime analysis, to which we now turn.

Productivity of the food regime concept

There are several new departures, forthcoming in a special issue of *Agriculture and Human Values* on updating food regime analysis, important to address. To begin, Pritchard addresses, and argues against, the idea of a third food regime thus far. He suggests that a key question for food regimes scholars is 'how to theorise agriculture's incorporation into the WTO?' – translating this issue into 'the question of whether agriculture's incorporation into the WTO should be understood as facilitating a free market "third" food regime, or if it represents a state-centered carryover of the crisis of the second food regime, which it is incapable of resolving.' And his answer is that the collapse of the Doha Round negotiations in July 2008 'makes it possible, for the first time, to offer a conclusive assessment to this question' – his assessment being that the WTO is a 'carryover from the politics of the crisis of the second food regime, rather than representing a putative successor' (Pritchard forthcoming). This assessment is consistent with his claim that 'The essential feature of the food regimes approach is that it is best used as a tool of hindsight. It can help order and organise the messy reality of contemporary global food politics, but its applications are necessarily contingent upon an unfolding and unknowable future' (Pritchard 2007, 8). And this claim, of course, revisits the question of the status of the food regime concept.

One way of addressing this puzzle is to query the separation, and juxtaposition, of 'states' and 'markets' framing Pritchard's case. Fundamental to previous conceptions of food regimes has been the observation that food circuits are politically constructed, and in fact institutionalised in one way or another. All markets are political

---

[30]Another interpretation is the reluctance of the authors to commit to a definitive characterisation of a 'third food regime', either because the terms of reference are changing themselves, or because of different emphases.

constructs, and so a 'free market' regime is that in name only, and the important point is to see how and why that discourse is deployed and institutionalised. As Pritchard notes, the WTO has concentrated the competitive and sovereign relations of a quite unequal state system, not merely being the 'organisational vehicle of globally powerful sovereign states', nor being a 'supra-national entity with independent agency to implement free market agriculture' (forthcoming). The WTO certainly did evolve to manage the crisis following the collapse of the 'second food regime', and in that sense did not represent a clean-break successor regime. On the other hand, given the paucity of cases, and previous observations that the contradictory relations in the first regime were conditions for resolution in the second, this might be extended to an understanding of the constitution of a 'third regime'. Pritchard's position, however, is that 'the WTO encapsulated but did not move beyond the crisis of the second food regime' (forthcoming).

Arguably, the existence of the WTO is one thing, and its paralysis expresses (and reinforces) the serious asymmetries of the state system, but its protocols have at the same time enabled a massive corporate consolidation and assault on smallholder agriculture behind the backs, as it were, of the negotiators (distracted by the trade paradigm from the social and ecological contradictions attending such arrangements).[31] By identifying the demise of the WTO, as the centerpiece of a possible third food regime, Pritchard draws attention to its contradictory political role in a world where agro-export intensification has empowered a set of states in the global South known as the Group of 20 (G-20). That is, power relations have altered as a consequence of the combined liberalising effects of the Structural Adjustment and WTO regimes. What is noteworthy here is that behind the backs of the trade institutions agribusiness power has deepened through integrating the global food system, altering the geography of power in the state system itself, increasingly at odds with the initial power structure informing the WTO regime itself. Pritchard's notion of a 'hangover' is insightful, and helps to underline the seismic shifts underway in the global food system. His intervention suggests the need for a double vision: viewing the food regime as constituted through the state/market relationship, nevertheless recognising that the scope and modalities of 'market rule' are always changing, reflecting geo-political, competitive, technological, social and legitimacy concerns. The question remains whether and to what extent a private food regime is embedded in and behind the institutional trappings of the multilateral system.

This is the implication of the argument by Pechlaner and Otero (2008) for a third, 'neoliberal food regime', insofar as trade liberalisation and corporate-friendly intellectual property rights (IPR) protocols link neoliberal regulation and biotechnology. Using the NAFTA framework as their case, these authors argue that while 'the anticipated third regime is still finding its stasis' (Pechlaner and Otero 2008, 2), its impact depends ultimately on 'neoregulation' (of liberalisation and IPR) by states interpreting trade agreements, as well as the strength of resistance. For Pechlaner and Otero, the 'inter-relationship between regulatory change and genetic engineering are integral to the emerging third food regime', in which biotechnology is the 'central technology for capitalist agriculture' (2008, 2). Again, this is a significant dimension of

---

[31]Deploying Cutler's concept of the 'private regime' (2001), Peine (2009) argues from the vantage point of the Brazilian soybean agro-exporting economy, that a privatised agribusiness regime has emerged through the backdoor, via a series of appeals to the Dispute Settlement Board of the WTO to advance corporate/state trade agendas.

food regime analysis, grounding the 'abstraction from ecology and entitlements associated with a world agriculture... prefigured in the biological and socio-economic blueprints of the gene revolution and the WTO's Agreement on Agriculture' (McMichael 2005, 282). Through their region-specific case study, Pechlaner and Otero are suggesting in fact that this blueprint will take different forms insofar as it depends ultimately on (multilaterally-situated) state decisions in the construction of food relations. Mindful of the groundswell of opposition to this regime among Mexican *campesinos*, the authors conclude that these decisions 'will be highly influenced by the force of local resistance. Consequently, we argue that, despite the prevailing trends, sufficient local resistance to the technology could modify, or even derail, the technology's role in individual nations, and, accordingly, in the unfolding food regime as a whole' (Pechlaner and Otero 2008, 2). While the technology dimension is a critical component, the contestation of agricultural biotechnology is, arguably, integral to the corporate food regime, insofar as it is organised in relation to the destruction of indigenous and local knowledges, and livelihoods. In other words, a food regime's structure is dialectical, rather than unilinear.

Burch and Lawrence, who have linked the rise of the retailing sector, and its focus on 'own brands' as a competitive strategy vis-à-vis food manufacturers, to an argument about the appearance of a 'third food regime' (2005, and see 2007), resituate food regime history in context of the transformation of financial relations. Specifying a 'financialised food regime', Burch and Lawrence (2009) claim:

> What is new, though, is the role played by a number of financial institutions and instruments that have the capacity to re-organise various stages of the agri-food supply chain, and to alter the terms and conditions under which other actors in the chain can operate. In the case of the private equity company, for example, we see a fraction of capital which views the agri-food company – whether it is a third-party auditor, an input supplier, a farm operator, a food manufacturer or a retailer – as a bundle of resources which provide opportunities for a quick profit, which may or may not involve a restructuring, but which will eventually return the enterprise to the share market and then move on to another bundle of resources.

Further, Burch and Lawrence argue that financialisation becomes endemic to the food industry, from supermarkets establishing their own financial services in partnerships with banks, acting like private equity companies, 'realising shareholder value by exploiting corporate assets which were previously seen as passive investments', to food manufacturing companies generating 'rental income from licensing of brand names, or sub-contracting the production of internationally famous products, such as Coca-Cola, while charging local producers monopolistic prices for the supply of necessary ingredients and other intermediate inputs' and new income streams generated (and patented) by nutraceuticals and functional foods produced by food/chemical companies (forthcoming).[32] Clearly analysis of the

---

[32]Complementing this scenario, financialisation can be understood as a measure of general political de-regulation, and because 'agro-industrialisation is being rapidly globalised through the mobility of financial capital, and its ability to rapidly concentrate, centralise and coordinate global agribusiness operations' (McMichael 2005, 288). Vía Campesina (2004, 2) has noted that 'now capital is not content to buy labour and hold land as private property, but it also wants to turn knowledge, technology, farm technologies and seeds into private property as part of a strategy of unification of agrofood systems across the world.'

restructuring of the food industry – in this case its retailing arm – is a significant dimension of food regime analysis. The focus on financialisation is timely, and possibly portends the further centralisation of the corporate food sector as the global financial crisis unfolds.

Dixon (2009) advances an altogether quite different, but nonetheless critical, nutritional perspective on food regimes. Her argument is framed by the rise and fall of the 'nutrition transition', as a benchmark of modernity and positive national development. The transition, from plant-based diets towards consumption of animal protein, oils and fats, processed sugars and processed carbohydrates, is typically associated with rising affluence. From this basic association has arisen a policy focus on nutritionalisation of the food supply (greater dietary diversity and available energy leading to positive public health outcomes). Within this movement class diets have distributed healthy diets to affluent consumers, and highly processed high calorie foods for poorer populations, the resulting explosion of mal-nutrition (associated with obesity) paralleling a persistent under-nutrition for a considerable portion of humanity. Dixon identifies these latter phenomena as the crisis phase of the nutrition transition, with 'diseases of affluence' appearing alongside global regions of hunger. Underpinning this crisis is a 'cultural economy' involving nutritionalisation of modern food systems, based in a science of the 'metabolic fate of food' as a form of governance, that is, 'the co-option of nutrition science to extract surplus value and authority relations from food, and is most transparent when framing corporate strategies and public policies in terms of nutritional disease and health-wealth advancement' (Dixon 2009). Dixon breaks new ground in tracing nutritionalisation of food systems as an 'unbroken socio-technical and knowledge revolution' from the identification of the calorie in the late-nineteenth century, that is, the 'capacity to quantify human energy introduced "scientific eating" into public policy and legitimised the agrofood import-export complexes that underpinned the first and second food regimes' (2009). Her contribution to food regime analysis alerts us to the increasingly contested nature of the 'search for nutritional and diet-based ontological security' in a world of shrinking dietary diversity and natural resources, a legitimacy crisis of nutritional science (authority) and corporate nutritionalisation (vs. viable cuisines or cultural diets) as unwanted side-effects mount.

Finally, Campbell (forthcoming) develops the environmental dimension of food regime analysis, elaborating Friedmann's ecological sensibility, and her use of Polanyi to address 'the destructive power of distanciated and socially disembedded food relations'.[33] As Campbell (forthcoming) sees Friedmann's arguments:

> Two key relations (echoing much of her earlier work) emerged as lying at the heart of unsustainable relations in the two historical food regimes: *distance* (between production and consumption) and *durability* (of key food commodities like wheat). Her argument was that a 'sustainable' food regime needed to subvert these dynamics and create sites for re-embedding food in local settings. The positive outcome that could be achieved through subversion of distance and durability was to enable a turn towards locality and seasonality: thus re-embedding food within locally and ecologically-appropriate food systems.

Taking his cue from Friedmann's previous work on ecology (1993b, 2000, 2003), Campbell introduces the concepts of 'ecologies at a distance', and 'ecological

---

[33]This is a theme returned to in Friedmann and McNair (2008).

feedback', into food regime analysis, noting that prior food regime relationships invisibilised their ecological impacts. And in the contemporary, more explicitly global/integrated, era these impacts have begun to feed back into the cultural dynamics of food regimes, as environmental and public health concerns mount. Campbell views food regime cultures as markers of legitimacy and stability, especially from the ecological point of view of environmentalism: 'The durable cultural logic of each was characterised by the ability to disguise what Marx had, during [the] very period, described as an irreparable metabolic rift that increasingly disrupted the interaction between human beings and nature' (forthcoming). To illustrate, for the second food regime, he notes how cultural framing of pesticides within a technological optimism began unraveling with the critique stemming from Rachel Carson's *Silent Spring* (1962). Developing Friedmann's identification of social movement power to legitimise or challenge regime cultures, he focuses on contradictory tendencies in the current world food order, especially a 'food from nowhere regime'. The latter, built on the cultural legitimacy of 'cheap food' has 'an emerging acute problem of cultural legitimacy', stemming from declining trust in science, environmental mobilisation, communication of 'previously invisible relations typical of "food from nowhere", risk politics and food scares, retailer power and explicit consumer preferences, and a perceived nutrition crisis associated with convenience foods' (Campbell, forthcoming). Accordingly, Campbell claims:

> It is within this cluster of cultural dynamics that a cluster of food relations that can be termed Food from Somewhere has emerged. Just as the Food from Nowhere regime is concentrated in the cheaper end of the food market and rooted within a set of cultural framings that emphasise cheapness, convenience, attractive transformation through processing and rendering invisible the origins of food products, affluent consumers in Western societies are attaching cultural status to foods which they perceive to be opposite; that are attractively socially and ecologically-embedded.

And Campbell (forthcoming) concludes:

> while Foods from Somewhere do provide one site of opportunity for changing some key food relations and ecologies, the social legitimacy of this new form of food relations does rely on the ongoing existence of the opposite, more regressive, pole of world food relations. Resolving this tension is central to any attempt to continue opening up spaces for future, more sustainable, global-scale food relations.

Campbell's legacy, building on the environmental thread spun originally by Friedmann, is not only to re-ground food regime analysis explicitly in political ecology, but also to underline the current tension between abstraction and situation of food cultures. This focus resonates in the continuing tension between the food sovereignty movement claims that smallholders 'feed the world and cool the planet', and the development industry's attachment to a 'food security' rhetoric for 'feeding the world' by incorporating Southern farmers into a refashioned agriculture organised by transnational 'value chains' – as detailed in the World Bank's *World Development Report* (2008). In addition, by emphasising the notion of 'ecological feedback', Campbell reinforces Weis' substantive contribution to the ecological contradictions of the 'global food economy' with its food miles, mounting toxicity and the 'ecological hoofprint'. Weiss (2007) emphasises the environmental impact of 'meatification', stressing that 'moving away from meat-centred consumption patterns is an elemental part of reducing humanity's collective space in the biosphere and leaving room for other species into this century, with well-balanced plant-centred diets also holding the additional promise of an array of public health benefits' (2007, 171). This is an argument not just for a political ecological perspective on food regimes, but also for an ethical perspective, which I shall address in the final section.

Food regimes, the metabolic rift, and an epistemic rift

Friedmann's and Campbell's respective references to the metabolic rift is foundational. The food regime is ultimately anchored in the 'metabolic rift', Marx's term for the separation of social production from its natural biological base.[34] The 'metabolic rift' expresses the subordination of agriculture to capitalist production relations, that is, the progressive transformation of agricultural inputs (organic resources to inorganic commodities), reducing nutrient recycling in and through the soil and water, and introducing new agronomic methods dependent upon chemicals and bioengineered seeds and genetic materials produced under industrial conditions. As Moore (2000) notes, the metabolic rift underlies the historic spatial separation between countryside and city, as agriculture industrialises. As a fundamental consequence of this rift, fossil fuel dependence exerts a determinate constraint on the viability of industrial agriculture in the future.

Moore's treatment of the metabolic rift articulates the social division of labour and its world-scale, and imperial implications (otherwise known as the 'ecological footprint').[35] The mediation of the urban/rural spatial relation by commodity circuits, rather than cycles of waste and regeneration of natural processes, deepens the metabolic rift. Historically, the world was reordered along these lines initially via the colonial division of labour, anchored in monocultures producing tropical products for metropolitan industrial and personal consumption. With the development of chemical agriculture and biotechnology, the growing abstraction of agriculture as an 'input-output process that has a beginning and an end' (Duncan 1996, 123) means that rather than a complex embedded in, and regenerating, local biological cycles, agriculture can in principle be relocated to specific locales anywhere on the planet as the 'intrinsic qualities of the land matter less' (1996, 122). In effect, agro-industrialisation increasingly replicates the spatial mobility of manufacturing systems, including the sub-division of constituent processes into global commodity chains (such as the animal protein complex). And, to the extent that it is premised on the incorporation (contract farming for agribusiness) and/or dispossession of smallholders, it has a labour reserve at its disposal.

'Petro-farming' (Walker 2004) deepens the metabolic rift, by extending inputs of inorganic fertiliser, pesticides, herbicides along with mechanisation, increasing farm demand for carbon-emitting fuels and inputs, in addition to releasing soil carbon to the atmosphere along with even more damaging nitrous oxide from fertiliser use, and

---

[34]It captures 'the material estrangement of human beings within capitalist society from the natural conditions which formed the basis for their existence' (Foster 2000, 163).
[35]Thus Marx wrote in *Capital*, volume 1 (1967, 860): 'England has indirectly exported the soil of Ireland, without even allowing its cultivators the means for replacing the constituents of the exhausted soil.'

from livestock waste in factory farming. The agro-industrial model, aided by states enclosing common and peasant lands for agro-industrial estates, has deepened its global presence via a second, private phase of the Green Revolution, targeting feed crops, livestock, fruits and vegetables, and now agro-fuels. Represented as the vehicle to feed the world, and stimulating agro-export revenues in indebted states, agribusiness displaces those agro-ecological systems, including slow food systems, which could reverse the metabolic rift, as they use 6–10 times less energy than industrial agriculture, restore soils, and reduce emissions up to 15 percent, not to mention sustaining small-scale producer livelihoods.[36]

The metabolic rift underlies both the material and epistemic relations of capitalism (McMichael forthcoming). In separating agriculture from its natural foundations, the metabolic rift informs the episteme through which we analyse the value relations of commodity production. The abstraction of agriculture, and therefore the foundations of social production, means that value relations organise agriculture, and it comes to be understood in these terms. This has become readily apparent today, in context of the combined crises of food, energy and climate change. The so-called 'biofuels rush' renders agriculture indistinguishable from energy production in a context where peak oil is making its presence felt in world prices. The inflation of food prices in turn expresses an integrating fuel-food complex, as alternative energy sources displace food, with fuel, crops. Here, palm oil 'now used widely in food products ranging from instant noodles to biscuits and ice cream, has become so integrated into energy markets that its price moves in tandem with crude oil price' (Greenfield 2007, 4). Further, with rising oil prices, 'food is worth more as petrol than it is on the table, even if the subsidies are removed' (Goodall 2008). The fungibility of investment choices, in plant-based food or fuel, emphasises the extent to which value relations govern current food relations.

As suggested above, value relations enable us to situate food politics historically, and to underline how capital undermines agriculture and its ecological base and hydrological and atmospheric cycles. But a value relations perspective ultimately limits our understanding of alternatives. We are constrained to 'see like capital', our understanding of the processes and consequences of agro-industrialisation being governed by its application of the economic calculus to environment relations. To the extent that food regime analysis deploys the lens of value relations, it discounts the ecological calculus, whereby the social reproduction of alternative food cultures depends on restorative ecological practices beyond a market episteme. Arguably, the ecological calculus is emerging as an organising episteme of the mushrooming counter-movements – from food sovereignty through Slow Food to Fair Trade. Vía Campesina, in particular, and the food sovereignty movement in general, recognise the epistemic shift that is necessary to reverse the metabolic rift, by revaluing agro-ecology and a 'carbon-rich' future, where a human-scale agriculture performs the life-task of feeding those marginalised by corporate foods, sequestering atmospheric carbon and rebuilding depleted soils across this planet. This epistemic shift represents an ethical intervention by which the economic calculus of capitalist food regimes is replaced by an ecological calculus.[37] The latter is gaining legitimacy, as the

---

[36]Pretty *et al.* (2006), Apfelbaum and Kimble (2007).
[37]Martinez-Alier (2002, 147) characterises Vía Campesina's ecological vision in these terms.

global food system's contradictions, limits and injustices are revealed more clearly in the compounding of the energy, food and climate crises. The food sovereignty movement, in politicising the current food order (McMichael 2008a), draws attention to the severe shortcomings of commodifying food, and its ecological foundations, across the world, and in so doing offers a new ethic that would inform a decentered and democratic 'food regime'.[38]

This ethic seeks to recover the centrality of agriculture[39] as a foundation, rather than the receding baseline role it has played in development narratives. It means envisioning ways in which social life can be reconstituted around alternative principles that respect the ecological relations through which social reproduction occurs. Community-supported agricultures, Slow Food praesidia (Fonte 2006), vegetarianism, and chapters of the transnational food sovereignty movement express this ecological sensibility.

Conclusion

The food regime concept is a key to unlock not only structured moments and transitions in the history of capitalist food relations, but also the history of capitalism itself. That is the food regime is an important optic on the multiple determinations embodied in the food commodity, as a *genus* fundamental to capitalist history. As such, the food regime concept allows us to refocus from the commodity as object to the commodity as relation, with definite geo-political, social, ecological, and nutritional relations at significant historical moments. In this essay I have traced the evolution of food regime analysis from a rather stylised periodisation of moments of hegemony in the global order to a refocusing on moments of transition, and the various social forces involved in constructing and reconstructing food regimes. Using a genealogical approach, we have encountered versions of food regime analysis focusing on whether and to what extent there is a current food regime associated with the neo-liberal moment, and its institutional supports, sometimes through case studies of commodities (Pritchard 1996, 1998), or technologies (Pechlaner and Otero 2008). Since the authority of the concept is a public good, what we may make of the debates about the salience or existence of a 'third' food regime is that they express different vantage points in understanding the structuring relations in, and multiple dimensions of, the agrofood system.

A principal distinction I have made is between identifying food regime moments (stable periods of accumulation and associated transitional periods), and using food regime analysis to identify significant relationships and contradictions in capitalist processes across time and space. Further, rather than view the food regime as a bounded moment or period, it can alternatively (or complementarily) express uneven and/or particular structuring processes in food relationships associated with the world-history of capitalism. The value relations method lends itself to this kind of '*longue dureé*', or generic, property of the food regime. We might, then, either

---

[38]This organising principle is central to the current debate over the future of food and ecology, as suggested by the different, but complementary, accounts by Friedmann (2005) and McMichael (2005).

[39]Duncan (1996, 2007) makes the extended, historical case for this notion.

154   *Philip McMichael*

imagine the food regime as a constant presence in the capitalist historical landscape, and/or examine mutations as continual efforts to regulate and resolve what are ultimately quite contradictory and volatile relationships, given food's central role in social reproduction, and therefore in reproducing changing forms and relations of power.

Ultimately, as a historical construct, the food regime has ethical potential: regarding how we live on the earth, and how we live together.[40] In this sense it stands as a point of departure in specifying the political and ecological relations of food in the history of capital. But identifying key food regime contradictions raises epistemological questions concerning the value calculus, challenging the path-dependent account encouraged by analysis through the lens of capital. Whether 'ecology at a distance', food sovereignty, or agro-ecology, these perspectives insist on addressing what are considered 'externalities' in the economic calculus, embracing a holistic understanding of agriculture that dispenses with the society/nature binary, and politicises food system cultures. Finally, historicising food regime politics has the potential to transcend the increasingly discredited episteme of capital accumulation and advocate agricultural reorganisation according to socially and ecologically sustainable practices. This is the centrality of the food regime in the twenty-first century.

References

Altieri, M. 2008. Small farms as a planetary ecological asset: five key reasons why we should support the revitalisation of small farms in the Global South. *Food First*. Available from: http://www.foodfirst.org/en/node/2115 [Accessed 26 March 2009].
Apfelbaum, S.I. and J. Kimble. 2007. A dirty, more natural way to fight climate change. *Ithaca Journal*, 6 December, p. 9.
Araghi, F. 1995. Global de-peasantisation, 1945–1990. *The Sociological Quarterly*, 36(2), 337–68.
Araghi, F. 2003. Food regimes and the production of value: some methodological issues. *The Journal of Peasant Studies*, 30(2), 41–70.
Araghi, F. 2008. The invisible hand and the visible foot: peasants, dispossession and globalisation. *In:* A.H. Akram-Lodhi and C. Kay, eds. *Peasants and globalisation. Political economy, rural transformation and the agrarian question*. London & New York: Routledge, pp. 111–47.
Barndt, D. 2008. *Tangled routes. Women, work and globalisation on the tomato trail*. Aurora, ON: Garamond Press.
Bernstein, H. 2008. Agrarian questions from transition to globalisation. *In:* A.H. Akram-Lodhi and C. Kay, eds. *Peasants and globalisation. Political economy, rural transformation and the agrarian question*. London: Routledge, pp. 214–38.
Bonnano, A., *et al.* 1994. *From Columbus to Conagra: the globalisation of agriculture and food*. Lawrence, KS: University of Kansas Press.
Borras, S.M. Jr., M. Edelman and C. Kay, eds. 2008. *Transnational agrarian movements confronting globalisation*. Chichester: Wiley-Blackwell.
Bové, J. and F. Dufour. 2001. *The world is not for sale*. London: Verso.
Burbach, R. and P. Flynn. 1980. *Agribusiness in the Americas*. New York, NY: Monthly Review Press.
Burch, D. and G. Lawrence. 2005. Supermarket own brands, supply chains and the transformation of the agrofood system. *International Journal of the Sociology of Agriculture and Food*, 13(1), 1–18.

[40]Such ethical perspectives constitute an important thread through Friedmann's work (e.g., 1993, 2003), and feature in Duncan (1996, 2007) and in the final, visionary chapter of Weis (2007).

Burch, D. and G. Lawrence, eds. 2007. *Supermarkets and agri-food supply chains. Transformations in the production and consumption of foods*. Cheltenham: Edward Elgar.

Burch, D. and G. Lawrence. 2009. Towards a third food regime: behind the transformation. *Agriculture and Human Values*, 26(4).

Burros, M. 2008. Supermarket chains narrow their sights. *New York Times*, 6 August, F1.

Buttel, F.H. 2001. Reflections on late-twentieth century agrarian political economy. *Sociologia Ruralis*, 41(2), 11–36.

Campbell, H. Forthcoming. The challenge of corporate environmentalism: social legitimacy, ecological feedbacks and the 'food from somewhere' regime. *Agriculture and Human Values*.

Campbell, H. and B. Coombes. 1999. 'Green protectionism' and organic food exporting from New Zealand: crisis experiments in the breakdown of Fordist trade and agricultural policies. *Rural Sociology*, 64(2), 302–19.

Carson, R. 1962 *Silent spring*. Boston, MA: Houghton Mifflin.

Cutler, A.C. 2001. Critical reflections on the Westphalian assumptions of international law and organisation: a crisis of legitimacy. *Review of International Studies*, 27(2), 133–50.

Davis, M. 2001. *Late Victorian holocausts. El Nino famines and the making of the Third World*. London: Verso.

Davis, M. 2006. *Planet of slums*. London: Verso.

Desmarais, A.-A. 2007. *La Vía Campesina. Globalisation and the power of peasants*. Halifax: Fernwood Publishing.

DeWalt, B. 1985. Mexico's second green revolution: food for feed. *Mexican Studies/Estudios Mexicanos*, 1(1), 29–60.

Dixon, J. 2002. *The changing chicken: chooks, cooks and culinary culture*. Sydney: University of New South Wales Press.

Dixon, J. (2009). From the imperial to the empty calorie: how nutrition relations underpin food regime transitions. *Agriculture and Human Values*, 26(4).

Dolan, C. 2004. On farm and packhouse: employment at the bottom of a global value chain. *Rural Sociology*, 69(1), 99–126.

Dolan, C. and J. Humphrey 2000. Governance and trade in fresh vegetables: the impact of UK supermarkets on the African horticulture industry. *Journal of Development Studies*, 37(2), 147–76.

Duncan, C. 1996. *The centrality of agriculture. Between humankind and the rest of nature*. Montreal: McGill-Queen's University Press.

Duncan, C. 2007. The practical equivalent of war? Or, using rapid massive climate change to ease the great transition towards a new sustainable anthropocentrism. Available from: http://globetrotter.berkeley.edu/GreenGovernance/Rapid%20Climate%20Change. pdf [Accessed 26 March 2009].

Fold, N. 2002. Lead firms and competition in 'bi-polar' commodity chains: grinders and branders in the global cocoa-chocolate industry. *Journal of Agrarian Change*, 2(2), 228–47.

Fold, N. and B. Pritchard. 2005. *Cross-continental food chains*. London: Routledge.

Fonte, M. 2006. Slow foods presidia: what do small producers do with big retailers?. *In:* T. Marsden and J. Murdoch, eds. *Between the local and the global: confronting complexity in the contemporary agri-food sector, research in rural sociology and development*. Oxford: Elsevier.

Foster, J.B. 2000. *Marx's ecology. Materialism and nature*. New York, NY: Monthly Review Press.

Friedland, W. 1994. The global fresh fruit and vegetable system: an industrial organisation analysis. *In:* P. McMichael, ed. *The global restructuring of agro-food systems*. Ithaca, NY: Cornell University Press, pp. 173–89.

Friedmann, H. 1978. World market, state and family farm: social bases of household production in an era of wage-labour. *Comparative Studies in Society and History*, 20(4), 545–86.

Friedmann, H. 1982. The political economy of food: the rise and fall of the postwar international food order. *American Journal of Sociology*, 88S, 248–86.

Friedmann, H. 1987. International regimes of food and agriculture since 1870. *In:* T. Shanin, ed. *Peasants and peasant societies*. Oxford: Basil Blackwell, pp. 258–76.

Friedmann, H. 1991. Changes in the international division of labour: agri-food complexes and export agriculture. *In:* W. Friedland, L. Busch, F.H. Buttel and A.P. Rudy, eds. *Towards and new political economy of agriculture.* Boulder, CO: Westview Press.

Friedmann, H. 1993a. The political economy of food: a global crisis. *New Left Review*, 197, 29–57.

Friedmann, H. 1993b. After Midas' feast. *In:* P. Allen, ed. *Food for the future: conditions and contradictions of sustainability.* New York, NY: John Wiley & Sons.

Friedmann, H. 1994. Distance and durability: shaky foundations of the world food economy. *In:* P. McMichael, ed. *The global restructuring of agro-food systems.* Ithaca, NY: Cornell University Press.

Friedmann, H. 2000. What on earth is the modern world-system? Food-getting and territory in the modern era and beyond. *Journal of World-System Research*, VI(2), 480–515.

Friedmann, H. 2003. Eating in the gardens of Gaia: envisioning polycultural communities. *In:* J. Adams, ed. *Fighting for the farm: rural America transformed.* Philadelphia, PA: University of Pennsylvania Press.

Friedmann, H. 2005. From colonialism to green capitalism: social movements and the emergence of food regimes. *In:* F.H. Buttel and P. McMichael, eds. *New directions in the sociology of global development. Research in rural sociology and development*, Vol. 11. Oxford: Elsevier, pp. 229–67.

Friedmann, H. and P. McMichael. 1989. Agriculture and the state system: the rise and fall of national agricultures, 1870 to the present. *Sociologia Ruralis*, 29(2), 93–117.

Friedmann, H. and A. McNair. 2008. Whose rules rule? Contested projects to certify 'local production for distant consumers'. *Journal of Agrarian Change*, 8(2–3), 408–34.

Frith, M. 2006. Ethical foods boom tops £2bn a year and keeps growing. *The Independent*, 13 October, p. 21.

George, S. 1977. *How the other half dies. The real reasons for world hunger.* Montclair, NJ: Allenheld, Osmun & Co.

Goodall, C. 2008. Burning food: why oil is the real villain in the food crisis. *The Guardian*, 30 May.

Goodman, D. and M. Watts. 1994. Reconfiguring the rural or fording the divide? Capitalist restructuring and the global agro-food system. *The Journal of Peasant Studies*, 22(1), 1–49.

Goodman, D. and M. Watts, eds. 1997. *Globalising food. Agrarian questions and global restructuring.* London: Routledge.

Goodman, D., B. Sorj and J. Wilkinson. 1987. *From farming to biotechnology. A theory of agro-industrial development.* Oxford: Basil Blackwell.

Gore, A. 2006. *An inconvenient truth.* New York, NY: Rodale.

Goss, J. and D. Burch. 2001. From agricultural modernisation to agri-food globalisation: the waning of national development in Thailand. *Third World Quarterly*, 22(6), 969–86.

Greenfield, H. 2007. Rising commodity prices & food production: the impact on food & beverage workers. International Union of Food, Agricultural, Hotel, Restaurant, Catering, Tobacco and Allied Workers' Associations (IUF), December.

Halperin, S. 2005. Trans-local and trans-regional socio-economic structures in global development: a 'horizontal' perspective. *In:* F.H. Buttel and P. McMichael, eds. *New directions in the sociology of global development.* Oxford: Elsevier Press, pp. 19–56.

Hollis, W.L. and F.L. Tullist, eds. 1986. *Food, the state, and international political economy.* Omaha, NB: University of Nebraska Press.

International Assessment of Agricultural Knowledge, Science and Technology for Development (IAASTD) 2008. Executive summary of the synthesis report. Available from: www.agassessment.org/docs/SR_Exec_Sum_280508_English.pdf [Accessed 26 March 2009].

Jaffee, D. 2007. *Brewing justice: fair trade coffee, sustainability and survival.* Berkeley, CA: University of California Press.

Kloppenburg, J.R. Jr. 1988. *First the seed. The political economy of plant biotechnology, 1492–2000.* Cambridge: Cambridge University Press.

Kneen, B. 2002. *Invisible giant. Cargill and its transnational strategies.* London: Pluto Press.

Lang, T. and M. Heasman. 2004. *Food wars. The global battle for mouths, minds and markets.* London: Earthscan.

Lappé, F.M. 1971. *Diet for a small planet.* New York, NY: Ballentine.

Lappé, F.M., J. Collins and P. Rosset. 1998. *World hunger: twelve myths.* New York, NY: Grove.

Le Heron, R. 1993. *Globalised agriculture.* Oxford: Pergamon.

Llambi, L. 1994. Comparative advantages and disadvantages in Latin American nontraditional fruit and vegetable exports. *In:* P. McMichael, ed. *The global restructuring of agro-food systems.* Ithaca, NY: Cornell University Press, pp. 190–213.

Madeley, J. 2000. *Hungry for trade.* London: Zed Books.

Martinez-Alier, J. 2002. *The environmentalism of the poor. A study of ecological conflicts and valuation.* Cheltenham: Edward Elgar.

Marx, K. 1967. *Capital*, Volume 1. Moscow: Progress Publishers.

McMichael, A.J., J.W. Powles, C.D. Butler and R. Uauy. 2007. Food, livestock production, energy, climate change, and health. *The Lancet*, 13 September.

McMichael, P. 1984. *Settlers and the agrarian question. Foundations of capitalism in colonial Australia.* Cambridge: Cambridge University Press.

McMichael, P. 1992. Tensions between national and international control of the world food order: contours of a new food regime. *Sociological Perspectives*, 35(2), 343–65.

McMichael, P., ed. 1994. *The global restructuring of agro-food systems.* Ithaca, NY: Cornell University Press.

McMichael, P., ed. 1995. *Food and agrarian orders in the world economy.* Westport, CT: Praeger.

McMichael, P. 1996. *Development and social change. A global perspective.* Thousand Oaks, CA: Pine Forge Press.

McMichael, P. 2000. A global interpretation of the rise of the East Asian food import complex. *World Development*, 28(3), 409–24.

McMichael, P. 2002. La restructuration globale des systems agro-alimentaires. *Mondes en Developpment*, 30(117), 45–54.

McMichael, P. 2005. Global development and the corporate food regime. *In:* F.H. Buttel and P. McMichael, eds. *New directions in the sociology of global development.* Oxford: Elsevier Press.

McMichael, P. 2008a. Peasants make their own history, but not just as they please... *Journal of Agrarian Change*, 8(2–3), 205–28.

McMichael, P. 2008b. Food sovereignty, social reproduction and the agrarian question. *In:* A.H. Akram-Lodhi and C. Kay, eds. *Peasants and globalisation. Political economy, rural transformation and the agrarian question.* London: Routledge, pp. 288–311.

McMichael, P. 2009a. Contemporary contradictions of the global development project: geopolitics, global ecology and the 'development climate'. *Third World Quarterly*, 30(1), 247–62.

McMichael, P. 2009b. Banking on agriculture: a review of the *World development report 2008*. *Journal of Agrarian Change*, 9(2), 205–28.

McMichael, P. Forthcoming. In the short run are we all dead? A perspective on the development climate. *Review.*

McMichael, P. and H. Friedmann. 2007. Situating the 'retailing revolution'. *In:* D. Burch and G. Lawrence, eds. *Supermarkets and agrofood supply chains. Transformations in the production and consumption of foods.* Cheltenham: Edward Elgar.

Mintz, S. 1986. *Sweetness and power. The place of sugar in modern history.* New York, NY: Vintage.

Moore, J.W. 2000. Environmental crises and the metabolic rift in world-historical perspective. *Organisation & Environment*, 13(2), 123–57.

Moran, W., *et al.* 1996. Family farmers, real regulation, and the experience of food regimes. *Journal of Rural Studies*, 12(3), 245–58.

Morgan, D. 1980. *Merchants of grain.* New York, NY: Penguin.

Panitch, L. and S. Gindin. 2004. Global capitalism and American empire. *In:* L. Panitch and C. Leys, eds. *The new imperial challenge: socialist register 2004.* London: Merlin Press.

Patel, R. 2006. International agrarian restructuring and the practical ethics of peasant movement solidarity. *Journal of Asian and African Studies*, 41(1–2), 71–93.

Patel, R. 2007. *Stuffed and starved. Markets, power and the hidden battle over the world's food system*. London: Portobello Books.

Pechlaner, G. and G. Otero. 2008. The third food regime: neoliberal globalism and agricultural biotechnology in North America. *Sociologia Ruralis*, 48(4), 1–21.

Peine, E. 2009. *The private state of agribusiness*. PhD Dissertation, Development Sociology, Cornell University.

Perkins, J.H. 1997. *Geopolitics and the green revolution: wheat, genes and the Cold War*. New York, NY: Cambridge University Press.

Pistorius, R. and J. van Wijk. 1999. *The exploitation of plant genetic information. Political strategies in crop development*. Oxon: CABI Publishing [place in Oxon].

Pretty, J., *et al.* 2006. Resource conserving agriculture increases yields in developing countries. *Environmental Science & Technology*, 40(4), 1114–9.

Pritchard, B. 1996. Shifts in food regimes, regulation and producer cooperatives: insights from the Australian and US dairy industries. *Environment and Planning A*, 28(5), 857–75.

Pritchard, B. 1998. The emergent contours of the third food regime: evidence from Australian dairy and wheat sectors. *Economic Geography*, 74(1), 64–74.

Pritchard, B. 2007. Food regimes. *In:* R. Kitchin and N. Thrift, eds. *The international encyclopedia of human geography*. Amsterdam: Elsevier.

Pritchard, B. Forthcoming. The long hangover from the second food regime: a world-historical interpretation of the collapse of the WTO Doha Round. *Agriculture and Human Values*.

Pritchard, B. and D. Burch. 2003. *Agro-food globalisation in perspective. International restructuring in the processing tomato industry*. Aldershot: Ashgate.

Raynolds, L.T. 1994. The restructuring of Third World agro-exports: changing production relations in the Dominican Republic. *In:* P. McMichael, ed. *The global restructuring of agro-food systems*. Ithaca, NY: Cornell University Press, pp. 214–37.

Raynolds, L.T., *et al.* 1993. The 'new' internationalisation of agriculture: a reformulation. *World Development*, 21(7), 1101–21.

Raynolds, L.T., D. Murray and J. Wilkinson, eds. 2007. *Fair trade: the challenges of transforming globalisation*. Abingdon: Routledge.

Reardon, T., *et al.* 2003. The rise of supermarkets in Africa, Asia and Latin America. *American Journal of Agricultural Economics*, 85(5), 1140–6.

Rifkin, J. 1992. *Beyond beef. The rise and fall of the cattle culture*. New York, NY: Penguin.

Roberts, J. 2008. *The end of food*. New York, NY: Houghton Mifflin.

Rostow, W.W. 1960. *The stages of economic growth*. Cambridge: Cambridge University Press.

Sachs, J. 2008. *Commonwealth*. New York, NY: Penguin.

Thomson, R. 1987. *Green gold: bananas and dependency in the eastern Caribbean*. London: Latin American Bureau.

Tubiana, L. 1989. World trade in agricultural products: from global regulation to market fragmentation. *In:* D. Goodman and M. Redclift, eds. *The international farm crisis*. New York, NY: St. Martin's Press.

Vía Campesina 2001. Our world is not for sale. Priority to people's food sovereignty. Bulletin, 1 November. Available from: http://www.viacampesina.org/welcome_english.php3 [Accessed 15 June 2005].

Vía Campesina 2004. The domination of capital over agriculture. Bulletin 4, 18 June. Available from: http://www.viacampesina.org/welcome_english.php3 [Accessed 15 June 2005].

Walker, R. 2004. *The conquest of bread. A hundred and fifty years of agribusiness in California*. New York, NY: New Press.

Weiss, T. 2007. *The global food economy. The battle for the future of farming*. London: Zed Books.

Winders, B. 2009. *The politics of food supply. US agricultural policy in the world economy*. New Haven, CT: Yale University Press.

World Bank 2008. *World development report 2008: agriculture for development*. Washington, DC: The World Bank.

Livelihoods perspectives and rural development

Ian Scoones

Livelihoods perspectives have been central to rural development thinking and practice in the past decade. But where do such perspectives come from, what are their conceptual roots, and what influences have shaped the way they have emerged? This paper offers an historical review of key moments in debates about rural livelihoods, identifying the tensions, ambiguities and challenges of such approaches. A number of core challenges are identified, centred on the need to inject a more thorough-going political analysis into the centre of livelihoods perspectives. This will enhance the capacity of livelihoods perspectives to address key lacunae in recent discussions, including questions of knowledge, politics, scale and dynamics.

Introduction

Livelihoods perspectives have been central to rural development thinking and practice in the past decade. But where do such perspectives come from, what are their conceptual roots, and what influences have shaped the way they have emerged? This paper responds to these questions with an historical review of key moments in debates about rural livelihoods, identifying the tensions, ambiguities and challenges of such approaches. A complex archaeology of ideas and practices is revealed which demonstrates the hybrid nature of such concepts, bridging perspectives across different fields of rural development scholarship and practice. Yet, in arguing that livelihoods perspectives are important for integrating insights and interventions beyond disciplinary or sectoral boundaries, the paper also touches on some of the limitations, dangers and challenges. In particular, the paper highlights the problems arising from a simplistic application of synthetic frameworks which have come to dominate certain aspects of applied development discussion and practice over the past decade. Looking to the future the paper identifies a number of core challenges, centred on the need to inject a more thorough-going political analysis into the centre of livelihoods perspectives. This, the paper argues, will enhance the capacity of livelihoods perspectives to address key lacunae in recent discussions, including questions of knowledge, politics, scale and dynamics.

I would like to thank Jun Borras, Robert Chambers and Ingrid Nyborg for comments on an earlier draft of this paper. I would also like to thank the participants of the ESRC Seminar on 'A critical re-evaluation of the history of the development and evolution of SLAs', hosted by Livelihoods Connect at IDS, University of Sussex on 13 October 2008, where some of these ideas were shared.

Any basic search of literature or development project material will uncover numerous mentions to livelihoods approaches, perspectives, methods and frameworks. A mobile and flexible term, 'livelihoods' can be attached to all sorts of other words to construct whole fields of development enquiry and practice. These relate to locales (rural or urban livelihoods), occupations (farming, pastoral or fishing livelihoods), social difference (gendered, age-defined livelihoods), directions (livelihood pathways, trajectories), dynamic patterns (sustainable or resilient livelihoods) and many more.

Livelihoods perspectives start with how different people in different places live. A variety of definitions are offered in the literature, including, for example, 'the means of gaining a living' (Chambers 1995, vi) or 'a combination of the resources used and the activities undertaken in order to live'.[1] A descriptive analysis portrays a complex web of activities and interactions that emphasises the diversity of ways people make a living. This may cut across the boundaries of more conventional approaches to looking at rural development which focus on defined activities: agriculture, wage employment, farm labour, small-scale enterprise and so on. But in reality people combine different activities in a complex *bricolage* or portfolio of activities. Outcomes of course vary, and how different strategies affect livelihood pathways or trajectories is an important concern for livelihoods analysis. This dynamic, longitudinal analysis emphasises such terms as coping, adaptation, improvement, diversification and transformation. Analyses at the individual level can in turn aggregate up to complex livelihood strategies and pathways at household, village or even district levels.

Diversity is the watchword, and livelihoods approaches have challenged fundamentally single-sector approaches to solving complex rural development problems. The appeal is simple: look at the real world, and try and understand things from local perspectives. Responses that follow should articulate with such realities and not try and impose artificial categories and divides on complex realities. Belonging to no discipline in particular, livelihoods approaches can allow a bridging of divides, allowing different people to work together – particularly across the natural and social sciences. Being focused on understanding complex, local realities livelihoods approaches are an ideal entry point for participatory approaches to inquiry, with negotiated learning between local people and outsiders.

Following the strong advocacy for sustainable livelihoods approaches in development from the 1990s (Chambers and Conway 1992 and later Scoones 1998, Carney 1998, 2002, Ashley and Carney 1999), many development agencies started to advocate livelihoods approaches as central to their programming, and even organisational structures. Yet the simple, rather obvious, argument for a livelihoods perspective, as discussed further below, is not so easy to translate into practice, with inherited organisational forms, disciplinary biases and funding structures constructed around other assumptions and ways of thinking.

Over the last decade or so 'livelihoods' has thus emerged as a boundary term (Gieryn 1999), something that brings disparate perspectives together, allows conversations over disciplinary and professional divides and provides an institutional bridging function linking people, professions and practices in new ways. But several questions arise. Where did these perspectives come from? What

---

[1] http://www.livelihoods.org/info/dlg/GLOSS/Gloss3.htm#1 (glossary for distance learning guide).

brought people together around such perspectives at a particular historical moment? And what tensions, conflicts and dissonances arise?

A brief archaeology of ideas and approaches

Despite the claims of some genealogies of livelihoods thinking, such perspectives did not suddenly emerge on the scene in 1992 with the influential Chambers and Conway paper. Far from it: there is a rich and important history that goes back another 50 or more years where a cross-disciplinary livelihoods perspective has profoundly influenced rural development thinking and practice. One early example is the work of the Rhodes-Livingstone Institute in what is today Zambia. This involved collaborations of ecologists, anthropologists, agriculturalists and economists looking at changing rural systems and their development challenges (Fardon 1990). While not labelled as such this work was quintessential livelihoods analysis – integrative, locally-embedded, cross-sectoral and informed by a deep field engagement and a commitment to action.

Yet such perspectives did not come to dominate development thinking in the coming decades. As theories of modernisation came to influence development discourse, more mono-disciplinary perspectives ruled the roost. Policy advice was increasingly influenced by professional economists, rather than the rural development generalists and field-based administrators of the past. With the framing in terms of predictive models, of supply and demand, inputs and outputs, both micro and macro economics in different ways, offered a framing which suited the perceived needs of the time. The post-World War II institutions of development – the World Bank, the UN system, the bilateral development agencies, as well as national governments in newly independent countries across the world – reflected the hegemony of this framing of policy, linking economics with specialist technical disciplines from the natural, medical and engineering sciences. This pushed alternative sources of social science expertise, and particularly cross-disciplinary livelihoods perspectives, to the side. While, alternative, radical Marxist perspectives engaged at the macro-level on the political and economic relations of capitalism in post-colonial formations, they rarely delved into the particular, micro-level contextual realities on the ground.

Of course this was not universally true, and there were some important contributions of both economists and Marxist scholars, particularly in the fields of agricultural economics and geography, who offered a more nuanced view. The village studies tradition, dominated by economists, but not exclusively so, was an important, empirically-based alternative to other economic analyses of rural situations (Lipton and Moore 1972). A classic series of studies in India, for example, looked at the diverse impacts of the Green Revolution (Farmer 1977, Walker and Ryan 1990). In many respects these were livelihood studies, although with a focus on the micro-economics of farm production and patterns of household accumulation. In developing the distinctive actor-oriented approach of the Wageningen School, Norman Long was referring to livelihood strategies in his studies in Zambia at this time (Long 1984, see De Haan and Zoomers 2005). In the same period, from a different theoretical tradition, field studies such as the classic examination of rural change in northern Nigeria by Michael Watts (1983), *Silent Violence*, offered important insights into the contested patterns of livelihood change.

These studies provided important inspirations to wider bodies of work that followed. Building on the village studies work, household and farming systems studies of different sorts became an important part of development research in the 1980s (Moock 1986), particularly with a focus on intra-household dynamics (Guyer and Peters 1987). Farming systems research was encouraged in a range of countries, with the aim of getting a more integrated, systems perspective on farm problems. Later, agro-ecosystem analysis (Conway 1985) and rapid and participatory rural appraisal approaches (Chambers 2008) were added to the repertoire, expanding the range of methods and styles of field engagement.

Studies focusing on livelihood and environmental change were also an important strand of work. A concern for dynamic ecologies, history and longitudinal change, gender and social differentiation and cultural contexts meant that geographers, social anthropologists and socio-economists offered a series of influential rich-picture analyses of rural settings in this period.[2] This defined the field of environment and development, as well as wider concerns with livelihoods under stress, with the emphasis on coping strategies and livelihood adaptation.

This line of work overlapped substantially with studies that emerged from Marxist political geography, but had, in some respects, another intellectual trajectory which came to be labelled as political ecology (Blaikie and Brookfield 1987, Robbins 2003, Forsyth 2003). At root, political ecology focuses on the intersections of structural, political forces and ecological dynamics, although there are many different strands and variations. The commitment to local-level fieldwork, with understandings embedded in the complex realities of diverse livelihoods, but linking to more macro-structural issues, are all important characteristics.

The environment and development movement of the 1980s and 1990s threw up in particular concerns about linking a focus on poverty reduction and development with longer-term environmental shocks and stresses. The term 'sustainability' entered the lexicon in a big way following the publication of the Brundtland report in 1987 (WCED 1987) and became a central policy concern with the UN Conference on Environment and Development in Rio in 1992 (Scoones 2007). The sustainable development agenda combined, often in a very uneasy way, concerns with livelihoods and the priorities of local people, the central feature of Agenda 21, and global concerns with environmental issues, enshrined in conventions on climate change, biodiversity and desertification. In cross-disciplinary academic research, these issues have in turn been explored in studies of socio-ecological systems, resilience and sustainability science (Folke *et al.* 2002, Clarke and Dickson 2003).

Thus all these approaches – village studies, household economics and gender analyses, farming systems research, agro-ecosystem analysis, rapid and participatory appraisal, studies of socio-environmental change, political ecology, sustainability science and resilience studies (and many other strands and variants) – have offered diverse insights into the way complex, rural livelihoods intersect with political, economic and environmental processes from a wide range of disciplinary perspectives, drawing from both the natural and social sciences. Each has different emphases and disciplinary foci, and each has engaged in rural development policy and practice in different ways, with more or less influence. Where, then,

---

[2]For example, Richards (1985), Mortimore (1989), Davies (1996), Fairhead and Leach (1996), Scoones (1996), among many others.

do 'livelihood perspectives' – and particularly 'sustainable rural livelihood approaches' – fit into this complex and variegated history?

Sustainable rural livelihoods: a policy story

The connection of the three words 'sustainable', 'rural' and 'livelihoods' as a term denoting a particular approach was possibly first made in 1986 in a hotel in Geneva during the discussion around the Food 2000 report for the Bruntdland Commission.[3] Involving M.S. Swaminathan, Robert Chambers and others, the report laid out a vision for a people-oriented development that had as its starting point the rural realities of poor people (Swaminathan *et al.* 1987). This was a strong theme in Chambers' writing, and especially in his massively influential book, *Rural Development: Putting the Last First* (Chambers 1983). About the same time, through the initiative of Richard Sandbrook, sustainable livelihoods became a focus for a conference organised by the International Institute for Environment and Development in 1987 (Conroy and Litvinoff 1988), and was the subject of Chambers' (1987) overview paper.

But it was not until 1992, when Chambers and Conway produced a working paper for the Institute of Development Studies that a now much used definition of sustainable livelihoods emerged. This stated:

> A livelihood comprises the capabilities, assets (including both material and social resources) and activities for a means of living. A livelihood is sustainable when it can cope with and recover from stresses and shocks, maintain or enhance its capabilities and assets, while not undermining the natural resource base.[4]

This paper is now seen as the starting point of what came to be known later in the 1990s as the 'sustainable livelihoods approach'. At the time its aims were less ambitious, and emerged out of on-going conversations between the two authors who saw important links between their respective concerns with 'putting the last first' in development practice and agro-ecosystem analysis and the wider challenges of sustainable development. The paper was widely read at the time, but it did not go much further, and had little immediate purchase on mainstream development thinking.

Arguments about local knowledge and priorities and systemic concerns with sustainability issues did not have much traction in the hard-nosed debates about economic reform and neo-liberal policy of that period. Despite numerous books and papers, the neo-liberal turn from the 1980s had extinguished effective debate on alternatives. Debates about livelihoods, employment and poverty emerged around the 1995 World Summit for Social Development in Copenhagen,[5] but a livelihoods angle remained at the margins of the mainstream, with debates framed in terms of employment. Of course strands of the participation argument for local involvement and a livelihoods focus were incorporated into the neo-liberal paradigm, along with

---

[3]Robert Chambers (personal communication, October 2008), although, as he points out, there are various other earlier antecedents, including a paper for a 1975 Commonwealth Ministerial Meeting entitled 'Policies for Future Rural Livelihoods'.
[4]As adapted by Scoones (1998), Carney *et al.* (1999) and others.
[5]http://www.un.org/esa/socdev/wssd/.

narratives about the retreat of the state and demand-oriented policy; yet, for some, this became part of a 'new tyranny' (Cooke and Kothari 2001). In the same way, sustainability debates became part-and-parcel of market-oriented solutions and top-down, instrumental global environmental governance (Berkhout *et al.* 2003). The wider concerns about complex livelihoods, environmental dynamics and poverty-focused development, however, remained on the side-lines.

But all this changed in the latter part of the 1990s and into the 2000s, when the formulaic solutions of the Washington Consensus began to be challenged – both on the streets, such as in the 'battle of Seattle' at the World Trade Organisation Ministerial Conference of 1999, in the debates generated by global social movements around the World Social Fora (from 2001 in Porto Alegre), in academic debate, including in economics (from Stiglitz onwards), and in countries whose economies had not rebounded with the magic medicine of neo-liberal reform and whose state capacities had been decimated along the way. More parochially, for those hooked into UK-focused debates about development, a key moment came in 1997 with the arrival of a new Labour government, with a development ministry, the Department for International Development (DfID), a vocal and committed minister, Clare Short, and a White Paper that committed explicitly to a poverty and livelihoods focus (see Solesbury 2003).[6]

In particular, in its opening section, the White Paper mentioned the promotion of 'sustainable rural livelihoods' as a core development priority. Indeed, the UK government had already commissioned work in this area, with several research programmes underway, including one coordinated by the Institute of Development Studies (IDS) at the University of Sussex, with work in Bangladesh, Ethiopia and Mali. This multi-disciplinary research team had been developing an approach which attempted to analyse livelihood change in a comparative way, and had developed a diagrammatic checklist to link elements of the field enquiry (Scoones 1998). In addition to interacting with work being pioneered by the International Institute for Sustainable Development (Rennie and Singh 1996) and the Society for International Development (Almaric 1998), this drew substantially on parallel IDS work on 'environmental entitlements' which, building on the classic work of Sen (1981), emphasised the mediating role of institutions in defining access to resources, rather than simply production and abundance (Leach *et al.* 1997).

Like the IDS sustainable livelihoods work, this was an attempt to draw economist colleagues into a discussion about questions of access and the organisational and institutional dimensions of rural development and environmental change. Drawing on work by North (1990) among others, these approaches used the language of institutional economics, combined with considerations of environmental dynamics (especially from the 'new ecology' perspective) (see Scoones 1999) and social, political and cultural contexts, drawing on social anthropology and political ecology. It chimed very much with the work of Bebbington (1999) who developed a 'capitals and capabilities' framework for looking at rural livelihoods and poverty in the Andes, again drawing on Sen's classic work.

In the notionally trans-disciplinary subject area of development, making sense to economists is a must. With economists only recently having discovered institutions – or at least a particular individualistic, rational-actor version – in the form of new institutional economics and social relations and culture, defined in terms of 'social

---

[6]http://www.dfid.gov.uk/Pubs/files/whitepaper1997.pdf.

capital', following Putnam *et al.* (1993), a moment had opened up to generate some productive conversation, even if largely on disciplinary economics' terms. Thus, both the environmental entitlements approach (Leach *et al.* 1997, 1999) and its more popular cousin, the sustainable livelihoods framework (Scoones 1998, Carney 1998) emphasised the economic attributes of livelihoods as mediated by social-institutional processes. The sustainable livelihoods framework in particular linked inputs (designated with the term 'capitals' or 'assets') and outputs (livelihood strategies), connected in turn to outcomes, which combined familiar territory (of poverty lines and employment levels) with wider framings (of well-being and sustainability) (see Figure 1).

This all echoed discussion around the meanings and definitions of poverty, which was beginning to accommodate broader, more inclusive perspectives on well-being and livelihoods (Baulch 1996). The input-output-outcome elements of the livelihoods framework were of course easily recognised by economists, and were amenable to quantitative analysis and the application of numerous long questionnaires. Some livelihoods analysis has unfortunately never moved much beyond this, missing out on wider social and institutional dimensions.

In particular, the focus on 'capitals' and the 'asset pentagon'[7] kept the discussion firmly in the territory of economic analysis. There was of course important discussion about how assets could be combined, substituted and switched, with different portfolios emerging over time for different people in different places, and linking changes in natural capital ('the environment') with social and economic dimensions was an important step forward. A broader view of assets was also

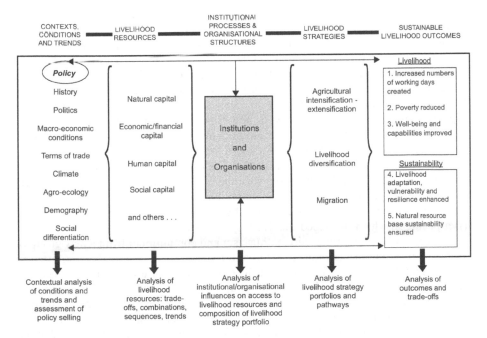

Figure 1. Sustainable livelihoods framework: a checklist (Scoones 1998).

---

[7]A core feature of the DfID version of the framework (see Carney *et al.* 1999).

advocated. Bebbington (1999, 22), for example, saw assets as 'vehicles for instrumental action (making a living), hermeneutic action (making living meaningful) and emancipatory action (challenging the structures under which one makes a living)'. However, perhaps predictably, it was the more instrumental, economic focus that remained at the core of the discussion, and defined much subsequent action on the ground.

In some respects the focus on the 'asset pentagon' and the use of the 'capitals' metaphor was an unfortunate diversion. Other work on sustainable livelihoods had emphasised other features. For example, the IDS studies[8] stressed in particular the idea of institutions and organisations as mediating livelihood strategies and pathways. These were socio-cultural and political processes which explained how and why diverse asset inputs linked to strategies and outcomes. They were subject to power and politics and were where questions of rights, access and governance were centred. Thus a different explanatory angle, with a different disciplinary emphasis, was being offered within the same framework; one that emphasised complex processes requiring in-depth qualitative understandings of power, politics and institutions, and so a very different type of field research.

One explanation for the down-playing of this dimension of sustainable livelihoods analysis over time was the way a framework being used as a checklist for a multi-disciplinary field enquiry in three countries became something much bigger, with many more claims and associations attached to it. The move from diagrammatic checklist to framework – or more precisely the Sustainable Livelihoods Framework, with capital letters, or the Sustainable Livelihoods Approach, with an acronym, SLA, happened in the course of 1998. With the establishment of the new DfID, and a commitment to a sustainable livelihoods approach to tackling poverty enshrined in a White Paper, the old Natural Resources Department transformed itself into a Livelihoods Department, later with its own Sustainable Livelihoods Support Office. An advisory committee was established, led by Diana Carney then of the Overseas Development Institute in London. The committee consisted of DfID staff, from a range of departments, as well as outsiders from the research and NGO community. The committee deliberated on the way forward – how would a 'sustainable livelihoods approach' become operational? And how could a substantial amount of new development funds be channelled to livelihoods-focused poverty reduction? A simple, integrating approach was needed that would tie people into this conversation, and become a way of explaining – and making happen – the idea. At one meeting in London, the IDS checklist diagram was shared, and then transformed by more imaginative people with better skills in computer graphics to what became the DfID framework: essentially the same diagram, but with different nomenclature, and the asset pentagon which described the five 'capital' assets.

This was an exciting time, with enthusiasm and commitment from a new group of people with often a quite radical vision, and a government seemingly committed to doing something about it. This was not the old world of natural resources specialists (archetypically concerned with soils not people) and economists (with their interest in growth and trickle down), but a new, integrated perspective centred on normative, political commitments to banish poverty – and later supported by widespread public campaigns, at least in the UK, from Jubilee 2000 to Make Poverty History.

---

[8]See Carswell *et al.* (1999), Brock and Coulibaly (1999), Shankland (2000), Scoones and Wolmer (2002).

Of course the social development advisors in DfID pointed out (correctly) that they had been advocating livelihoods approaches, sensitive to local needs and cultural contexts forever. Others argued that this was 'just new wine in old bottles' – a reinvention of the failed integrated rural development paradigm of the 1970s. But advocates of a sustainable livelihoods approach argued strongly that this time it was different. The mistakes of old-style, area-based development were not going to be made again, and social and cultural issues would not just enter as part of a *post-hoc* 'told you so' evaluation process, but would be right at the core of the development endeavour.

With money and politics behind an idea – and now an attractive and well-marketed framework, with guidance sheets, an on-line distance learning guide and a growing methods toolbox, shared through the web-based network, Livelihoods Connect[9] – the concept could travel, gaining momentum – and large doses of misapplication and misunderstanding along the way. The first stop on this journey was the DfID Natural Resource Advisors conference of 1998. Framework ideas had already been widely shared, and the concepts and practices were debated intensely with numerous case studies presented (Carney *et al.* 1999). There were of course strong detractors, but many realised the opportunities of opening up debates – as well as the implications for funding flows. The NGO community was important too, bringing fresh ideas and field experiences for elaborating a livelihoods approach from Oxfam, CARE and others. The United Nations Food and Agriculture Programme (FAO) too became interested, as did the United Nations Development Programme (UNDP), creating a diverse array of livelihoods approaches (Carney *et al.* 1999).

In the coming years there was a snowballing of interest, with the flames fanned by effective promotion and communications activities. A whole professional cadre of livelihoods advisors was built up in DfID and other organisations, and soon comparative assessments of different approaches across agencies emerged, high-lighting the differences in interpretation and application of different versions of 'the SL framework' (Hussein 2002). Livelihoods approaches now seemed to be applied to everything: livestock, fisheries, forestry, agriculture, health, urban development and more. A veritable avalanche of papers emerged, all claiming the sustainable livelihoods brand.[10] As the approach became more centrally part of development programming, attempts were made to link it with operational indicators (Hoon *et al.* 1997), monitoring and evaluation (Adato and Meinzen Dick 2002), sector strategies (Gilling *et al.* 2001) and poverty reduction strategy papers (Norton and Foster 2001). But perhaps the more interesting applications were areas where clearly cross-cutting themes could be opened up by a livelihoods perspective. Thus HIV/AIDS discussions were recast from a health to a livelihoods focus (Loevinshon and Gillespie 2003), diversification of livelihoods, migration and non-farm rural income was put at the centre of the rural development agenda (Tacoli 1998, De Haan 1999, Ellis 2000) and complex emergencies, conflict and disaster responses were now seen through a livelihoods lens (Cannon *et al.* 2003, Longley and Maxwell 2003).

---

[9]www.livelihoods.org.
[10]Applications were across sectoral areas – from water (Nicol 2000) to forestry (Warner 2000), natural resource management (Pound 2003), animal genetic resources (Anderson 2003), agriculture (Carswell 1987) to urban development (Farrington *et al.* 2002), river basin management (Cleaver and Franks 2005) and fisheries (Allison and Ellis 2001).

One of the recurrent criticisms of livelihood approaches is that they ignore politics and power. But this is not strictly true. Livelihoods approaches encompass a broad church, and there has been some important work that has elaborated what is meant, in different variants of different frameworks, by 'transforming structures and process', 'policies, institutions and processes', 'mediating institutions and organisations', 'sustainable livelihoods governance' or 'drivers of change' (cf. Davies and Hossain 1987, Hyden 1998, Hobley and Shields 2000, DfID 2004). These reflections have addressed the social and political structures and processes that influence livelihood choices. Power, politics and social difference – and the governance implications of these – have been central to these concerns (Scoones and Wolmer 2003). Unfortunately, though, such debates remained at the margins. While different people made the case for the importance of such political dimensions, dominant concerns were elsewhere – largely focused on a fairly instrumental poverty reduction agenda, framed by economics.

The various frameworks did not help either. Clearly an argument could be made that 'power was everywhere' – from contexts, to constructions and access to capitals, as mediating institutions and social relations, guiding underlying choices of strategies and influencing options and outcomes. Some tried to make politics more explicit, adding 'political capital' to the list of assets, and emphasising that social capital implied attention to power relations. But, as the critiques of a 'capitals' approach – and particularly a focus on social capital – have elaborated, such additions do not really deal with the complex intersections of the structural bases of power – in political interests, competing discourses and embedded practices – diminishing such complexity to a lowest common denominator metric (Harriss 1997). Thus, the regular pleas to pay attention to power and politics often fell on deaf ears, and an instrumental application proceeded as normal, but with a livelihoods label.

The 'community of practice' associated with sustainable livelihoods approaches in this period certainly had a strong normative commitment to poverty reduction and bottom-up, participatory approaches. The branded approaches began to be associated not just with analytical tools (frameworks and checklists), but normative positions. The DfID guidance sheets were quite explicit:

> Firstly, the approach is 'people-centred', in that the making of policy is based on understanding the realities of struggle of poor people themselves, on the principle of their participation in determining priorities for practical intervention, and on their need to influence the institutional structures and processes that govern their lives. Secondly, it is 'holistic' in that it is 'non-sectoral' and it recognises multiple influences, multiple actors, multiple strategies and multiple outcomes. Thirdly, it is 'dynamic' in that it attempts to understand change, complex cause-and-effect relationships and 'iterative chains of events'. Fourthly, it starts with analysis of strengths rather than of needs, and seeks to build on everyone's inherent potential. Fifthly, it attempts to 'bridge the gap' between macro- and micro-levels. Sixthly, it is committed explicitly to several different dimensions of sustainability: environmental, economic, social and institutional.[11]

A coalition of players was built up committed to this style of development. This cut across government, multilateral and NGO players who saw themselves in some way bound together by such a perspective. Wider social movements and local groups, as

---

[11]Quoted at http://www.chronicpoverty.org/toolbox/Livelihoods.php; see DfID guidance sheets at www.livelihoods.org/info/guidance_sheets_pdfs/sect8glo.p.

well as government officials across developing countries, were also active, as this shift in positioning of the aid industry was coincident with their values and politics. Others took a more instrumental stand, as livelihoods thinking became a guarantee of a consultancy or funded aid project, and a proliferation of training courses and advisory services were now being offered from all sorts of sources and of varying quality.

The decline and fall of livelihoods perspectives?

Where have debates about livelihoods and their sustainability ended up in 2009? For some, the destination is a development aid backwater, having lost both the political and financial momentum of being at the centre of influence. One reading of the story is a period of strategic opportunism followed by inevitable disappointment; of dilution and diversion, as ideas become part of the mainstream in large organisations. But there is another, more positive, reading. The rise of livelihood perspectives in rural development thinking and practice from the 1990s did make a difference. Aid money was spent in different ways, new people with different values and skills were hired, and, for once, even if grossly inadequately, local contexts were better understood and poor, marginalised people were involved in plans and decisions (Neely *et al.* 2004). The intersection of academic debate and practical action provided numerous insights and lessons (not all positive by any means) and, in the process, new articulations of livelihoods approaches were elaborated, linking livelihoods to debates on rights, governance and agrarian change, for example.

So why are livelihoods perspectives seemingly not as prominent today compared to a decade ago? Four recurrent failings of livelihoods perspectives can be highlighted. The first relates to the lack of engagement with processes of economic globalisation. To illustrate this, a return to the policy story is required. In the UK context, the 2000 White Paper focused on macro-economic and governance questions, and became known as 'the revenge of the economists'.[12] Despite the accommodation of economic thinking in the sustainable livelihoods framework, it was not enough. Livelihoods approaches were often dismissed as too complex, and so not compatible with real-world challenges and decision-making processes. Idealism, complexity, naïvety, lack of political nous and incompatibility with existing sectorally-based organisations were all accusations made. Other bigger, macro-economic, global-scale questions were, it was argued, more important, and a project-focused, micro-scale approach was not appropriate to the new aid modalities of direct budget support and the Paris agenda (Clarke and Carney 2008).

And such critics had a point. Livelihoods approaches, coming as they did from a complex disciplinary parentage that emphasised the local, have not been very good at dealing with big shifts in the state of global markets and politics. In the frameworks, these were dumped in a box labelled 'contexts'. But what happens when contexts are the most important factor, over-riding the micro-negotiations around access to assets and the finely-tuned strategies of differentiated actors? While the economists in the development agencies were arguing for a growth agenda, based on 'sound macro-economic principles', political economists were also ready to point out

---

[12]http://www.dfid.gov.uk/Pubs/files/whitepaper2000.pdf. By contrast to the 25 mentions of the word livelihood(s) in the 1997 White Paper, just three years on this paper had only three.

the dangers of naïve localism and idealistic liberal analyses that ignore the structural forces of class and capital.

The second failing relates to the lack of attention to power and politics and the failure to link livelihoods and governance debates in development. Of course there were attempts to engage, including work on livelihoods and decentralisation (Manor 2000, SLSA 2003a, Ribot and Larsen 2005), rights-based approaches (Moser and Norton 2001, Conway *et al.* 2002, SLSA 2003b) and linking wider questions of agrarian change (Lahiff 2003). But these efforts failed to have much purchase. In many ways, livelihoods debates had generated their own business, creating livelihoods for consultants, trainers, NGO practitioners and researchers engaged in local-level development. This largely practitioner community often failed to connect with those concerned with state politics, governance regimes and the emergent discussions around agrarian futures among the social movements. It had in many respects got stuck, both intellectually and practically. The weak and sometimes confusing and contradictory theorisation of politics and power, meant that an intellectual articulation with both mainstream political science governance debates and more radical agrarian change discussions was missing.

Another strand of development thinking which really came to the fore in the late 1990s and early 2000s, and was equally focused at the macro, global level, was the need to deal urgently with climate change. Were sustainable livelihood approaches up to this challenge, perhaps *the* big issue of the twenty-first century, one that development could not ignore? Despite the use of the word 'sustainable', the third failing has been the lack of rigorous attempts to deal with long-term secular change in environmental conditions. With more and more data confirming the likely impacts of climate change, particularly in parts of the world where poverty and livelihoods-oriented development has been focused, the danger was evident that livelihoods approaches, as originally conceived, were just ignoring the big picture: fiddling while Rome burned.

In livelihoods discourse 'sustainability' tended to refer to coping with immediate shocks and stresses, where local capacities and knowledge, if effectively supported, might be enough. The iconic cases of mobile pastoralists (Scoones 1995) or adaptive dryland farmers (Mortimore 1989) were well known. But were such local strategies enough? Many thought not, and new climate change adaptation studies emerged which focused on adaptation to long-term change (Adger *et al.* 2003). As discussed below, a central future challenge must be integrating livelihoods thinking and understandings of local contexts and responses with concerns for global environmental change.

Finally, a fourth area that livelihood studies failed to grapple with were debates about long-term shifts in rural economies and wider questions about agrarian change. A rich description of livelihood complexity in the present was one thing, but what were future livelihoods going to look like – in 10, 20 or 50 years? Perhaps local-level adaptation ameliorates poverty at the margins, but does it address more fundamental transformations in livelihood pathways into the future? These issues of course have been raised by many working firmly in the livelihoods tradition, including research on livelihood diversification (Ellis 2000) and 'de-agrarianisation' (Bryceson 1996) in Africa.

These four failures to engage – with processes of economic globalisation, with debates about politics and governance, with the challenges of environmental sustainability and with fundamental transformatory shifts in rural economies – have meant that the research and policy focus has shifted away from the contextual, trans-disciplinary and cross-sectoral insights from livelihood perspectives, often back to a predictable default of macro-economic analyses. One response might be: fair enough, livelihoods perspectives were never meant to do more than this, and different approaches are needed for these new problems. Horses for courses. Another view, however, is that what livelihoods perspectives offer, these other perspectives often miss out on, with potentially damaging consequences. Instead, the argument goes, what is needed is a re-energising of livelihoods perspectives with new foci and priorities to meet these new challenges. This is the theme of the final section of this paper.

Re-energising livelihoods perspectives: new foci, new priorities?

Livelihood perspectives offer, I have argued, a unique starting point for an integrated analysis of complex, highly dynamic rural contexts. Drawing on diverse disciplinary perspectives and cutting across sectoral boundaries, livelihoods perspectives provide an essential counter to the monovalant approaches that have dominated development enquiry and practice. With more complexity, more diversity and more uncertainty about possible rural futures such an embedded approach is, I contend, essential. Yet livelihoods approaches have been accused of being good methods in search of a theory (O'Laughlin 2004). Does a re-energised livelihoods perspective need a new meta-theory to carry it forward? As discussed below, although a more explicit attention to the theorisation of key concepts, with especial attention to the understanding of power and politics is clearly required, a more pluralist, hybrid vision is probably more appropriate if a solid, field-based, grounded empirical stance is to remain at the core. But in order to be responsive to new contexts a number of challenges lie ahead. I identify four: the need to articulate livelihoods perspectives with concerns of knowledge, politics, scale and dynamics. Each offers opportunities for extending, expanding and enriching livelihoods perspectives from a variety of different perspectives.

Knowledge

In the last decade livelihoods debates have emerged in a particular discursive space in the development debate. Providing a 'boundary terminology', they have been able to break down divides, build bridges and transform the focus of debates and implementation practice in some fundamental ways. Livelihoods thinking has often carried with it some explicit normative commitments around a set of widely-shared principles – people matter, contexts are important, a focus on capacities and capabilities, rather than needs, and a normative emphasis on poverty and marginality. Such efforts have constructed new methods, frameworks, institutions and funding streams and, with these, new alliances and networks, or what Hajer (1995) would term a discourse coalition.[13]

Through processes of discursive framing – creating typologies and categories, defining inclusions and exclusions – this has forged a politics of livelihoods

---

[13]The discourse and associated coalition remains however largely Anglophone. Sustainable livelihoods language and concepts have proven very difficult to translate into other languages – and sometimes fit uncomfortably with other culturally-defined intellectual traditions.

knowledge. 'Livelihood' is a seemingly neutral, descriptive word – about making a living – yet livelihood perspectives have been adopted widely, appearing in outputs from the World Bank to the most radical social movement. But what are the power relationships underlying this new discourse, and how do they in turn shape action? The underlying politics of livelihoods knowledge-making has been rarely discussed, and if so only obliquely. But when terms emerge which gain power and influence in constructing and shaping debates, it is worth reflecting on livelihoods perspectives as discourse, as well as methods and analytical tools.

Three dimensions are relevant. First, is the deployment of normative assumptions. Very often in discussion of livelihoods – and particularly sustainable livelihoods – a set of ideas about bottom-up, locally-led, participatory development dovetails with livelihoods analysis. But what is left out by this particular normative framing? For example, rights, justice and struggles for equality are sometimes obscured by more instrumentalist perspectives, coincident with conventional planned development and neo-liberal governance framings. Yet questions of values are central. Arce (2003), for example, offers the case of coca farming in Bolivia, asking whose livelihoods count – and to what and whose ends? Second, the livelihoods literature is replete with classifications and typologies, often contrasting ideal types with alternatives with pejorative ascriptions. But who is to say that, for example, subsistence farmers, poachers, border jumpers or sex workers are pursuing inappropriate livelihoods in need of rescue, discipline or transformation? Third, are questions of directionality and ideas of 'progress' in development. What does the framing of livelihood analysis say about whether things are heading towards positive or negative ends? What is assumed to be a 'good' or a 'bad' livelihood? What needs transformation through the disciplining practices of 'development'? These questions often remain unaddressed or only implicitly treated.

For example, the World Bank's 2008 *World Development Report* on agriculture focused on the importance of livelihoods, characterised by different strategies – based on farming (market-oriented and subsistence), labour, migration and diversification – and three different types of economy: agriculture-based, transforming and urbanised (World Bank 2007, 76). A strong narrative line suggests that progress (development) is about moving through a series of assumed evolutionary stages, with transitions which can be facilitated through a range of interventions in technologies, markets, support institutions and policies, as illustrated by the success stories of Brazil, China, India and Indonesia (cf. World Bank 2007, 5, figure 2). As with other narratives about agricultural change, with an implicit evolutionary argument about progress and modernisation (cf. debates about 'mixed farming', for example, Scoones and Wolmer 2002), the assumption is that the end point, with agriculture as a business, driven by entrepreneurship and vibrant markets, linked to a burgeoning urban economy, is the ideal to strive for.[14] Such framings of course present a normative version of 'good' and 'bad' livelihoods and so 'good' and 'bad' rural futures, defining 'progress' in a particular way. While accepting diverse, complex livelihoods as an empirical reality (certainly an advance from many other

---

[14]By contrast, the International Assessment of Agricultural Knowledge, Science and Technology for Development (IAASTD 2008), presented a very different, and much contested, narrative about progress, and directions for the future. Here, more complex, livelihood concerns were put centre-stage, with principles of equity, access and sustainability guiding the normative framing.

analyses), the assumption is that these are starting points for a future trajectory to something better.

When emanating from influential institutions and cast in a rational-technical framing, as with the World Bank's *World Development Report*, such statements carry with them major consequences. The institutional power behind ideas creates a particular politics of knowledge in the development field, and the role of the World Bank and other donor agencies are key (Broad 2006). Such dominant framings are, in turn, reinforced by educational and training institutions, as scientific knowledge, policy and development practice become co-constructed. Unpacking, questioning, challenging and recasting such perspectives is vital. Livelihoods analysis, by the World Bank or any other actor, is not a neutral exercise; knowledge production is always conditioned by values, politics and institutional histories and commitments (Keeley and Scoones 2003).

Therefore, although livelihoods analysis frameworks and methods definitely offer a way of uncovering complexity and diversity in ways that has often not been revealed before, the important question is: what happens next? Which option is best, and for whom? How do different framings get negotiated? How does knowledge for action get defined? The politics of knowledge and framing often gets kept under wraps. Livelihoods analysis is presented as a rigorous and rational process, yet inevitably it is pursued with many buried assumptions and commitments. While such analysis may be good at opening up inputs to debate, offering descriptive insight into local complexity, it is less good at defining outputs, which often get narrowed down. The problem is that livelihoods analysis can be made to serve multiple purposes and ends. As a malleable concept which opens up such rich diversity in empirical description, it can equally be squashed down into the narrow instrumentalism of log-frames and planning formats, or get deployed by particular political commitments, dominated in recent years by neo-liberal reform.

In order to avoid such closing down, and maintain a process of appraisal, assessment and intervention which remains open, attention to the processes through which livelihoods knowledge is negotiated and used is required (cf. Stirling 2008). With knowledge politics around framings and normative commitments more explicit, opportunities to deliberate upon the political choices inherent in livelihoods analyses potentially emerge. Rather than relying on a bland listing of principles or, worse, keeping such questions of values and politics off the agenda with a naïve plea to rationality, a focus on inclusive deliberation around livelihood framings and directions of change can come to the fore.

Politics

Politics and power thus must be central to livelihood perspectives for rural development. Politics is not just 'context', but a focus for analysis in and of itself. It is not just a matter of adding another 'capital' to the assets pentagon (Baumann 2000), with all the flawed assumptions of equivalence and substitutability inherent. While, as discussed earlier, some excellent work has been carried out on local-level power dynamics and institutional and organisational politics, the attention to power and politics must, of course, move beyond the local level to examine wider structures of inequality. Basic questions of political economy and history matter: the nature of the state, the influence of private capital and terms of trade, alongside other wider structural forces, influence livelihoods in particular places. This is conditioned by histories of places and peoples, and their wider interactions with colonialism, state-making and globalisation.

All this is, in many senses, blindingly obvious. But an unhelpful divide often persists in livelihoods analyses between micro-level, locale-specific perspectives, emphasising agency and action, and broader, macro-level structural analysis. Both speak of politics and power, but in very different ways. This is down in large part to disciplinary proclivities, separated out along the classic structure-agency axis of the social sciences. Yet, livelihood perspectives must look simultaneously at both structure and agency and the diverse micro- and macro-political processes that define opportunities and constraints. While Giddens' concept of 'structuration' (1984) is rather cumbersome, the basic argument for recursive links across scales and between structural conditions and human action is essential. Although developed to some degree in some earlier precursors of the livelihoods frameworks (cf. Bebbington 1999, Leach *et al.* 1999), such basic analytical moves have not been central to livelihoods analysis, with a preference often towards locality and agency, black-boxing wider structural features.

This is a problem which needs to be addressed. It is one of the reasons that, in some respects, livelihoods perspectives have been side-lined in debates about governance and the politics of globalisation. The 1992 book, *Rural Livelihoods: Crises and Responses* (Bernstein *et al.* 1992) is probably the most comprehensive attempt to integrate livelihoods perspectives with these more structural political economy concerns. There are also other rich strands of scholarship to draw on, which would allow livelihoods analysis to put politics centre stage. However, these have sometimes got lost in the micro-economic reformulations of livelihoods analysis. Thus, the long-standing work on agro-food systems (Goodman and Watts 1997, McMichael 1994) and agrarian change (Bernstein and Byers 2001), for example, provide important insights, while political ecology explicitly explores links between the local level and broader political-economic structures (Peet and Watts 1996). In the same way feminist scholarship is keenly aware of links between personal and bodily questions and broader structural forces defining power relations in diverse livelihood settings (Kabeer 1994).

Attention to how livelihoods are structured by relations of class, caste, gender, ethnicity, religion and cultural identity are central. Understanding of agrarian structures requires, as Bernstein *et al.* (1992, 24) point out, asking the basic questions: who owns what, who does what, who gets what and what do they do with it? Social relations inevitably govern the distribution of property (including land), patterns of work and divisions of labour, the distribution of income and the dynamics of consumption and accumulation. As with gender and other dimensions of social difference, questions of class must be central to any livelihoods analysis. But, as O'Laughlin (2004, 387) argues:

> Class, not as an institutional context variable, but as a relational concept, is absent from the discourse of livelihoods. Accordingly, political space is very limited – focusing mainly on 'empowering' the poor, without being clear about how this process takes place or who might be 'disempowered' for it to occur.

A more explicit theorisation of politics, power and social difference is thus required. Livelihoods analysis is still required to unpick the complex threads and contextual specificity, but it must be located, as O'Laughlin argues, in a relational understanding of power and politics which identifies how political spaces are opened up and closed down.

So, how can an attention to politics and power be put at the heart of livelihoods perspectives? Some would say it already is. Much livelihoods analysis centres on the basic question of how different people gain access to assets for the pursuit of livelihoods. This must necessarily encompass questions of power and politics. Institutions – the rules of game governing access – are of course mediated by power relations. And struggles over access involve both individual efforts and collective action through organised politics, involving alliances, movements or party politics. The livelihoods 'tool box' is not short of methods for looking at this type of political process operating across scales.[15] But, as discussed, in the overly instrumental work driven by development imperatives these are often not used – or only in a light, descriptive way.[16] In sum, there is an urgent need to bring politics back in to livelihoods perspectives. As Sue Unsworth (2001, 7) argues:

> Poverty reduction requires a longer term, more strategic understanding of the social and political realities of power, and confronts us with ethical choices and trade-offs which are much more complex ... A more historical, less technical way of looking at things can provide a sense of perspective.

Thus to enrich livelihood perspectives further, there is a need to be more informed by an explicit theoretical concern with the way class, gender and capitalist relations operate (O'Laughlin 2004), asking up-front who gains and loses and why, embedded in an analysis informed by theories of power and political economy and so an understanding of processes of marginalisation, dispossession, accumulation and differentiation.

Scale

One of the claims of livelihoods perspectives is that they link the micro with the macro. As already discussed, this is often more of an ambition than a reality. One of the persistent failings of livelihoods approaches has been the failure to address wider, global processes and their impingement on livelihood concerns at the local level. Livelihoods perspectives have thus often failed to engage with debates about globalisation, for example, ceding the terrain to macro-economics, notoriously under-informed about local-level complexities.

As global transformations continue apace, attention to scale issues must be central to the reinvigoration of livelihoods perspectives. Again, while there have been failings and absences, there have been some important contributions which can be drawn upon and made more central to livelihoods approaches for the future. An important collection of papers edited by Tony Bebbington and Simon Batterbury (2001, 370) emphasised the significance of what they termed transnational livelihoods and the 'analytical value of grounding political ecologies

---

[15]See for example, Murray (2001, 2002); www.livelihoods.org; www.policy-powertools.org/; www.chronicpoverty.org/toolbox/Livelihoods.php.
[16]There is a good argument for 'optimal diplomatic omission' in order to gain access to formal agendas and open up policy spaces – and livelihoods perspectives, with their all-embracing coverage and trans-disciplinary approach, are a good route to this – but this is no excuse for a lack of underlying political analysis to inform such engagements.

of globalisation in notions of livelihood, scale, place and network'. With cases examining migration, remittance flows and rural social movements, the importance of looking at linking solid, place-based analysis with broader scales, including trans-national connections, is emphasised. Looking beyond the local to wider landscapes is of course central to geographical analysis, and the notion of 'scape' has been extended to look at the patterns of practices of globalisation (Appadurai 1996). To meet these challenges, Bebbington and Batterbury (2001, 377) argue for:

> A broader enterprise in which political ecology, cultural geography, development studies and environmental politics are all involved, even if they have differing entry points. This broader enterprise is one that struggles to understand the ways in which peoples, places and environments are related and mutually constituted, and the ways in which these constitutions are affected by processes of globalisation.

A variety of approaches lend themselves to this sort of analysis. Network approaches (Castells 1996), flow analysis (Spaargaren *et al.* 2006) and value chain, commodity system or filiere approaches (Kaplinsky and Morris 2001) have become important lenses in different areas for looking at processes of change across scales. Yet there has been poor articulation with livelihoods approaches. Some initiatives stand out, however. For example, there have been some excellent multi-sited, comparative, scaled studies linking local-level analysis to broader processes of change (e.g. Warren *et al.* 2001). There have also been attempts to link approach to, for example, understanding trade regimes and livelihoods (Stevens *et al.* 2003) or combining value chain and livelihoods assessments (Kanji *et al.* 2005). These are all efforts to build on if scale questions – linking the micro to the macro and *vice versa* – are to be addressed.

The challenge for the future is to develop livelihoods analyses which examine networks, linkages, connections, flows and chains across scales, but remain firmly rooted in place and context. But this must go beyond a mechanistic description of links and connections. Such approaches must also illuminate the social and political processes of exchange, extraction, exploitation and empowerment, and so explore the multiple contingent consequences of globalisation on rural livelihoods. They must ask how particular forms of globalisation and associated processes of production and exchange – historically from colonialism to contemporary neo-liberal economics – create both processes of marginalisation and opportunity. In such a view 'the global' and 'the local' are not separated – either physically or analytically – but intimately intertwined through connections, linkages, relations and dynamics between diverse locales. Livelihoods analysis must thus expose the inevitably highly variegated experiences of globalisation, and so the implications of multiple transformations and diverse livelihood pathways.

Dynamics

Another challenge for livelihoods perspectives is to deal with long-term change. The term *sustainable* livelihoods implies that livelihoods are stable, durable, resilient and robust in the face of both external shocks and internal stresses. But what stresses and what shocks are important? How is sustainability assessed? And how are future generations' livelihoods made part of the equation? This has been a weak element in much livelihoods analysis, despite earlier pleas.[17] The focus instead has often been on coping and short-term adaptation, drawing on a rich heritage of vulnerability analysis (cf. Swift 1989), rather than attention to systemic transformation due to long-run secular changes.

For example, in a study from rural Zimbabwe, Frost *et al.* (2007) present a highly pessimistic vision of livelihood sustainability. They argue forcefully that livelihoods interventions in the study area have made no difference, and that people are stuck in a more fundamental trap which palliative, and very expensive, measures are not geared up to deal with. But such single time-frame analyses may miss out on longer-term dynamics and the potentials for more radical transformations. Historical analyses of livelihood change highlight how long-term shifts in livelihood strategies emerge (Mortimore 2003, Wiggins 2000). People's initiative and local knowledge enhances resilience to shocks and stresses. In long-run livelihood change, specific dynamic drivers, operating over decades, are highlighted as important. These include demography (Tiffen *et al.* 1994), regional economic shifts and urbanisation (Tiffen 2003), migration (Batterbury 2001), land-use (Fairhead and Leach 1996) and climate (Adger *et al.* 2003).

Without attention to these long-run, slow variables in dynamic change, a snap-shot view describing desperate coping may miss slow transformations for the better – as people intensify production, improve environmental conditions, invest or migrate out. But, in the same way, a rosy picture of local, adaptive coping to immediate pressures, based on local capacities and knowledge, may miss out on long-term shifts which will, in time, undermine livelihoods in more fundamental ways. Long-term temperature rises may make agriculture impossible, shifts in terms of trade may undermine the competitiveness of local production or migration of labour to urban areas may eliminate certain livelihood options in the long-term.

Sustainability and resilience thus cannot always emerge through local adaptation in conditions of extreme vulnerability. Instead, more dramatic reconfigurations of livelihoods may have to occur in response to long-run change. This is highlighted in particular by the challenge of climate change. Livelihoods language has certainly been incorporated into thinking about climate adaptation, linking climate change to development objectives (Lemos *et al.* 2007, Boyd *et al.* 2008). But much of this has been rather instrumental, merely dressing up standard rural development interventions in climate adaptation clothing. Bringing perspectives on livelihoods into climate change responses requires more than this, with a more careful unpacking of the inter-relationships between vulnerability and resilience perspectives (Nelson *et al.* 2007).

Livelihoods analysis that identifies different future strategies or pathways provides one way of thinking about longer-term change. Dorward *et al.* (2005), for example, distinguish between 'hanging in', 'stepping up' and 'stepping out'. Different people, because of their current asset base and livelihood options, are likely, given future trends, to end up just coping, moving to new livelihood options

---

[17]Chambers and Conway (1992, 26), for example, urged for a consideration of 'net sustainable livelihoods' defined as 'the number of environmentally and socially sustainable livelihoods that provide a living in a context less their negative effects on the benefits and sustainability of the totality of other livelihoods everywhere'. They also explicitly argued that the interests of unborn generations be included in discussions about contemporary development.

or getting out completely. In the same way Pender (2004) identifies future livelihood pathways for the highlands of Central America and East Africa based on comparative advantages – in agricultural potential, market access, infrastructure provision and population densities, among other variables. Thus for different sites, future pathways are envisaged – and so different types of intervention are required – if livelihood options are to be enhanced. On the basis of detailed livelihood analyses of mixed crop-livestock systems in Ethiopia, Mali and Zimbabwe, Scoones and Wolmer (2002) identify eight different livelihood pathways, conditioned by patterns of social difference and institutional processes, with different people's options channelled down particular pathways, reinforced by policy processes, institutional pressures and external support.

These examples thus identify multiple future options – or pathways – some positive, some negative; some supported by external intervention and policy, some not. But how sustainable are such pathways, given the possible, but always uncertain, future shocks and stresses, and long-term drivers of change? Here other literatures may help enhance livelihoods thinking, and bring debates about sustainability more firmly back into discussions. First, are approaches focused on the analysis of the resilience of socio-ecological systems (Folke *et al.* 2002, Gunderson and Holling 2002, Walker and Salt 2006). These identify the importance of looking at interactions between slow and fast variables and cross-scale interactions between them, and the interactions these have on resilience – defined as the amount of change a system can undergo while maintaining its core properties. While emerging from ecology and a concern for complex, non-linear dynamics of ecosystems, resilience thinking has increasingly been applied to interactions between ecological and social systems across scales (Berkes *et al.* 1998). As with 'sustainability science' (Clarke and Dickson 2003), the central concern is with sustaining 'life support systems', and the capacity of natural systems to provide for livelihoods into the future, given likely stresses and shocks. While well developed for ecological and engineering systems, the extension of resilience concepts to social-economic-cultural-political systems is definitely 'work in progress', but an area with increasing attention and innovation.[18]

A second area, with similar concerns but with different origins, is work on transitions in socio-technical systems (Geels and Schot 2007, Smith and Stirling 2008). Emerging from science and technology studies, such approaches explore how interacting social and technical systems move towards more sustainable configurations. This may not be through gradual, incremental shifts, but through more radical transitions, where new social, economic and technological systems unfold. This may, in particular, emerge from 'niches', where experiments with alternatives occur at a small scale at the margins, only to become mainstream at a later date when conditions change and opportunities arise (Smith 2006).

Livelihoods perspectives could be significantly enhanced by some interaction with these literatures, converging as they do on key concerns for rural livelihoods – including adaptive capacity/capability, institutional flexibility and diversity of responses, as key ingredients of sustainability.

---

[18]See in particular the work of the Resilience Alliance (www.resalliance.org).

Conclusion

Livelihoods perspectives offer an important lens for looking at complex rural development questions. As argued by Scoones and Wolmer (2003, 5):

> A sustainable livelihoods approach has encouraged ... a deeper and critical reflection. This arises in particular from looking at the consequence of development efforts from a local-level perspective, making the links from the micro-level, situated particularities of poor people's livelihoods to wider-level institutional and policy framings at district, provincial, national and even international levels. Such reflections therefore put into sharp relief the importance of complex institutional and governance arrangements, and the key relationships between livelihoods, power and politics.

But in order to have continued relevance and application, livelihoods perspectives must address more searchingly and concretely questions across the four themes highlighted above: knowledge, politics, scale and dynamics. These are challenging agendas, both intellectually and practically. For those convinced that livelihoods perspectives must remain central to development, this is a wake-up call. The vibrant and energetic 'community of practice' of the late 1990s has taken its eye off the ball. A certain complacency, fuelled by generous funding flows, a comfortable localism and organisational inertia has meant that some of the big, emerging issues of rapid globalisation, disruptive environmental change and fundamental shifts in rural economies have not been addressed. Innovative thinking and practical experimentation has not yet reshaped livelihood perspectives to meet these challenges in radically new ways.

But, more positively, around the four themes outlined above a new livelihoods agenda opens up. This does not mean abandoning a basic commitment to locally-embedded contexts, place-based analysis and poor people's perspectives; nor does it mean slavishly responding to the framings provided by dominant disciplines such as economics. But there is an urgent need to rethink, retool and reengage, and draw productively from other areas of enquiry and experience to enrich and reinvigorate livelihoods perspectives for new contemporary challenges. A re-energised livelihoods perspective thus requires, first, a basic recognition of cross-scale dynamic change and, second, a more central place for considerations of knowledge, power, values and political change. The themes of knowledge, scale, politics and dynamics, I argue, offer an exciting and challenging agenda of research and practice to enrich livelihood perspectives for rural development into the future.

References

Adato, M. and R. Meinzen-Dick. 2002. Assessing the impact of agricultural research on poverty using the sustainable livelihoods framework. *Environment and production technology division discussion paper no. 89*. Washington, DC: International Food Policy Research Institute.

Adger, W.N., *et al*. 2003. Adaptation to climate change in developing countries. *Progress in Development Studies*, 3(3), 179–95.

Allison, E. and F. Ellis. 2001. The livelihoods approach and management of small-scale fisheries. *Marine Policy*, 25(2), 377–88.

Almaric, F. 1998. *The sustainable livelihoods approach. General report of the sustainable livelihoods project, 1995–1997*. Rome: SID.

Anderson, S. 2003. Animal genetic resources and sustainable livelihoods. *Ecological Economics*, 45(3), 331–9.

180   *Ian Scoones*

Appadurai, A. 1996. *Modernity at large: cultural dimensions of globalisation*. Minneapolis, MN: University of Minnesota Press.
Arce, A. 2003. Value contestations in development interventions: community development and sustainable livelihoods approaches. *Community Development Journal*, 38(3), 199–212.
Ashley, C. and D. Carney. 1999. *Sustainable livelihoods: lessons from early experience*. London: DFID.
Batterbury, S. 2001. Landscapes of diversity: a local political ecology of livelihood diversification in South-Western Niger. *Ecumene*, 8(4), 437–64.
Baulch, R. 1996. Neglected trade-offs in poverty measurement. *IDS Bulletin*, 27(1), 36–43.
Baumann, P. 2000. Sustainable livelihoods and political capital: arguments and evidence from decentralisation and natural resource management in India. *ODI working paper*, 136. London: ODI.
Bebbington, A. 1999. Capitals and capabilities: a framework for analysing peasant viability, rural livelihoods and poverty. *World Development*, 27(12), 2012–44.
Bebbington, A. and S. Batterbury, eds. 2001. Transnational livelihoods and landscapes: political ecologies of globalisation. *Ecumene*, 8(4), 369–464.
Berkes, F., C. Folke and J. Colding. 1998. *Social and ecological systems: management practices and social mechanisms for building resilience*. Cambridge: Cambridge University Press.
Berkhout, F., M. Leach and I. Scoones, eds. 2003. *Negotiating environmental change: new perspectives from social science*. Cheltenham: Edward Elgar.
Bernstein, H. and T. Byres. 2001. From peasant studies to agrarian change. *Journal of Agrarian Change*, 1(1), 11–56.
Bernstein, H., B. Crow and H. Johnson, eds. 1992. *Rural livelihoods: crises and responses*. Oxford: Oxford University Press.
Blaikie, P. and H. Brookfield. 1987. *Land degradation and society*. London: Methuen and Company.
Boyd, E., *et al.* 2008. Resilience and 'climatising' development: examples and policy implications. *Development*, 51(3), 390–6.
Broad, R. 2006. Research, knowledge, and the art of 'paradigm maintenance': the World Bank's development economics vice-presidency (DEC). *Review of International Political Economy*, 13(3), 387–419.
Brock, K. and N. Coulibaly. 1999. Sustainable rural livelihoods in Mali. *IDS research report*, 35. Brighton: IDS.
Bryceson, D. 1996. Deagrarianisation and rural employment in Sub-Saharan Africa: a sectoral perspective. *World Development*, 24(1), 97–111.
Cannon, T., J. Twigg and J. Rowell. 2003. *Social vulnerability, sustainable livelihoods and disasters*. London: DFID.
Carney, D., ed. 1998. *Sustainable rural livelihoods: what contribution can we make?*. London: DFID.
Carney, D. 2002. *Sustainable livelihoods approaches: progress and possibilities for change*. London: DFID.
Carney, D., *et al.* 1999. Livelihood approaches compared: a brief comparison of the livelihoods approaches of the UK Department for International Development (DFID), CARE, Oxfam and the UNDP. A brief review of the fundamental principles behind the sustainable livelihood approach of donor agencies. *Livelihoods connect*. London: DFID.
Carswell, G. 1997. Agricultural intensification and rural sustainable livelihoods. *IDS working paper*, 64. Brighton: IDS.
Carswell, G., *et al.* 1999. Sustainable livelihoods in Southern Ethiopia. *IDS research report*, 44. Brighton: IDS.
Castells, M. 1996. *The rise of the network society, vol. 1 of the information age: economy, society and culture*. Oxford: Blackwell Publishers.
Chambers, R. 1983. *Rural development: putting the last first*. London: Longman.
Chambers, R. 1987. Sustainable rural livelihoods. A strategy for people, environment and development. An overview paper for Only One Earth: Conference on Sustainable Development, 28–30 April 1987. London: IIED.
Chambers, R. 1995. Poverty and livelihoods: whose reality counts?. *ID discussion paper*, 347. Brighton: IDS.
Chambers, R. 2008. *Revolutions in development inquiry*. London: Earthscan.

Chambers, R. and G. Conway. 1992. Sustainable rural livelihoods: practical concepts for the 21st century. *IDS discussion paper*, 296. Brighton: IDS.

Clarke, J. and D. Carney. 2008. Sustainable livelihoods approaches – what have we learned?. Background paper, ESRC Livelihoods Seminar, 13 October. *Livelihoods Connect.* Brighton: IDS.

Clarke, W. and N. Dickson. 2003. Sustainability science: the emerging research program. *Proceedings of the National Academy of Science*, 100(14), 8059–61.

Cleaver, F. and T. Franks. 2005. *Institutions elude design: river basin management and sustainable livelihoods.* Alternative Water Forum. Bradford Centre for International Development, Bradford.

Conroy, C. and M. Litvinoff, eds. 1988. *The greening of aid. Sustainable livelihoods in practice.* London: Earthscan Publications.

Conway, G. 1985. Agroecosystems analysis. *Agricultural Administration*, 20(1), 31–55.

Conway, T., *et al.* 2002. Rights and livelihoods approaches: exploring policy dimensions. *Natural Resource Perspectives*, 78. London: ODI.

Cooke, B. and U. Kothari, eds. 2001. *Participation: the new tyranny?.* London: Zed Books.

Davies, S. 1996. *Adaptable livelihoods. Coping with food insecurity in the Malian Sahel.* London: MacMillan.

Davies, S. and N. Hossain. 1987. Livelihood adaptation, public action and civil society: a review of the literature. *IDS working paper*, 57. Brighton: IDS.

De Haan, A. 1999. Livelihoods and poverty: the role of migration – a critical review of the migration literature. *Journal of Development Studies*, 36(2), 1–47.

De Haan, L. and A. Zoomers. 2005. Exploring the frontier of livelihoods research. *Development and Change*, 36(1), 27–47.

DfID 2004. Drivers of change. Public Information Note, September 2004. London: DfID. Available from: http://www.gsdrc.org/docs/open/DOC59.pdf [Accessed 17 March 2009].

Dorward, A., *et al.* 2005. *A guide to indicators and methods for assessing the contribution of livestock keeping to livelihoods of the poor.* London: Department of Agricultural Sciences, Imperial College.

Ellis, F. 2000. *Rural livelihoods and diversity in developing countries.* Oxford: Oxford University Press.

Fairhead, J. and M. Leach. 1996. *Misreading the African landscape: society and ecology in a forest-savanna mosaic.* Cambridge: Cambridge University Press.

Fardon, R., ed. 1990. *Localising strategies: regional traditions of ethnographic writing.* Edinburgh: Scottish Academic Press.

Farmer, B. 1977. *Green revolution.* London: MacMillan.

Farrington, J., T. Ramasut and J. Walker. 2002. Sustainable livelihoods approaches in urban areas: general lessons, with illustrations from Indian examples. *ODI working paper*, 162. London: ODI.

Folke, C., *et al.* 2002. Resilience and sustainable development: building adaptive capacity in a world of transformations. *AMBIO: A Journal of the Human Environment*, 31(5), 437–40.

Forsyth, T. 2003. *Critical political ecology: the politics of environmental science.* London: Routledge.

Frost, P., *et al.* 2007. In search of improved rural livelihoods in semi-arid regions through local management of natural resources: lessons from case studies in Zimbabwe. *World Development*, 35(11), 1961–74.

Geels, F. and J. Schot. 2007. Typology of sociotechnical transition pathways. *Research Policy*, 36(3), 399–417.

Giddens, A. 1984. *The constitution of society.* Cambridge: Polity Press.

Gieryn, T. 1999. *Cultural boundaries of science: credibility on the line.* Chicago, IL: University of Chicago Press.

Gilling, J., S. Jones and A. Duncan. 2001. Sector approaches, sustainable livelihoods and rural poverty reduction. *Development Policy Review*, 19(3), 303–19.

Goodman, D. and M. Watts. 1997. *Globalising food: agrarian questions and global restructuring.* London: Routledge.

Gunderson, L. and C. Holling, eds. 2002. *Panarchy: understanding transformations in human and natural systems.* Washington, DC: Island Press.

Guyer, J. and P. Peters. 1987. Conceptualising the household: issues of theory and policy in Africa. *Development and Change*, 18(2), 197–214.

Hajer, M. 1995. *The politics of environmental discourse: ecological modernisation and the policy process*. Oxford: Oxford University Press.

Harriss, J., ed. 1997. Policy arena: 'missing link' or analytically missing? The concept of social capital. *Journal of International Development*, 9(7), 919–71.

Hobley, M. and D. Shields. 2000. *The reality of trying to transform structures and processes: forestry in rural livelihoods*. London: ODI.

Hoon, P., N. Singh and S. Wanmali. 1997. Sustainable livelihoods: concepts, principles and approaches to indicator development. Available from: http://www.sustainable-livelihoods.com/pdf/sustainablelivelihoodsc-1.pdf [Accessed 17 March 2009].

Hussein, K. 2002. *Livelihoods approaches compared: a multi-agency review of current practice*. London: ODI.

Hyden, G. 1998. Governance and sustainable livelihoods. Paper for the Workshop on Sustainable Livelihoods and Sustainable Development, 1–3 October 1998, jointly organised by UNDP and the Center for African Studies, University of Florida, Gainesville.

IAASTD 2008. *International assessment of agricultural knowledge, science and technology for development*. Washington, DC: World Bank.

Kabeer, N. 1994. *Reversed realities: gender hierarchies in development thought*. London: Verso Press.

Kanji, N., J. MacGregor and C. Tacoli. 2005. *Understanding market-based livelihoods in a globalising world: combining approaches and methods*. London: IIED.

Kaplinksy, R. and M. Morris. 2001. Handbook for value chain research. Report prepared for IDRC, Canada.

Keeley, J. and I. Scoones. 2003. *Understanding environmental policy processes. Cases from Africa*. London: Earthscan Publications.

Lahiff, E. 2003. Land and livelihoods: the politics of land reform in Southern Africa. *IDS Bulletin*, 34(3), 54–63.

Leach, M., R. Mearns and I. Scoones. 1997. Environmental entitlements: a framework for understanding the institutional dynamics of environmental change. *IDS Discussion Paper*, 359. Brighton: IDS.

Leach, M., R. Mearns and I. Scoones. 1999. Environmental entitlements: dynamics and institutions in community-based natural resource management. *World Development*, 27(2), 225–47.

Lemos, M., *et al.* 2007. Adapting development and developing adaptation. *Ecology and Society*, 12(2), 26.

Lipton, M. and M. Moore. 1972. *The methodology of village studies in less developed countries*. Brighton: IDS, University of Sussex.

Loevinsohn, M. and S. Gillespie. 2003. HIV/AIDS, food security and rural livelihoods: understanding and responding. Discussion paper – food consumption and nutrition division. Washington, DC: IFPRI.

Long, N. 1984. *Family and work in rural societies. Perspectives on non-wage labour*. London: Tavistock.

Longley, C. and D. Maxwell. 2003. *Livelihoods, chronic conflict and humanitarian response: a synthesis of current practice*. London: ODI.

Manor, J. 2000. *Decentralisation and sustainable livelihoods*. Brighton: IDS.

McMichael, P. 1994. *The global restructuring of agro-food systems*. Ithaca, NY: Cornell University Press.

Moock, J., ed. 1986. *Understanding Africa's rural household and farming systems*. Boulder, CO: Westview Press.

Mortimore, M. 1989. *Adapting to drought, farmers, famines and desertification in West Africa*. Cambridge: Cambridge University Press.

Mortimore, M. 2003. Long-term change in African drylands: can recent history point towards development pathways?. *Oxford Development Studies*, 31(4), 503–18.

Moser, C. 1998. The asset vulnerability framework: reassessing urban poverty reduction strategies. *World Development*, 26(1), 1–19.

Moser, C. and A. Norton. 2001. *To claim our rights. Livelihood security, human rights and sustainable development*. London: ODI.

Murray, C. 2001. *Livelihood research: some conceptual and methodological issues.* Manchester: Chronic Poverty Research Centre.

Murray, C. 2002. Livelihoods research: transcending boundaries of time and space. *Journal of Southern African Studies*, Special issue: Changing livelihoods, 28(3), 489–509.

Neely, C., K. Sutherland and J. Johnson. 2004. Do sustainable livelihoods approaches have a positive impact on the rural poor? A look at twelve case studies. *Livelihoods Support Programme Paper*, 16. Rome: FAO.

Nelson, D., W.N. Adger and K. Brown. 2007. Adaptation to environmental change: contributions of a resilience framework. *Annual Review of Environment and Resources*, 32, 345–73.

Nicol, A. 2000. Adopting a sustainable livelihoods approach to water projects: implications for policy and practice. *ODI working paper*, 133. London: ODI.

North, D. 1990. *Institutions, institutional change and economic performance.* Cambridge: Cambridge University Press.

Norton, A. and M. Foster. 2001. *The potential of using sustainable livelihoods approaches in poverty reduction strategy papers.* London: ODI.

O'Laughlin, B. 2004. Book reviews. *Development and Change*, 35(2), 385–403.

Peet, R. and M. Watts. 1996. *Liberation ecologies: environment, development, social movements.* London: Routledge.

Pender, J. 2004. Development pathways for hillsides and highlands: some lessons from Central America and East Africa. *Food Policy*, 29(4), 339–67.

Pound, B., *et al.* 2003. *Managing natural resources for sustainable livelihoods: uniting science and participation.* Ottawa: IDRC.

Putnam, R., R. Leornardi and R. Nanetti. 1993. *Making democracy work: civic traditions in modern Italy.* Princeton, NJ: Princeton University Press.

Rennie, J. and N. Singh. 1996. *Participatory research for sustainable livelihoods: a guidebook for field projects.* Ottawa: IISD.

Ribot, J. and A. Larson, eds. 2005. *Democratic decentralisation through a natural resource lens.* London: Routledge.

Richards, P. 1985. *Indigenous agricultural revolution: ecology and food crops in West Africa.* London: Hutchinson.

Robbins, P. 2003. *Political ecology: a critical introduction.* Oxford: Blackwell.

Scoones, I., ed. 1995. *Living with uncertainty. New directions in pastoral development in Africa.* London: IT Publications.

Scoones, I. 1996. *Hazards and opportunities farming livelihoods in dryland Africa: lessons from Zimbabwe.* London: Zed Press.

Scoones, I. 1998. Sustainable rural livelihoods: a framework for analysis. *IDS working paper*, 72. Brighton: IDS.

Scoones, I. 1999. New ecology and the social sciences: what prospects for a fruitful engagement?. *Annual Review of Anthropology*, 28, 479–507.

Scoones, I. 2007. Sustainability. *Development in Practice*, 17(4), 589–96.

Scoones, I. and W. Wolmer, eds. 2002. *Pathways of change in Africa: crops, livestock and livelihoods in Mali, Ethiopia and Zimbabwe.* Oxford: James Currey.

Scoones, I. and W. Wolmer, eds. 2003. Livelihoods in crisis? New perspectives on governance and rural development in Southern Africa. *IDS Bulletin*, 34(3).

Scoones, I., *et al.* 2007. Dynamic systems and the challenge of sustainability, *STEPS working paper*, 1. Brighton: STEPS Centre.

Sen, A. 1981. *Poverty and famines. An essay on entitlement and deprivation.* Oxford: Oxford University Press.

Shankland, A. 2000. Analysing policy for sustainable livelihoods. *IDS Research Paper*, 49. Brighton: IDS.

SLSA Team 2003a. Decentralisations in practice in Southern Africa. *IDS Bulletin*, 34(3), 79–96.

SLSA Team 2003b. Rights talk and rights practice: challenges for Southern Africa. *IDS Bulletin*, 34(3), 97–111.

Smith, A. 2006. Green niches in sustainable development: the case of organic food in the United Kingdom. *Environment and Planning C: Government and Policy*, 24(3), 439–58.

Smith, A. and A. Stirling. 2008. Social-ecological resilience and socio-technical transitions: critical issues for sustainability governance, *STEPS working paper*, 8. Brighton: STEPS Centre.

Solesbury, W. 2003. *Sustainable livelihoods: a case study of the evolution of DFID policy.* London: ODI.

Spaargaren, G., A. Mol and F. Buttel, eds. 2006. *Governing environmental flows. Global challenges to social theory.* Cambridge, MA: MIT Press.

Stevens, C., S. Devereux and J. Kennan. 2003. *International trade, livelihoods and food security in developing countries.* Brighton: IDS.

Stirling, A. 2008. Opening up and closing down: power, participation and pluralism in the social appraisal of technology. *Science Technology and Human Values,* 33(2), 262–94.

Swaminathan, M.S. 1987. *Food 2000: global policies for sustainable agriculture, report to the World Commission on Environment and Development.* London: Zed Press.

Swift, J. 1989. Why are rural people vulnerable to famine?. *IDS Bulletin,* 20(2), 8–15.

Tacoli, C. 1998. *Rural–urban linkages and sustainable rural livelihoods.* London: DFID Natural Resources Department.

Tiffen, M. 2003. Transition in Sub-Saharan Africa: agriculture, urbanisation and income growth. *World Development,* 31(8), 1343–66.

Tiffen, M., M. Mortimore and F. Gichuki. 1994. *More people, less erosion. Environmental recovery in Kenya.* Chichester: John Wiley.

Unsworth, S. 2001. *Understanding pro-poor change: a discussion paper.* London: DFID.

Walker, B. and D. Salt. 2006. *Resilience thinking. Sustaining ecosystems and people in a changing world.* Washington, DC: Island Press.

Walker, T. and J. Ryan. 1990. *Village and household economies in India's semi-arid tropics.* Baltimore, MD: Johns Hopkins University Press.

Warner, K. 2000. *Forestry and sustainable livelihoods.* Rome: FAO.

Warren, A., S. Batterbury and H. Osbahr. 2001. Soil erosion in the West African Sahel: a review and an application of a 'local political ecology'. *Global Environmental Change,* 11(1), 79–95.

Watts, M. 1983. *Silent violence: food, famine and peasantry in northern Nigeria.* Berkeley, CA: University of California Press.

WCED 1987. *Our common future. The report of the World Commission on Environment and Development.* Oxford: Oxford University Press.

Wiggins, S. 2000. Interpreting changes from the 1970s to the 1990s in African agriculture through village studies. *World Development,* 22(4), 631–62.

World Bank 2007. *World development report 2008: agriculture for development.* Washington, DC: World Bank.

Engendering the political economy of agrarian change

Shahra Razavi

Over the past thirty odd years, the analysis of agrarian social relations, institutions, and movements has benefited from the insights offered by feminist scholars whose intellectual project has been to bring into the political economy of agrarian change the pervasiveness of gender relations and their interconnections with broader processes of social change. The implications of this re-thinking are potentially radical. Gender analysis has interrogated some of the dominant orthodoxies in agrarian studies: in conceptualising households and their connections to broader economic and political structures; in deepening the analysis of rural markets as social and political constructions with highly unequalising tendencies; and in better understanding both the role and the limitations of different institutional arrangements (involving states, markets, and 'communities') for the management of local resources. However, the complexities of this research have been sanitised and distorted by neoclassical economists and powerful development organisations that speak the same language, as well as by some advocates keen to get their messages heard as they push for policy change, or alternatively ignored and sidelined by some of the political economists of agrarian change. In this paper I attempt to show some of the key contributions of feminist scholarship to agrarian studies, the extent to which these have been taken up by mainstream debates – whether neoclassical or political economy – and where there is scope for more empirical and theoretical work.

Introduction

Over the past thirty odd years, the analysis of agrarian social relations, institutions, and movements has benefited from the insights offered by feminist scholars whose intellectual project has been to bring into the political economy of agrarian change the pervasiveness of gender relations and their interconnections with broader processes of social change. The implications of this re-thinking are potentially radical. Gender analysis has interrogated some of the dominant orthodoxies in agrarian studies: in conceptualising peasant households and their connections to broader economic and political structures; in deepening the analysis of rural markets as social and political constructions with highly un-equalising tendencies; and in better understanding both the role and the limitations of different institutional arrangements (involving states, markets, and 'communities') for the management of resources. However, the complexities of this research, as the paper argues, have been filtered out by neoclassical economists and powerful development organisations, as

I would like to thank Haroon Akram-Lodhi, Debbie Budlender, Lucia da Corta, and Cecile Jackson for comments on earlier versions of this paper. I alone remain responsible for any remaining errors and omissions.

well as by some gender policy advocates keen to get their messages heard. Nor have political economists of agrarian change made a serious effort to engage with it, although as female labour has become increasingly commodified (and hence visible to political economists) an area of mutual interest has developed around the question of labour in the rural informal economy, where social relations of gender are widely recognised (even by 'ungendered' political economists!) to be working in an exclusionary, hierarchical and exploitative manner.

Section 1 of this paper considers one of the most fertile areas of feminist research – the work on domestic institutions (households, families), their internal workings, as well as their connections to broader economic and political structures and processes – which emerged in the early 1980s in response to both the limitations of neoclassical economics as well as the 'silences' of political economy of agrarian change. The section goes on to argue that while aspects of the feminist critique of the unitary household model were selectively adopted and elaborated upon (some would say distorted) by neoclassically-inclined researchers and policy actors, their work had the salutary effect of exposing the severe limitations of methodological individualism for analysing gender relations, thereby prompting a feminist re-thinking of the household to which we turn at the end of this section.

If neoclassical economists are guilty of distorting gender relations, then political economists of agrarian change must be faulted for ignoring it. Working with a largely unitary household model, the brief Marxist interest in (uncommodified) 'domestic labour' fizzled out when criticised for its economism and functionalist modes of argumentation, and 'reproduction' continued to be seen by agrarian political economists as a gender-neutral process – flying in the face of feminist analysis which aims to make visible the uncommodified work (carried out through relations of family, kinship and friendship) involved in reproducing labour power on a day-to-day basis and across generations. This was also the period when feminists were beginning to develop their critique of neoliberal restructuring which was seen as an attempt by capital (and the state) to shift the burden of reproduction and care of the labour force onto the shoulders of women (and girls) whose unpaid labour was (wrongly) assumed to be infinitely elastic and the functioning of households (also wrongly) considered to be something that could be taken-for-granted. For all the references to 'reproduction', the political economy of agrarian change never seriously considered the relations between the largely feminised unpaid reproductive sphere and the more visible labour and commodities that entered the circuits of accumulation.

Gender, or rather female labour, was however visible in political economy research that documented different 'classes of rural labour' (Bernstein 2009, 55–81, this collection), the terms of their incorporation into markets partially determined by their domestic relations and responsibilities. This is explored in Section 2 of the paper which turns to the broader structures and institutions of the agrarian economy, most notably labour markets and land tenure institutions, permeated as they are by gender relations as well as other relations of power and inequality. This section raises a number of contentious questions, about the centrality that has been given to land rights as the basis for global policy prescriptions for female poverty and women's empowerment; it also asks how women's contractual inferiority in labour markets (broadly defined) can best be understood.

Section 3 turns to an emerging consensus across the political spectrum, that sees 'community' institutions – be they village councils, 'traditional' authorities, or 'indigenous' institutions and local states in processes of decentralisation – as the more suitable managers of rural resources and as the more socially embedded locus of decision-making (as opposed to the more distant and 'elite-dominated', if not 'predatory', state). As this section shows, feminist analysis raises questions about the particular forms these institutions take, their implications for *equality* of participation and benefit sharing. There are concerns that 'participation' can effectively translate into an additional unpaid work burden for particular categories of people who are deemed to be more suitable stewards of the family and the environment.

The new political economy and the domestic arena

Domestic structures – the world of families and households – appear as taken-for-granted by much of the new political economy, just as they were by nineteenth century political economists (Elson 1998). As Elson goes on to argue 'to understand the ordering (and disordering) of societies' (i.e. processes of social change), the analytical focus was, and remains, on political and economic processes and the detailed relations between them, with little sign of interest in how 'households are organised (or disorganised) both internally and in relation to economic and political structures' (1998, 189). Esping-Andersen, in a similar fashion, refers to 'the blindness of virtually all comparative political economy to the world of families' (1999, 11).[1] Figure 1, taken from Elson (1998), depicts the 'domestic sector' and its interconnections with the 'private sector' and the 'public sector' in a simple format by showing the circuits that connect them and through which flow goods, services, labour (with physical, technical and social capacities) and values (be they commercial, regulatory or provisioning). While domestic structures can produce

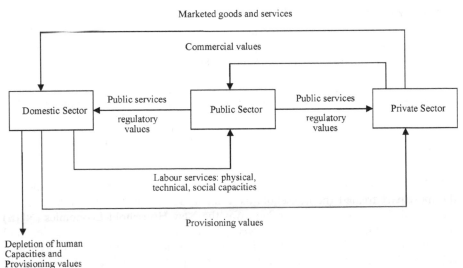

Figure 1.   The circular flow of output of goods and services.
Source: Elson (1998).

---

[1]Esping-Andersen's remarks are a self-critique, responding to feminist criticisms of his earlier work on welfare regimes which neglected the role of families and of women's unpaid work in social provisioning.

able-bodied and socialised workers/citizens on a daily and intergenerational basis who in turn contribute to the workings of the other sectors (both private/market and public/state), there is nothing automatic about this. Domestic structures too rely and depend on the flow of goods, services and values from other sectors. When these inputs (from the public and private sectors) are not sufficiently nourishing, as I hope to show in this section, then human capacities and provisioning values will be destroyed and they will drain away from the circular flow, as shown in the Figure.

Bringing the domestic arena into the analytical framework thus demands not only that we scrutinise what goes on inside domestic institutions – cooperation and conflict, pooling/sharing and inequality in resource allocation and in the division of unpaid work necessary to sustain the members of the household – but also that we tease out the interconnections between domestic structures and broader economic and political processes. My contention in this section is that within agrarian political economy more effort has been made by feminists to date in explicating the former set of issues – the internal workings of families and households – while the analysis of changes in household and family forms over time, and the systemic relations between the domestic, economic and political structures have been more difficult to pursue. My second contention is that the unpaid (non-commodified) provisioning of household members has been invisible to political economists of agrarian change and continues to be invisible in research on livelihoods. But my third and main contention is that while neoclassical economists and policy organisations allied to them have shown more interest in households and in selected aspects of feminist research, their methodological grounding and political projects have distorted the more complex, contextualised and nuanced aspects of feminist analysis.

Rethinking the unitary 'peasant' household

'The field of peasant studies generally implicitly equates "the peasant" with male household heads, which actually excludes the majority of the peasant population from the socio-economic analysis.' (White 1986, 60).

> it would be idle to see the capitalisation of peasant agriculture in isolation from the study of these conflicts of age and sex. To do so would be the hallmark neither of good economic history nor of Marxist analysis, but would be the opposite of both. ... But nor should recognition of the struggle between the generations and the sexes be taken to mean, *a priori*, that class concepts cease to be relevant to Russian agrarian history. (Harrison 1977, 151)

While both neoclassical economists and the founding fathers[2] of peasant studies for the most part treated the household as a 'black box', one strand of thinking that recognised the intra-household sphere was the New Household Economics (NHE) pioneered by Gary Becker (1981). NHE, however, analysed the production and consumption activities of the household with neoclassical analytical tools which relied on heroic assumptions (altruism on the part of the head of household), produced circular arguments (women specialise in homemaking because they earn

---

[2]The reference here is to the influential work of Alexander Vasilevich Chayanov, the Russian agricultural economist, who is considered by most as the father of peasant studies. His work and legacy are dealt with at length in the contributions by Shanin and Bernstein in this collection. For a feminist analysis of Chayanov's work see Deere (1995).

less in the market, and they earn less in the market because of their household responsibilities), and had a tendency to dissolve all differences between an idealised market sphere (the abstract market of economic textbooks) and the social sphere.[3]

The unitary household model has come under increasing fire over the past couple of decades by feminists from diverse disciplinary currents, and of late by neo-classically inclined microeconomists and modelers using so-called Nash-bargaining models.[4] For nearly three decades the feminist literature has drawn attention to the unequal distribution of resources and power within households along gender and generational lines. It has also documented women's greater attachment to the welfare of children, evident in their spending priorities for example (Bruce and Dwyer 1988). Some of these findings, as we argue in more detail below, seem to have been selectively taken up by policy institutions: anti-poverty programs, whether in the form of micro-credit or conditional cash transfers, increasingly target women on the grounds that they will spend the resources under their control in ways that will enhance family and child welfare, and policy documents on land tenure institutions and their reform frequently refer to gender-based constraints on women's ownership and access to land as an impediment to market efficiency and poverty reduction (World Bank 2003). These are not small achievements, but they raise the question as to why 'bowdlerised, impoverished or, for some, just plain wrong representations about gender issues have become embedded in development' (Cornwall *et al.* 2007, 2). I will turn to this question later in this section.

Much of the early feminist critique of agrarian political economy, which appeared in the pages of *The Journal of Peasant Studies* (JPS), was precisely aimed at driving the important message home that the peasant household was not an internally homogeneous institution, and that the 'household head' 'did not necessarily have the same interests' as other members of the household (White 1986, 60). It was further argued that these differing interests had important implications for the analysis of 'peasant resistance' (White 1986). In the move from a system of primarily male-controlled 'family farms' to a cooperative form of agriculture (in Vietnam) there were cases where within the same household wives were enthusiastic about the idea of collective work, while the husband was more resistant (White 1986). Conversely, as collective forms of production were weakened under the Chinese reforms in the early 1980s, and households given a more prominent productive role, a certain re-ordering of intrahousehold and interhousehold relations took place, including the increasing prevalence of three-generation households and 'aggregate families', along with the intensified use of family labour (Croll 1987). Implicit in these accounts was the understanding that male access to, or ownership of, land for household production effectively deepened the exploitation of women's labour through heavier workloads in the form of generally unpaid family labour (Jackson 1996). Patriarchy, understood as the rule of elder men over women and junior men, was not only acknowledged but also used by Harrison (1977) to explain practices of household partition and social mobility among the Russian peasant households.

---

[3]Fine (2004) refers to this as the risk of 'economic imperialism', that is, of economics colonising the other social sciences by imposing a neoclassical mode of analysis on all social phenomenon.
[4]The feminist literature on intra-household gender inequalities is extensive; prominent contributions include Agarwal (1990), Folbre (1986), Hart (1995), Kabeer (1994) and Whitehead (1981). For a critique of collective models of the household see in particular Hart (1995).

The accuracy of the unified household model, and the assumption of altruism on the part of the household head, was most dramatically shaken by the emerging evidence, especially from South Asia, of gender inequalities in well-being. Gender bias in basic demographic indicators (such as sex ratios, mortality and life expectancy) showed the younger females (especially those in the 0–5 age group) to be most vulnerable, and some of this was explained by significant female health disadvantages, caused by disadvantage in access to treatment and food (Agarwal 1984, Harriss 1990).[5] Explorations of intra-household inequalities in well-being (captured through sex ratios, nutritional indicators, and other 'softer' indicators) and in the distribution of burdens in contexts of seasonal scarcity and pervasive calamity such as during droughts and famines also showed the burden of coping to fall disproportionately on female members within poor households (Agarwal 1990). But the picture was far from uniform. Detailed case study material produced contradictory results: among the adult working poor, for example, *male* nutritional disadvantage was not uncommon (Harriss 1990, Kynch and McGuire 1994, Jackson and Palmer-Jones 1999) and vulnerability of younger females could not always be explained by underfeeding. As one exhaustive review of the evidence on intrafamily food allocation for South Asia concluded:

> It can be objected that in the context of the reciprocally causing relations of patriarchy, it is counter-intuitive that there be no general sex bias in food intake. This chapter has shown, however, that discrimination in feeding does not automatically imply discrimination in nutrition, and that discrimination in nutrition does not automatically imply disadvantage in welfare. [Adult] Females [on average] need absolutely less of most nutrients than do males. It is also clear that there are class, caste and regional differences in the operation of patriarchy. Furthermore, discrimination may exist and yet not be picked up even in welfare indicators. That average life expectancy for example has probably reached gender parity in India says little about sex bias under patriarchy. (Harriss 1990, 410)[6]

Intrahousehold gender inequalities in well-being were explained by the relative 'fall-back position' of women and men (captured in terms of relative wages) by those using bargaining models of the household (Sen 1990), or in terms of parents' maximising behaviour by those still loyal to the unitary household model (Rosenzweig and Schultz 1982).[7] The former, however, was considered to be too simplistic (Agarwal 1990) and the latter deeply misconstrued (for confusing income

---

[5]The contrast between north-west India (where female disadvantage was most marked) and south-east India (where such disadvantage was either absent or less acute) was a recurring theme. The literature on this issue is vast: the reader may refer to the extensive references found in Agarwal (1984, 1990) and Harriss (1990).

[6]By the end of the 1990s in fact the gender gap in life expectancy in *favour* of women was even more substantial, while at the same time the sex ratio among children in the 0–6 age group had deteriorated and was showing even greater female disadvantage than in the early 1990s (Deaton and Dréze 2002). Discriminatory patterns of access to health care are also likely to have intensified in the 1990s in the context of a substantial increase in private health care provision at the expense of public health care, especially where the former services have been geographically inaccessible and unaffordable for poor rural households (Jackson and Rao 2009).

[7]Rosenzweig and Schultz (1982) attempted to reconcile the disjuncture between the assumption of intra-household welfare maximisation (held by NHE) and evidence of intra-household welfare inequalities by arguing that parents were behaving rationally by ensuring

maximisation with welfare maximisation) (Folbre 1984). Agarwal (1990) identified the importance of at least five factors that together shape the 'fall-back position' of different household members (rather than wages alone being the determining factor): private ownership and control over assets, especially land; access to employment and other income-generating activities; access to common property resources; access to external social support systems such as broader kinship networks and informal credit and patronage systems; and support from the state (such as through public employment programs) and from non-governmental organisations and grassroots organising.

Research findings from smallholder production in sub-Saharan Africa, especially West Africa, provided critics with a different kind of ammunition against the unitary household model. The pertinence of smallholder agricultural systems of sub-Saharan Africa to the gender critique of neoclassical models stemmed from the prevalence in the region of less corporate forms of householding involving the relative autonomy of mother-child units, compared to the more corporate male-headed households prevalent in the regions of 'classical patriarchy' (Caldwell 1978, Kandiyoti 1988). These features of smallholder agriculture had their roots in the pre-commodity economies and societies of the nineteenth century prior to modern transformations. Women's and men's access to resources (including land), for example, had been symmetrical in form, even if men's capacity to command and effective possession had been far more extensive than women's (Whitehead 1990, 438, Mackenzie 1990). Furthermore, women had a *dual* productive role: working both independently of other members of the household (i.e. having separate access to land and other resources for farming or other economic activities), and also contributing to household production as unremunerated family labour (Whitehead 1990, 438). The other distinctive feature was that resources of husbands and wives were not merged into a single conjugal fund (Whitehead 1981). The responsibilities for providing for children's well-being were often divided between father and mother – a feature which required women to be economically independent to some degree, and which was also 'structurally linked to polygyny' (Whitehead 1990, 439).

While there are some elements of continuity between these historical antecedents and modern gender relations, the transformation of rural production systems, and the commodification and individualisation of productive resources, were experienced very differently by women and men. As agrarian economies were commercialised and rural class differentiation was intensified, women's independent farming came under increasing pressure, while many men were able to solidify their command over land, labour, and capital resources. It is within this context that several case studies captured intra-household conflicts between women and men over crop rights and over women's labour (Dey 1981). The story of Jahaly Pacharr in the Gambia (Carney 1988) was one among several such studies. In the context of this Gambian contract-farming irrigated rice scheme, women challenged their husbands' claims to

---

the welfare maximisation of those children with a greater likelihood of economic productivity. Although not sharing the same theoretical premises (of unitary household model), several other contributors to the Indian debate on sex ratios, also drew attention to the important divide between 'north' and 'south' Indian patterns of female disadvantage by attributing these differences to women's differing labour force participation rates (higher in the south and lower in the north-west) (Bardhan 1974) as well as differing marriage costs (typically high dowries of north-west) (Miller 1981). These economic theses of female disadvantage were however critically scrutinised by Agarwal (1984, 179–80).

their labour where they were not successful in receiving compensation from them, by withdrawing their labour from rice fields and engaging in other economic activities (such as through hiring themselves out as wage labourers or engaging in petty trade). The absence of incentives which could lure female labour into rice farming, it was claimed, was affecting the household's capacity to intensify its labour usage, which was a primary objective of the project, thereby undermining the viability of the project and indeed of Gambia's attempt to enhance national food security through rice import-substitution.

A somewhat different but related strand of thinking, which resonated with gender advocates in development agencies like the World Bank and the International Food Policy Research Institute (IFPRI), used microeconomic analytical tools to argue that the structure of male and female incentives in farm households in sub-Saharan Africa was leading to 'allocative inefficiencies' and a muted agricultural supply response. It was further claimed that if the allocation of productive resources were not skewed against women farmers, smallholders would be able to produce more, with positive implications for rural poverty.[8] Such arguments were abstracted from a set of empirical accounts of agricultural production in sub-Saharan Africa, most notably, Udry and Alderman's analysis of an ICRISAT agricultural survey in six communities in Burkina Faso (Udry and Alderman 1995). While attractive from an advocacy standpoint, given the synergies they suggested between gender inequality and adverse agricultural outputs, closer analyses of these studies found the policy conclusions that are drawn to be overstated (Whitehead 2001) and questionable as a strategy for reducing poverty in rural sub-Saharan Africa (O'Laughlin 2007).

Interestingly, there are some parallels between the above-mentioned neoclassical arguments for gender equality, and arguments made by some economists (labeled 'neo-populist' by Byres (2004)) for redistributive land reform as a mechanism for reducing poverty and enhancing growth and equity (Griffin *et al.* 2002). The argument for redistributive land reform, like the efficiency argument for gender equality, is based on the understanding that current resource allocation (to large landowners, male farmers) is inefficient and that the re-allocation of resources (to small farmers, female farmers) will improve static efficiency, enhance agricultural growth and reduce poverty. Both arguments abstract from a very small number of empirical studies (Berry and Cline's 1979 work is the reference for the 'inverse relationship' between farm size and productivity, Udry and Alderman's (1995) work is the reference for the inefficiency of gender inequality), and then generalise to a 'mythical uniform terrain' (O'Laughlin 2007, 40).

Even if one agrees with some of the empirical findings on which such claims are grounded (the efficiency of the small farmer, the gender inequalities in intrahouse-hold resource allocation) it is the way in which those facts are read and interpreted which is problematic. For example, the efficiency of the small farm (in specific contexts where it can be shown to operate) is surely a sign of distress and exploitation of family labour, rather than being indicative of its technical superiority. Likewise the inequalities in women's and men's access to certain resources (land, fertiliser) which are highlighted may be part of a larger and *joint* set of interests, even

---

[8]The policy documents making such arguments include Blackden and Bhanu (1999). The argument has been further generalised beyond sub-Saharan Africa in a key policy document, World Bank (2001). Earlier examples of this genre of analysis include Collier (1989) and Palmer (1991); for a critique see Lockwood (1992) and Razavi and Miller (1995).

if these are inequitably distributed (Whitehead and Kabeer 2001). A major flaw in the much-cited study by Udry is that the production of a single crop is treated in isolation from the farming and livelihood system as a whole (Whitehead 2001). For example the fact that men's crops receive more labour and fertiliser inputs could be part of a whole set of arrangements that privilege household heads, but these privileges pay off not just for themselves but for all household members (though not equally). 'In other words, it is important to take seriously household economic interdependence as well as economic separation' (Whitehead 2001, 18) – a theme to which I shall return at the end of this section. Last but not least, both sets of argument are static and ahistorical, ignoring the construction of class and gender inequalities as part of larger processes of accumulation and impoverishment in the context of capitalist transformation in the countryside (O'Laughlin 2007). This is inevitably linked to their methodological grounding in neoclassical economics.

The attraction of these case studies, not only to gender advocates but also to some analysts, is that at a superficial level they identify interconnections between intrahousehold gender relations and the broader economic objectives and policies (as depicted in Figure 1). And yet it is the narrow, static, and ahistorical manner in which gender inequality is understood and captured – not as inequalities in social relations of gender that are shaped by broader economic and political processes, but as a simple gender-disaggregation of crops and inputs – that makes these case studies problematic and the policy claims that are made in their name highly spurious and misleading.

Production and reproduction

Parallel to the neoclassical work on the household (in NHE), the 1960s and 1970s also witnessed an important set of debates within Marxist and feminist intellectual circles, subsequently referred to as the 'domestic labour debate'. Much of this debate was about how to conceptualise the domestic work of women and how it relates to the capitalist 'mode of production', so as to better understand the material basis of women's subordination.

However, to fit women's domestic work into Marxist analytical categories, designed for the analysis of paid commodity-producing labour, domestic labour was described as a 'client mode of production', somewhat similar to non-capitalist sectors such as subsistence production within peripheral social formations (Harrison 1973). Within the pages of *JPS* a parallel argument was elaborated by Veronica Bennholdt-Thomsen who suggested that housewives (presumably in the capitalist economies of the centre) and 'peasants (both men and women)' (1982, 241) in the periphery, both produce labour power for capital but without being compensated by capital (Bennholdt-Thomsen 1982). The argument that 'housewives' and 'peasants' produce 'surplus value', which was then transferred from the domestic to the capitalist sphere, was refuted by others for being inconsistent with the Marxist theory of value (Molyneux 1979).[9] There were also concerns that the recourse to functionalist arguments in conceptualising the relationship between capitalism and domestic

---

[9]Molyneux maintained that Harrison's argument was based on a false premise that treats as equivalent and, therefore, comparable the concrete labour in the domestic sphere and the abstract labour time of commodity production; thus, his argument could only hold if the 'law of value' was redefined.

labour had a tendency to economic reductionism (Molyneux 1979). It did not for example explain why it was women who did this work. This critique was in fact part of a much broader attempt by feminist analysts at the time to develop a theory of gender, which was integrated into and informed by the general analysis of changes in the global economy and, yet, which avoided crude analyses of gender relations made exclusively in terms of their function for capital and 'the reproduction of capitalist relations of production' (Pearson *et al.* 1981, x).

Curiously, Bennholdt-Thomsen (1982) was silent about the arduous domestic and care work that 'peasant women' undertook, in addition to their contribution to subsistence agricultural work, i.e. in preparing meals, fetching water and firewood, and caring for young children, frail elderly and sick family members – tasks that often consumed a considerable amount of time and energy in contexts with poorly developed infrastructure (such as sanitation, piped water, electricity). These issues received far more attention in the more grounded analyses of agrarian transformations which drew attention to the significant class and caste differentiations not only in domestic work – the 'double day' of the poor female agricultural labourer (Sharma 1985, 83) – but also in marital patterns and household structures, as well as in the degree of control that women of different class and caste households exercised over resources (common property resources, wages, land), and over their own sexuality and fertility (Sharma 1985, Agarwal 1990, 352–4).[10] Prefiguring what would later be labeled 'the patriarchal bargain' (Kandiyoti 1988) Sharma captured the harsh trade-offs that women in (Uttar Pradesh) North India often had to make between personal autonomy and material security within the given set of concrete constraints within which they operated:

> An understanding of the nature of women's oppression, and the importance of class differences in defining their subordination, makes it less easy to refer to 'false consciousness' as an explanation of why the poor would like to go into purdah or so strongly defend an institution – the family – which appears to be the locus of much of their oppression. (1985, 84)

The issue of 'reproduction' was also being addressed in feminist accounts of agricultural restructuring in socialist countries, where there was at least a rhetorical commitment on the part of the state to public and collective assistance in sharing the burden of domestic and care work. Despite the policies of collectivisation, the individual rural household had maintained some forms of production (albeit greatly reduced in scope), and because community services were still very unevenly distributed in rural areas, it remained the primary unit of consumption and reproduction (Croll 1987). By the early 1980s, however, there were warnings that new policy directions being adopted by the state were putting 'production' and 'reproduction' – the latter understood more narrowly in this case as biological reproduction – onto a collision course. The economic reforms initiated in the late 1970s (under the household responsibility system) were increasing the range and variety of productive demands (agriculture, sideline production) on the peasant household, leading to the intensified use of family labour and strategies for

---

[10]Several studies drew attention to women's domestic work, and particularly their specific responsibility for collecting fuel, fodder, and water under conditions of increasing deforestation and ecological deterioration (e.g. Agarwal 1984).

increasing the family labour pool, while the singular and universal 'one-child' focus in state population policy was working in the opposite direction (Croll 1983, 1987). The title of Croll's (1983) article – 'Production versus reproduction: a threat to China's development strategy' – was prescient as the full social ramifications of the 'one-child policy' fatally combined with strong son preference to produce one of the most dramatic instances of 'missing women' through sex-selective abortions and post-birth neglect (Klasen and Wink 2003).[11]

In sum, despite some of the limitations of the 'domestic labour debate' already alluded to, the main argument that unpaid domestic labour produces vital inputs for the economy in the form of a labour force that is available for work and a variety of other intangible social assets, is one that needs to be retained. Curiously though, much of the commentary on 'reproduction' in political economy research remains vague and un-gendered. O'Laughlin (2008) suggests that this vagueness in Marxist political economy (like liberal economics) stems from its fixation with the sphere of commodities where value is realised, and a 'certain dose of residual teleological evolutionism' (2008, 352) which sees commodification as a linear and irreversible process that will eventually lead to the withering away of non-commodified forms of work.

Not only have non-commodified forms of work expanded under current conditions of crisis and neoliberal restructuring (Gonzalez de la Rocha 1988), there are strong arguments for thinking that complete commodification (and de-familialisation) of unpaid care work is neither possible (as the experience of both socialist countries, and advanced social democratic welfare states show) nor necessarily desirable. Livelihoods depend not only on money wages, but also on the unpaid work that reproduces the labour force over time and across generations. Livelihood research (like political economy research), as Jackson and Rao (2009) have noted, pays lip service to these activities but there are still too few studies which collect data on this 'invisible' part of the rural economy, alongside descriptions and analyses of employment and farm production. A recent review of the literature on the gender-differentiated impacts of liberalisation policies in rural sub-Saharan Africa draws attention to the thin evidence base on women's increased work burdens both due to changes in agricultural production, and to the removal of subsidies on social services in education and health (Whitehead 2009).

Without good comparative studies over a wide range of circumstances, it will remain difficult to track the connections between productive and reproductive spheres, and to show how the unpaid work undertaken by women and girls within households and 'communities' expands when public social provisioning is curtailed or insufficient to meet needs. One concrete illustration of this expansion of unpaid care work can be gleaned from countries with high HIV/AIDS prevalence. In contexts where public health systems are unable to cope with the intensified need, in addition to the unpaid work that women and girls undertake for family and household members, states have come to rely on volunteer workers (among whom poor women predominate) mobilised by the not-for-profit organisations to assist the prime carers on a 'voluntary' basis without recognition or compensation, or for very basic stipends (Akintola 2004, Meena 2008).

---

[11]A care crisis has also been precipitated by the onset of rapid demographic ageing which is expected to persist for many decades to come (Peng and Phillips 2004).

Rural livelihood research also needs to be attentive to the role and impact of social policies, including both the provision of services and social protection measures (seen as an aspect of the 'social wage'), especially as many of the latter (social pensions, child and family allowances) are being extended to the rural population (in Brazil for example, while in South Africa social pensions were always provided in rural areas). The expansion of social protection measures can be understood as a response on the part of the state to anemic employment creation (even in contexts where growth rates have been respectable), which as Section 2 suggests, is a hallmark of the present crisis in global capitalism.

Rethinking the household

While gender analysis sees households as sites of struggle and inequality, there is far less consensus as to how the given inequalities and conflicts, as well as common interests and cooperative behaviour, should be understood. Agarwal (2003) is correct when she writes that the bargaining framework conceptualises intra-household interaction as containing 'both cooperation and conflict' (2003, 573), but the question that others (Jackson 2003, Whitehead and Kabeer 2001, Razavi 2003) have been raising is whether analyses that have been informed by the bargaining framework have sufficiently *captured and reflected* the *common* interests that all household members have in the overall economic success of their households, however unequally the resources and the burdens are divided? Do women work on their household plots because 'older men prevent them from leaving their farms in search of more productive employment' as Sender and Johnson (2004, 148) claim? What makes women and junior men stay inside the patriarchal household, even though they are allocated fewer resources, take on heavy work loads of largely 'unpaid labour' and enjoy less leisure time (Whitehead and Kabeer 2001)? Is it pure despotism on the part of the male household head that binds the household together? In the words of Sharma (1985, 84), why do poor women so strongly defend an institution 'which appears to be the locus of much of their oppression'?

Reflecting on the South African land reform program, Cherryl Walker (2003) admits that even where women have been listed as independent household heads and as beneficiaries in their own right, their access to land has been mediated overwhelmingly through their membership in patriarchal households. Skeptical of individual rights as the solution – an element that was incorporated in the land policy document, Land Redistribution for Agricultural Development (LRAD) issued in November 2000 – she presses for a deeper appreciation of the importance of household membership in poor women's lives, and thus the importance of ensuring women's rights to household resources. Had the LRAD framework with its emphasis on individual rights been in place from the start of the land reform program, very few, if any, of the women in the present beneficiary communities would have been able to access land through it – 'they are simply too poor, too isolated and too dependent on male authority to be able to establish individual rights to land' (Walker 2003, 143). Moreover many women beneficiaries endorse the household model implicit in the Department of Land Affairs' (DLA's) work, Walker claims, and some have struggled very hard to secure their household's interests. While a minority seemed interested in the idea of individual titles, de-linked from that of their husbands or families, few saw this as the solution to their problems. They were more interested in mechanisms that would secure and extend their rights to household resources, through joint titles, inheritance rights for their daughters and copies of title deeds (Walker 2003).

While intra-household inequalities in access to resources and conflicts over their distribution are well documented, this does not mean that a woman's level of well-being is unrelated to that of her husband, partner, father or brother. Women's and men's interests within marriage are both joint and separate, which is what makes gender struggles so complex (Jackson 2003). What many women seem to be saying in the South African case study above is that they want to redress the inequalities in gender relations that hamper their access to household resources and impinge negatively on their security, but not necessarily to opt out of the household in order to make it on their own in isolation.

The need to re-think the household also has to take into account the increasing prevalence, in regions such as Southern Africa perhaps most acutely, of households where women are living with their children and also often sustaining them without a significant economic contribution from the fathers of those children. In the process of capitalist development in agriculture in the region, while some women have been able to benefit as commercial farmers in their own right, or as wives of successful commercial farmers, much larger numbers of women have been concentrated in a stratum of rural households that lacks the resources to meet its own consumption needs and is seeking different kinds of wage work. One of the main characteristics of this growing stratum is that it contains a significant number of households headed and maintained by women (Whitehead 1990, O'Laughlin 1998).

In South Africa, for example, the majority of children live apart from their biological fathers (Budlender and Lund 2007). In July 2005, only just over a third (35 percent) of children (0 17 years) were resident with both their biological parents, two in five (39 percent) were living with their mother but not their father, while one in five (22 percent) did not have either biological parent living with them. Only three percent were living with their father but not their mother (Budlender and Lund 2007). The fact that a significant proportion of children do not reside in the same household as their biological fathers does not automatically mean that the latter do not contribute economically or otherwise to their well-being (although the evidence cited below suggests that in South Africa they do not contribute). As the anthropological literature has shown, households and families are not coterminous, and family members may be spread across several households (Moore 1994). Nor is it helpful to simply equate female headship with poverty, in the way that many policy organisations (both donors and NGOs) have been doing (in order to 'target' their assistance to the truly needy). The trajectories leading to female headship are clearly divergent, and the category of households labeled 'female-headed' a highly heterogeneous one (not to mention the ambiguities in how 'headship' is defined in the first place by both enumerators and respondents). Some of these conditions may constitute what can be reasonably thought of as poverty risk factors, such as households with young children maintained by women alone (Folbre 1990). But by aggregating these distinct categories of households generated through different social processes (e.g. migration, widowhood, divorce, having children outside marriage), and constructing a binary categorisation (female- and male-headed) it becomes impossible to interpret the evidence in a meaningful way (Razavi 1999).

Yet while women can and do in fact initiate both divorce and separation, and while they may enjoy and appreciate the greater autonomy of living on their own (Safa 1995 cited in Chant 2008), this should not blind us to the validity of their increasing primary economic responsibility and their growing vulnerability (O'Laughlin 1998). The patterns noted above for South Africa beg several questions: where are the missing fathers, and 'are they missed' (O'Laughlin 1998)? The fact that fathers are often absent probably means that many of the mothers must try to combine their care-giving roles with income-earning, which is clearly difficult in a situation of high unemployment (Budlender and Lund 2007). Data from the *2006 Household Survey* (Statistics South Africa 2007) suggest that dual-earner households constitute 16 percent of all households, households in which women are the single or dominant earner constitute 18 percent, households in which men are the single or dominant earner constitute 33 percent, while a significant 33 percent of households have no-one employed and presumably most survive on state social transfers (which have expanded massively over the last 15 years, reaching close to 25 percent of the population by 2007).

Hence, in over half of all households either men are absent (and few of these contribute financially to their children's upbringing)[12] or unable to make a financial contribution through paid forms of work suggesting an erosion of the male bread-winner model which relied for the most part on wage labour through migration – a feature shared by several countries in Southern Africa. The persistence of extensive structural unemployment – not only for men among whom unemployment stands at 23 percent, but also for women where it reaches a staggering 31 percent – raises fundamental questions about the development model that despite respectable levels of economic growth is so clearly inadequate in providing jobs and livelihoods.

In her detailed analysis of livelihoods in Botswana, O'Laughlin (1998) refers to the recurrence throughout Southern Africa, as a regional labour system, of a number of persistent structural commonalities: 'the dependence of rural livelihoods on disposable cash income, the sharpening polarisation of agrarian production, structural unemployment, the erosion of social support from kin and community, and the corresponding dependence on social transfers' (1998, 38). In Botswana the outflow of young men (and to a lesser extent young women) from rural areas into urban centres and beyond (especially as undocumented workers in South Africa), is reflected in the persistence of large numbers of households maintained by women. The fact that many women and men do not marry and establish common households, she underlines, 'is because they cannot, and not because they do not wish to do so' (1998, 24). In the context of long-term structural unemployment some men 'disappear analytically' because many poor men do not form households at all, and both rural poverty and the high incidence of women-headed households derive from the dominant model of accumulation that is exclusionary and polarising (O'Laughlin 1998).

This analysis of Southern Africa, which relies heavily on O'Laughlin's (1998) work, takes me back to Elson's diagram which appeared at the beginning of this paper (see Figure 1). We see here a dramatic illustration of why the 'domestic sector' cannot be taken for granted. A model of accumulation which produces persistent polarisation and long-term social exclusion is deeply disabling for families and households, undermining their ability to make provision for both their needs and the

---

[12]The 2008 *Income and expenditure survey* (Statistics South Africa 2008: 152) shows that of the total average household income of 74 589R, only 888R takes the form of 'alimony, palimony and other allowances' with a further 314R in 'other income from individuals' – a small fraction.

needs of the other 'sectors'. But it is disappointing that the resulting social disruptions are often accompanied by an ideological response that attributes the social malaise to the break-up of the (patriarchal) family (Moore 1994) and sometimes even blames women, who are managing their households as best they can even in the absence of decent jobs and livelihoods. This is why it is essential for the new political economy of agrarian change to recognise the interdependence between households, markets and states, and the fact that households and the unpaid (or un-commodified) work that sustains the labour force and capital accumulation, as being both responsive to, and dependent on, the workings of the wider political economy.

Gender and livelihoods: land, labour and the off-farm sector

In some regional contexts men may be missing from households and capital may no longer need 'the labour that it pulled from rural households over so many generations' (O'Laughlin 1998, 1), and elsewhere corporate farming may be seeking out a female workforce to produce agro export crops cheaply for an international market that is subject to competition, uncertainty and constantly fluctuating demand (Deere 2005), but gender hierarchies have not been erased from markets. Not only do female workers still encounter labour markets that are deeply segmented with their wages often being reported by bosses as 'pocket money' (Harriss-White 2000), but in the much-praised world of off-farm 'entrepreneurship' and self-employment too, women continue to be crowded into a limited range of activities with fewer entry barriers but also much lower returns (Whitehead and Kabeer 2001).

Gender inequalities within domestic institutions, as was elaborated in the previous section, are now increasingly recognised by many microeconomists. But the attentiveness to gender structures and hierarchies does not seem to extend to other social institutions, notably markets. Political economists (Polanyi 1957) and sociologists by contrast have long countered the abstract market of neoclassical textbooks by showing how 'real markets', as political (and social) constructs, are substantiated through the interaction of social groups and infused with social norms, hierarchies and exclusions (Hewitt de Alcántara 1993, Mackintosh 1990). Even though the 'old' political economy had relatively little to say about gender, a major preoccupation of feminist analyses of agrarian change has been to elaborate the gendered nature of agrarian markets and institutions and to show the inter-connections between gender power relations in the domestic sphere and in the broader institutional arena (most notably in land tenure systems, and in rural labour markets).

Land markets as gendered institutions

It is now widely recognised that the agrarian reforms implemented in the era of state-led redistributive efforts – the 'golden age' of land reform spanning from the 1910s through to the 1970s – were largely gender-blind. Gender analyses of land tenure institutions have exposed the complex of laws, customs, social norms, social relations and practices that conspire to exclude women from the ownership and control of property (especially immovable property like arable land) in many regions, most extensively researched in South Asia (Agarwal 1994) and Latin America (Deere and

Léon 2001). Agarwal's (1994) work, however, has been taken up to construct a rather simplistic view in some policy and advocacy circles that attributes in a blanket fashion rural women's poverty to their lack of access to land (World Bank 2001).[13]

There are continuing debates about the strategic significance and transformative potential of rights to land for women in South Asia,[14] and about the pertinence of that argument to other regional contexts, especially rural sub-Saharan Africa where women's farming is often constrained *not* because they are prevented from accessing land, but because they lack capital or regular non-agricultural income to hire labour, purchase inputs and access marketing channels (Whitehead 2001, O'Laughlin 2008). There are also larger questions about the linkages between land rights and diversified livelihood strategies (Hart 1996, Bernstein 2004) and the viability of a 'small farm' strategy (Byres 2004) as a route to poverty eradication. For reasons of space, below we touch on some of these issues, and only superficially.

Given the centrality of land markets, and of market-led agrarian reform, in the currently dominant liberal economic agenda, it is important to ask if market prescriptions are likely to accommodate women's land claims. In other words, how likely are low-income women to emerge as winners in the market-based land reform model and in land market transactions more broadly? Although the empirical base is far from comprehensive, a judicious reading of the existing evidence points to the severe limitations of land markets as a channel for women's inclusion (Razavi 2007). It is of course important not to homogenise women as a social group; there are always groups of women, for example urban women in formal employment or women in peri-urban areas who grow food for city markets, who may have accumulated enough resources to purchase land in their own name with full property rights (Lastarria-Cornhiel 1997, Mackenzie 1990). But for the vast majority of women smallholders, landless agricultural labourers and those doing different forms of informal work, market mechanisms are not likely to provide a channel for inclusion.

For sub-Saharan Africa Lastarria-Cornhiel's (1997) examination of the continent-wide evidence for the effects of land privatisation points to women as the largest group who have had little to gain from the trend toward privatised land tenure systems. In fact the transformation of African tenure systems have tended to further weaken women's already tenuous claims to land while other groups (community leaders and male household heads) have been able to strengthen their control over land to the detriment of women and some minority groups.[15] Like Platteau (1995), she argues that while previously a number of persons and community groups held different rights to a piece of land, with privatisation most of those rights were brought together and claimed by one person. In this process women have tended to lose out. The fact that women

---

[13]The issue of whether her work has been subverted by others or whether it methodologically lends itself to such advocacy is a contentious one.

[14]See in particular the exchange between Jackson (2003) and Agarwal (2003).

[15]Simple titling and land registration do not in themselves transform a customary tenure system into a freehold one – other changes are also needed for this to happen, namely the commercialisation of agriculture and the development of a land market. Hence, it is the general processes of privatisation and concentration that affect women's land and property rights negatively, rather than national land registration schemes *per se*.

enter the market system with no property, little cash income, minimal political power, and a family to maintain, works to their disadvantage.

The conclusion drawn by Whitehead and Tsikata's (2003) comprehensive review of the gender and land literature for sub-Saharan Africa, namely that in 'the development of private property regimes of any kind, sub-Saharan African women tend to lose the rights they once had ... either because their opportunities to buy land are very limited, or because local-level authorities practice gender discrimination' (2003 79) is sobering.[16] It has become even more important to underline this statement given the extent to which policy documents across the political spectrum advocate a blanket policy mix of private property rights and land-titling not only as a mechanism to encourage capital investment and foster a more efficient land market, but also as a solution to women's weak and tenuous place within land tenure institutions (World Bank 2003).

Have land markets been more inclusive of women in other regional contexts? For six countries in Latin America, Deere and Léon (2003) provide a useful database – something that is missing for many other countries and regions – on different modes of land acquisition (through inheritance, community, state, market, other) disaggregated by sex (see in particular Table 6.2 therein). Their evidence suggests that in the selected countries women tend to become landowners mainly through inheritance, while men do so through purchase in land markets (predictably, men are also privileged when land is acquired through the state).

Have women fared any better in so-called market-friendly land reform programs? Here reference will be made to the recent South African experience that attempted to meld a strong commitment to the goal of social justice, including gender equality, with the principles of market-led land reform. As is by now well-known, the intense negotiations around land reform led to a compromise solution or 'elite-pacted settlement': restitution and redistribution were endorsed but within the constraints of a market-led program informed by a 'willing seller/willing buyer' principle and protection for existing property rights.

There have been, as many predicted, severe constraints within this framework on effectively redistributing land, given the state's inability, within the market-friendly straightjacket, to acquire and redistribute productive land proactively and on a sufficiently large scale. Indeed by March 2005, less than 3.5 percent of the area designated as 'commercial farmland' had been redistributed. What is perhaps not always recognised is that a strictly demand-driven program also conflicts with the policy aim of reaching women (and other marginalised social groups), because it overlooks how power relations and divisions within communities structure 'demand' (Walker 2003). It effectively commits the state to respond to applications from social groups that are already constituted, in which it is likely that women's role will be a marginal and dependent one. Hence, in South Africa gender equality has been reflected in what Walker (2003) calls 'first tier' policy documents such as the White Paper on land. Yet at the level of implementation, as her detailed research unravels, the commitment to gender equality has been far less evident.

The above evidence, though far from conclusive, nevertheless seems to suggest that gender advocates should have serious reservations about land markets, both formal and informal, as a mechanism for women's inclusion. Yet women's rights

---

[16]The reader may refer to the extensive case study material referenced in this paper.

advocates do not seem to adopt a *uniformly* critical stance vis-à-vis land markets. In the context of recent policy efforts to liberalise land tenure institutions in Tanzania, for example, while some women's rights advocacy groups were deeply skeptical of the liberalisation agenda, given the adverse implications of private property regimes for resource-constrained women, others did not share this dim view of land markets (Manji 1998, Tsikata 2003). In fact, some of the most influential gender advocacy groups supported the liberalisation of land markets and land titling as creating opportunities for women to purchase land on their own account and have it registered in their own name to be inherited by their descendents (Tsikata 2003).

A fair probing of the positions taken by women's rights advocates in such debates would have to be cognisant of the concerns many of these groups have about how 'customary practices' have worked to undermine, rather than enhance, women's tenure security.[17] Reflecting a similar discontent, Tripp (2004) claims that in Uganda 'Purchasing land has, in effect, become a way of circumventing the traditional authorities' (2004, 4). Other women's rights advocates point out that liberalisation of land, whatever its risks and merits, is already underway and hence women should seek to gain a place in the emerging markets. This kind of position rather than being simply a reflection of middle class interests (Manji 1998), may be indicative of their pragmatism given the constraints within which they are operating.

Criticism of 'customary practices', however, should not lead to the oversimplified conclusion that land markets are a gender-neutral terrain, and that the main constraint women face is lack of information about their legal rights (hence, the advocacy for legal literacy). In other words land cannot be the 'magic bullet': there is much evidence from the region to show that land can lie fallow while women (and men) seek casual wage work in nearby plantations because they do not have the resources to work the land (Sender *et al.* 2006).

While the evidence to support the 'efficiency of small farms' and the alleged 'non-agricultural spin-offs' from small family farms is weak (Hart 1996), access to land that supports multiple livelihoods by underwriting the money wage – by providing some security in times of unemployment and in old age – is viable. Seen from this perspective, the significance of land is closely intertwined with the labour question, or the 'fragmentation of labour' as Bernstein (2004) refers to it.

It is within this context that we can also place some of the contemporary demands for land 'from below' – not only through spontaneous land occupations and collective mobilisations, but also through the kind of survivalist clamoring for land that Kandiyoti (2003) documents in the case of Uzbekistan. In the latter context, in a situation where collective enterprises have not been able to pay their workers' wages, rural households, and rural women in particular, have fallen back on household and subsidiary plots for self-subsistence and survival. But women's current land hunger, as Kandiyoti emphatically underscores, must be understood in the context of both a wish to reinstate the terms of their former social contract with collective enterprises (as *formal employees* which included a wide range of social benefits) and their despair given the lack of viable employment opportunities.

---

[17]We need far better empirical research to show the extent to which customary practices are actually working to women's disadvantage (as there are case studies pointing to more ambiguous outcomes).

Livelihoods and labour markets[18]

One striking feature of agrarian change and industrialisation in contemporary developing societies to which many observers have drawn attention is the growing prevalence of what is sometimes referred to as livelihood diversification, defined as 'the process by which rural families construct a diverse portfolio of activities and social support capabilities in their struggle for survival and in order to improve their standard of living' (Ellis 1998, 4, Bryceson 1999, see also contribution by Ian Scoones (2009, 171–96) in this collection).

For the sake of clarity it is important to underline that diversification out of agriculture is not itself the problem. On the contrary, the movement of labour out of agriculture and into higher income activities has been an important feature of economic transformation in many currently more developed countries, including countries in East Asia such as China where employment in Town and Village Enterprises (TVEs) and the setting up of side-line operations in the 1980s greatly boosted rural household incomes, with surpluses often reinvested back into agricultural technology (Croll 1987, Bramall and Jones 2000). As Christine White remarked many years ago, 'Peasants frequently have the ambition of becoming non-peasants (whether for themselves or for their children): to become members of higher-income or higher-status groups' (1986, 62). The dilemma which needs to be solved is *how* to transform an economy of 'poor petty commodity producers into "collective masters" of a more productive, prosperous and technologically advanced rural economic system' (White 1986, 62).

The problem therefore is when diversification out of agriculture takes the form of activities that are predominantly low-return and survivalist, which perpetuate poverty and (self-)exploitation. The fact that 'poorer countries today confront more formidable barriers to comprehensive industrialisation – and *a fortiori* to the generation of comparable levels of industrial employment – than did the advanced industrial countries in the past' (Bernstein 2004, 204), is an important part of what drives this process. For vast sections of the population, both female and male, this means a constant search for income through wage work and 'self-employment' that is often thinly disguised wage work, in or away from the village (Breman 1996). In developing countries with extensive informal economies, off-farm activities are for the most part informal, even if a small fraction of workers obtain jobs in the formal economy (as public sector employees, usually frontline service providers in rural clinics and schools).

Diversification therefore captures several different economic processes and its blanket use to describe all forms of non-farm employment is misleading. It is particularly important, from the point of view of thinking about poverty, to distinguish between diversification as a survival strategy and diversification that feeds into a process of accumulation (Whitehead and Kabeer 2001). However, even when off-farm activities feed into a process of capital accumulation, as in the case of informalised and sub-contracted labour for large enterprises (be they horticultural or industrial), the worker may receive a wage that is below a living wage or the national minimum wage (if one exists). Compared to self-employment in low-return activities

---

[18]*Labor markets* are understood here very broadly to include both *wage* labour markets as well as *quasi* labour markets (where workers sell a product or a service, but within a set of dependent relationships that limit their authority over the employment arrangement).

such as beer brewing or small-scale trading, in such cases there is at least some scope for workers to organise and demand regular payments, piece-rates that are equivalent to the minimum wage, or capital to improve their work spaces.

Evidence from developing countries as a general rule suggests that men tend to be overrepresented in the top segments of the informal economy, while women tend to be overrepresented in the bottom segment; and the relative shares of men and women in the intermediate segments vary across sectors and countries (Chen et al. 2005, Chen 2009). Available evidence also suggests that there are significant gaps in earnings within the informal economy: informal employers have the highest earnings on average; followed by their employees and informal employees of formal firms; then own account operators, casual wage workers, and industrial outworkers. Research findings also suggest that it is difficult to move up these segments due to structural barriers, cumulative disadvantage and an 'economic order' that rests on what Barbara Harriss-White (2003) calls 'social regulation' (as opposed to state regulation) which by its very nature is exclusive and discriminatory. Many workers, especially women, remain trapped in the overcrowded, lower-earning and more risky segments (Chen 2009).

To explain gender disadvantage in the off-farm segment, Whitehead and Kabeer (2001) draw attention to the 'low reserve price' of women's labour. Women's reserve price of labour is likely to be low 'where the income potential of their own production is low, where the income generating opportunities off farm are few, or give low returns to labour, and where there is urgent need' (2001, 16). In addition to the capital constraints that many studies have drawn attention to, the 'distress' character of women's labour market engagement – the fact that women often engage in casual wage labour or other income-generating activities off-farm when needs have to be met urgently (for example when a debt has to be repaid or medical expenses have to be covered) – has been identified as a key factor that leads to a low reserve price. This is a hypothesis that deserves more empirical verification, especially given the competing view in the literature that attributes women's confinement to all forms of 'flexible' labour in the different sectors of the economy (casual agricultural labour, maquilas, industrial home work) to their 'domestic role within the family' and the fact that they 'have been socialised to have the flexibility to combine production and reproduction' (Flores 1991, cited in Deere 2005, 32). The latter proposition, in my view, overstates the choices that landless and land-poor women have when they commodify their labour, while it plays down the structures of constraint under which female labour commodification happens – constraints which arise from the urgency of needs (within the domestic realm), from the low income potential of their own production (if they are smallholders for example) as well as the social construction of women's labour as 'unskilled' or 'less skilled' (regardless of the actual content and complexity of tasks involved) which permeates rural labour markets.

Macro data for India suggests striking differences between trends in female and male employment. From 1975 there has been a general trend for male labour, throughout rural India, to move from agricultural wage labour to rural non-agriculture employment (Kapadia and Lerche 1999). In the context of the general stagnation in agriculture and agricultural employment that characterised the 1990s, while male participation in the rural non-agricultural sector has further increased, female employment has more or less remained stable and confined to agriculture. Livelihood diversification in rural areas seems therefore to have offered opportunities to men – as small businessmen and as employees in better paid and higher status jobs, while for women it has involved extremely lowly paid work as a last resort when agricultural wage work is absent (Jackson and Rao 2009, Gayathri 2005).[19]

Additionally, it has been claimed, for some parts of India at least (most notably Andhra Pradesh and Uttar Pradesh), that the assertiveness of male labour and their ability to free themselves from 'attached labour' arrangements, has inserted women even further into the casual labour force and sometimes into forms of 'unfree labour relations' (da Corta and Venkateshwarlu 1999, Lerche 1999). Here too it seems women frequently engage in casual wage labour under conditions of 'distress', in order to meet households' needs when men are either unable to find 'suitable' work or because their wages are not handed over for the households' collective needs (da Corta and Venkateshwarlu 1999). Hence, reliance on wives and female kin, who now shoulder an increasing share of the household budget through casual labouring, has assisted poor Scheduled Caste men in their struggles for dignity and decent wages vis-à-vis rural employers.

The gendered-differentiated pattern of workers struggles in the production process (and their intersections with domestic relations) in this context contrast with those that Hart (1991) found in rural Malaysia in the late 1980s. There female agricultural labourers were far more able than men to define themselves as workers and to organise and resist landlord attempts at individualising their contracts, while their husbands were often caught up in relations of deference and patronage with rural party bosses who wielded considerable political and economic power, which included access to off-farm resources and jobs on which male workers had become heavily dependent in the context of extensive mechanisation of agriculture and labour displacement.

Bottom-up development: the role of social movements and 'communities' in processes of decentralisation

A long history links feminism to practices of local democracy, grassroots organising and social movements contesting power structures, 'exposing previously unchallenged bias, and rewriting political agendas' (Phillips 2002, 77). The fact that women have often been visibly present in social movements, whether in protests organised by environmental groups or land invasions undertaken by landless movements, has not, however, guaranteed that their gender interests will be furthered nor that gender relations will be transformed (Agarwal 1998). It took more than a decade of activism by feminists in the Brazilian Movimento dos Trabalhadores

---

[19]The promotion of corporate farming for export has been happening on a much smaller scale in India and sub-Saharan Africa (although Kenya, South Africa and Zimbabwe may be exceptions), than in Latin America. The evidence analysed by Deere (2005) for Latin America suggests that corporate farming of non-traditional export crops has increased absolute employment opportunities for women in agriculture, given the highly labour-intensive nature of production and packing (compared to traditional agricultural exports such as coffee and bananas). Occupational segregation by gender contributes to lower wages for women. Yet while wages in the horticultural export sector may be low and conditions of work problematic (lack of health and safety conditions being a major concern), wages tend to be better than those prevailing elsewhere in the rural economy.

Rurais Sem Terra (MST), for example, for women's rights to be strongly articulated by rural social movements in Brazil (Deere 2003). Nor has it been easy for female labourers to get gender-specific issues seriously taken up by rural labour organisations, trade unions and left parties who have often had what one observer referred to as the 'liquidationist approach to the woman question', namely dealing with working women's problems primarily as class problems (Omvedt 1978, 391).

At the same time, gender analyses of agrarian change and institutions have been weary of an emerging consensus since the mid-1980s across the political spectrum, that sees 'community institutions' – be they village councils, 'traditional' authorities, or 'indigenous' institutions – as the more suitable managers of rural resources and as the more socially embedded locus of decision-making (as opposed to the distant, elite-dominated, if not 'predatory' central state).[20] While gender analysts have seen this as a welcome shift away from the 'top-down State-dependent approach' to natural resource and land management (Agarwal 1998, 56), the questions they have been asking is what shape these community institutions should be taking (should 'traditional' authorities and institutions be revived as some advocates have argued?) and what would this imply for *equality* of participation and benefit sharing (Agarwal 1998, 56, Whitehead and Tsikata 2003)?

In sub-Saharan Africa, much land distribution and land access is governed by locally managed systems of 'customary' rights. In the 1980s the international financial institutions identified the absence of private property rights in land as a barrier to agricultural growth, and gave full support to privatisation, titling and registration of land. Yet subsequent research carried out by the World Bank and the Land Tenure Centre at the University of Wisconsin showed insignificant differences in the productivity and investment of lands held in freehold title compared with those held under customary tenure (Mighot-Adholla *et al.* 1994, cited in Whitehead and Tsikata 2003, 82). Received wisdom within the World Bank's Land Policy Division has thus been swinging in favour of 'building on customary tenures and existing institutions' (World Bank 2003).

At the other end of the political spectrum, for some of the more progressive donor agencies and policy advocates such as OXFAM, subsidiarity and devolution are key objectives in current land reform policy. Given the history of political abuse and processes of land alienation and 'land grabbing' facilitated by national political elites, they claim that it is best that decisions on land management and control be taken at the lowest levels possible, 'closer to home' in the words of the Shivji Commission in Tanzania (Tsikata 2003). Here 'the local' is seen as a site of resistance against the state (and international capital). This approach fits the general support that many of these organisations and international NGOs provide for participation, building of local capacities, and local-level democracy.

Yet from the perspective of gender equality the local state is in a very 'ambiguous position' (Beall 2005). It is the part of the state that is located closest to 'the people' as many of the advocates of decentralisation claim, facilitating its engagement with women as well. Nevertheless precisely because of its closeness to society, the local state can also become too closely intertwined with social institutions and existing power structures. There is very little discussion, however, as Whitehead and Tsikata (2003) show, by those who support decentralisation as to how the proposed local

---

[20]This has been happening in a neoliberal climate that has been largely anti-state, at least in rhetoric if not in practice.

level systems might work in practice, including their capacity to deliver more equitable (and especially gender-equitable) resource allocation. Similar concerns were voiced by women's rights advocates in the Tanzanian debates to which reference has already been made citing a study which found female-headed households to be excluded from access to clan lands (interestingly, the same study also found that purchasing land was a possibility rarely exercised because of the lack of resources). Women were reported to be generally unhappy with the local administration bodies 'for reasons of corruption, under-representation of women and bias against them arising from prejudices and ideologies which cast them as less reliable protectors of clan land than men' (Rwebangira *et al.* (no date) cited in Tsikata 2003, 172).

Parallel apprehensions about local authorities were echoed in a series of consultation meetings that the National Land Committee (NLC) and the Programme for Agrarian Studies at the University of Western Cape (South Africa) held with rural communities about the Communal Land Rights Bill during 2002 and 2003 (Claassens 2005). Here too, many women recounted the difficulties they faced in trying to secure land allocations from 'traditional' leaders, especially if the claimant was a single mother or a widow. Traditional leaders often justified their reluctance to allocate land to single mothers by reference to the danger of 'outside' men gaining rights in the community via marriage (Claassens 2005, 18–21).

In the case of environmental management in India, Agarwal (1998) drew attention to a similar consensus among a number of environmental scholars and activists, including 'ecofeminists', which not only romanticised the pre-colonial period (as one of ecological stability and social harmony) but also put their uncritical reading of that idealised past to prescriptive use, by for example suggesting the revitalisation of caste groupings and other village communities and institutions and giving them a more central role in the management of common property resources. Field-based evidence from different parts of India showed that a wide range of initiatives (some state-initiated, others autonomously developed by village councils and elders, or NGOs) were in fact under way and directly involved local communities in the management of degraded forest land. While in terms of regeneration many of these initiatives had achieved notable successes, the results were 'far from impressive in terms of gender equity in resource management and benefit sharing' (Agarwal 1998, 75). Women's virtual absence from many forest protection groups and hence their lack of involvement in framing the new rules of access meant that they were either excluded from the forests or their access was made more difficult (even though their needs for fuel and fodder were often pressing).

This is not to suggest that 'customary' practices (in the case of land management) and 'community participation' (in environmental management) uniformly work to women's disadvantage. There are many cases of women using different courts and other dispute-settlement forums and using arguments grounded in either 'customary' or modernist principles, whichever is to their advantage, to make their claims (Whitehead and Tsikata 2003, 95). There are also cases of forest management committees where women are numerically preponderant and also vocal, hence underlining the point that social constraints are not insurmountable (Agarwal 1998). But outcomes depend not only on power dynamics on the ground, but also the *kinds of structures* that are being put in place.

It is difficult to insert gender-equality concerns (or broader social equality concerns) to these processes where traditional institutions have a patriarchal character, as in the case of 'chieftancies' that are being re-invigorated in some African countries. Women's rights advocates in South Africa have had serious apprehensions about the place given to traditional authorities in rural local government. The concerns are rooted in the ANC's attempts since 1994 to accommodate some of the demands of the Inkatha Freedom Party (IFP) with its stronghold in rural KwaZulu-Natal, as well as the ANC's reluctance to alienate the traditional authorities given their perceived importance for delivering rural constituencies to the party (Beall 2005). Given the fact that the 'traditionalism' that is espoused by the IFP and many of its adherents in the tribal authorities is deeply patriarchal (and a product of colonial and apartheid policies), these political maneuverings have effectively blunted the ANC's commitments to gender equity in rural affairs (Walker 2003). In 2000 the new Minister for Agriculture and Land Affairs, in notable contrast to the Ministry's modernising agenda for the commercial sector, advocated building on existing local institutions and structures in the communal areas (Walker 2005).

February 2004 saw the stormy passage through parliament of the long-delayed Communal Land Rights Act (CLRA), to the dismay of its critics (which included the Commission on Gender Equality as well as land sector NGOs and community groups). The critics argued that the legislation would entrench the powers of undemocratic 'traditional authorities' over communal land, fail to secure the tenure rights of women living on this land, and ultimately undermine the significant role that common property resources play in the livelihood strategies of the rural poor (Walker 2005, 297).

This was followed by another Act, the Traditional Leadership and Governance Framework Act (TLGFA) which deemed existing tribal authorities to be traditional councils, provided they meet new composition requirements within a year. It is likely that a lasting legacy of these two Acts will be 'to bolster the power of traditional leaders relative to that of the holders and users of family or individual land rights' and 'harden the terrain' within which rural people, and especially rural women, struggle for change (Claassens 2005, 41). There are widespread concerns that the two Acts pre-empt rural people's right to choose their own representatives on the same basis as people living in other parts of the country, and that because women's 30 percent quota does not need to be elected it could lead to the selection by traditional leaders of acquiescent females to sit on traditional councils (Claasens 2005, 41).

What kind of political dynamics then are being unleashed by the re-turn to 'the customary', and the revival of 'traditional' authorities? As Whitehead and Tsikata (2003) rightly warn, these could have highly disempowering implications for rural African women and their claims on resources. The main problem, as they argue, is that women have too little political voice at all the decision-making levels that are implied by the land question: not only within formal law and government, but also within local level management systems. The evidence cited above for South Africa provides glimpses into the power dynamics at the local level and raises a cautionary note about the limitations of 'traditional' institutions in delivering socially equitable, and in particular gender-equitable, outcomes. There is, however, a dire need for more grounded long-term research that can provide solid evidence to substantiate (or perhaps refute) such findings.

Questions regarding traditional authorities aside, there are many warnings across the literature that 'the local' can indeed be a site of unequal rural social relations, with implications for women and other less powerful social groups. Although there can be many benefits to decentralisation, its advocates often neglect the prevalence of inequitable repressive local power structures (Barraclough 2005, 34). Many women's movement activists in developing countries have eschewed state-centric politics, questioning whether women's rights could be substantiated by states that were fundamentally undemocratic in character. States are clearly not (gender) neutral in designing and implementing policies, but they are not self-evidently patriarchal either. Yet it also needs to be recognised that states may be more permeable to women's interests than local patriarchal institutions in contexts of strong social conservatism.

Concluding remarks

In this paper I have tried to trace some of the key contributions of feminist scholars to the analysis of agrarian social relations, institutions, and livelihoods and to explain where these contributions have intersected with other currents of thinking and scholarship, both neoclassical and political economy, and where there is scope for more theoretical and empirical work. One of the key contentions in this paper is that in an era when liberal and neo-liberal thinking has become hegemonic within the social sciences (as well as in the political and policy fields) an impoverished and methodologically individualist reading of gender – one that abstracts 'female' and 'male' rational choice actors from their broader political and social contexts – has been selectively taken up and elaborated by policy think tanks and economists versed in neo-classical economics. Policy prescriptions which were hitherto aimed at getting resources (land, credit, employment) into the hands of the male breadwinners, have been relatively painlessly revised to cater to the new 'economic woman', who is not only as hard-working as her male counterpart but also more likely to 'invest' the resources she commands on the welfare of her children (by all accounts a good investment in the now dominant view of equality which redefines equality in terms of 'equality of opportunity' rather than the old 'equality of outcome'). This has in turn exposed the dangers and limitations of the liberal methodology for gender analysis, and prompted a timely re-thinking of 'the household' and of gender relations more broadly by feminists themselves.

While the appreciation of power relations, of historical change, and of capitalist processes of impoverishment and accumulation in political economy analyses of agrarian change would suggest that a social relations analysis of gender would find a more hospitable terrain there (da Corta 2008), the spaces for intellectual cross-fertilisation have not been as expansive as one would have assumed (or wished for). As this paper has repeatedly suggested, in a world where 'markets' and 'states' are seen as the key institutional arenas of the macro political economy where capital accumulation takes place, and where 'domestic' spaces and relations and the unpaid economy (that produces labour) have no analytical significance and space, women (and gender issues) have become analytically most visible when they have entered the workforce. It is around the 'fragmentation of labour' therefore that some of the more useful exchanges between political economy of agrarian change and gender studies have materialised.

However, some of the same forces that have fostered the expansion of informality in labour markets have also deepened reliance on noncommodified forms of work for care and sustenance at least among those classes who cannot afford the commodified equivalents (hired domestic/care workers, private purchase of health and education), and fostered changes in strategies governing kinship, marriage and the constitution and dissolution of households. While it would be misleading to see these complex social transformations as being simply derived from economic relations and imperatives in a functionalist manner, it would also be wrong to see them as separate spheres. For feminist analysis it remains as important today as it was in the early 1980s to see households and their relations as 'integrated into and informed by the general analysis of changes in the global economy' and, yet, in a way that avoids crude analyses of gender relations made exclusively in terms of their function for capital (Pearson *et al.* 1981, x). It is high time that the political economy of agrarian change acknowledged 'the other economy' (Donath 2000) of uncommodified work, domestic institutions and social relations.

References

Agarwal, B. 1984. Women, poverty and agricultural growth in India. *The Journal of Peasant Studies*, 13(4), 165–220.
Agarwal, B. 1990. Social security and the family: coping with seasonality and calamity in rural India. *The Journal of Peasant Studies*, 17(3), 341–412.
Agarwal, B. 1994. *A field of one's own: gender and land rights in South Asia*. Cambridge: Cambridge University Press.
Agarwal, B. 1998. Environmental management, equity and ecofeminism: debating India's experience. *The Journal of Peasant Studies*, 25(4), 55–95.
Agarwal, B. 2003. Women's land rights and the trap of neo-conservatism: a response to Jackson. *Journal of Agrarian Change*, 3(4), 571–85.
Akintola, O. 2004. *A gendered analysis of the burden of care on family and volunteer caregivers in Uganda and South Africa*. Working Paper, Health Economics and HIV/AIDS Research Division, University of KwZulu-Natal, Durban.
Bardhan, P. 1974. On life and death questions: poverty and child mortality. *Economic and Political Weekly*, 9(32–34), 1293–304.
Barraclough, S. 2005. In quest of sustainable development. *OC No. 4*. Geneva: UNRISD.
Beall, J. 2005. Decentralising government and centralising gender in Southern Africa: lessons from the South African experience. *Occasional Paper 8*. Geneva: UNRISD.
Becker, G.S. 1981. *A treatise on the family*. Cambridge, MA: Harvard University Press.
Bennholdt-Thomsen, V. 1982. Subsistence production and extended reproduction: a contribution to the discussion about the modes of production. *The Journal of Peasant Studies*, 9(4), 241–54.
Bernstein, H. 2004. 'Changing before our very eyes': agrarian questions and the politics of land in capitalism today. *Journal of Agrarian Change*, 4(1–2), 190–225.
Bernstein, H. 2009. V.I. Lenin and A.V. Chayanov: looking back, looking forward. *The Journal of Peasant Studies*, 36(1), 55–81.
Berry, R.A. and W.R. Cline. 1979. *Agrarian structure and productivity in developing countries*. Baltimore, MD: The Johns Hopkins Press.
Blackden, M. and C. Bhanu. 1999. *Gender, growth and poverty reduction: 1998 Africa poverty status report*. Washington, DC: World Bank for the Special Program of Assistance for Africa.
Bramall, C. and M.E. Jones. 2000. The fate of the Chinese peasantry since 1978. *In:* D. Bryceson, C. Kay, and J. Mooij, eds. *Disappearing peasantries? Rural labour in Africa, Asia and Latin America*. London: ITDG, pp. 262–78.
Breman, J. 1996. *Footloose labour: working in India's informal economy*. Cambridge: Cambridge University Press.

Bruce, J. and D. Dwyer. 1988. *A home divided: women and income in the Third World.* Stanford, CA: Stanford University Press.

Bryceson, D.F. 1999. Sub-Saharan Africa bewixt and between: rural livelihood practices and policies. *ASC working paper 43.* Leiden: Africa-Studiecentrum.

Budlender, D. and F. Lund. 2007. South Africa research report 1: setting the context. *Political and social economy of care project, mimeo.* Geneva: UNRISD.

Byres, T. 2004. Neo-classical populism 25 years on: déjà vu and déjà passé. Towards a critique. *Journal of Agrarian Change,* 4(1–2), 17–44.

Caldwell, J.C. 1978. A theory of fertility: from high platteau to stabilisation. *Population and Development Review,* 4(4), 553–77.

Carney, J. 1988. Struggles over crop rights and labour within contract farming households in a Gambian irrigated rice project. *The Journal of Peasant Studies,* 15(3), 334–49.

Chant, S. 2008. The 'feminisation of poverty' and the 'feminisation' of anti-poverty programmes: room for revision?. *Journal of Development Studies,* 44(2), 165–97.

Chen, M. 2009. Informalisation of labour markets: is formalisation the answer?. *In:* S. Razavi, ed. *The gendered impacts of liberalisation: towards 'embedded liberalism'?, Routledge/UNRISD research in gender and development.* New York, NY: Routledge, pp. 191–218.

Chen, M.A., *et al.* 2005. *Progress of the world's women 2005: women, work, and poverty.* New York, NY: UNIFEM.

Claassens, A. 2005. The communal land rights act and women: does the act remedy or entrench discrimination and the distortion of the customary?. *Occasional paper no. 28.* Cape Town: Programme for Land and Agrarian Studies, University of the Western Cape.

Collier, P. 1989. *Women and structural adjustment.* Oxford: Oxford University, Unit for the Study of African Economics.

Cornwall, A., E. Harrison and A. Whitehead. 2007. Gender myths and feminist fabled: the struggle for interpretive power in gender and development. *Development and Change,* 38(1), 1–20.

Croll, E. 1983. Production versus reproduction: a threat to China's development strategy. *World Development,* 11(6), 467–81.

Croll, E. 1987. New peasant family forms in rural China. *The Journal of Peasant Studies,* 14(4), 469–99.

Da Corta, L. 2008. *The political economy of agrarian change: dinosaur or phoenix?* Unpublished Mimeo, London.

Da Corta, L. and D. Venkateshwarlu. 1999. Unfree relations and the feminisation of agricultural labour in Andhra Pradesh, 1970–95. *The Journal of Peasant Studies,* 26(2–3), 71–139.

Deaton, A. and J. Dréze. 2002. Poverty and inequality in India: a reexamination. *Economic and Political Weekly,* 37(36), 3729–48.

Deere, C.D. 1995. What difference does gender make? Rethinking peasant studies. *Feminist Economics,* 1(1), 53–72.

Deere, C.D. 2003. Women's land rights and social movements in the Brazilian agrarian reform. *Journal of Agrarian Change,* 3(1–2), 257–88.

Deere, C.D. 2005. The feminisation of agriculture? Economic restructuring in rural Latin America. *Occasional paper no.1.* Geneva: United Nations Research Institute for Social Development.

Deere, C.D. and M. Léon. 2001. *Empowering women: land and property rights in Latin America.* Pittsburgh, PA: University of Pittsburgh Press.

Deere, C.D. and M. Léon. 2003. The gender asset gap: land in Latin America. *World Development,* 31(6), 925–47.

Dey, J. 1981. Gambian women: unequal partners in rice development projects. *In:* N. Nelson, ed. *African women in the development process.* London: Frank Cass.

Donath, S. 2000. The other economy: a suggestion for a distinctively feminist economics. *Feminist Economics,* 6(1), 115–23.

Ellis, F. 1998. Household strategies and rural livelihood diversification. *Journal of Development Studies,* 35(1), 1–38.

Elson, D. 1998. The economic, the political and the domestic: businesses, states, and households in the organisation of production. *New Political Economy,* 3(2), 189–208.

Esping-Andersen, G. 1999. *Social foundations of postindustrial economies*. Oxford: Oxford University Press.

Fine, B. 2004. Social policy and development: social capital as point of departure. *In:* T. Mkandawire, ed. *Social policy in a development context*. Geneva: UNRISD.

Folbre, N. 1984. Market opportunities, genetic endowments and intrafamily resource distribution: comment. *American Economic Review*, 74(3), 518–20.

Folbre, N. 1986. Hearts and spades: paradigms of household economics. *World Development*, 14(2), 245–55.

Folbre, N. 1990. *Mothers on their own: policy issues for developing countries*. New York, NY: Population Council and International Centre for Research on Women.

Gayathri, V. 2005. Gender, poverty and employment in India. *Journal of Social and Economic Development*, 7(1), 29–52.

Gonzalez de la Rocha, M. 1988. Economic crisis, domestic reorganisation and women's work in Guadalajara, Mexico. *Bulletin of Latin American Research*, 7(2), 207–23.

Griffin, K.A., R. Khan and A. Ickowitz. 2002. Poverty and the distribution of land. *Journal of Agrarian Change*, 2(3), 279–330.

Harrison, J. 1973. The political economy of housework. *Bulletin of the Conference of Socialist Economists*, Winter, 35–52.

Harrison, M. 1977. Resource allocation and agrarian class formation: the problem of social mobility among Russian peasant households, 1880–1930. *The Journal of Peasant Studies*, 4(2), 127–61.

Harriss-White, B. 1990. The intrafamily distribution of hunger in South Asia. *In:* J. Dreze and A. Sen, eds. *The political economy of hunger. Volume I: entitlement and well-being*. Oxford: Clarendon Press, pp. 351–424.

Harriss-White, B. 2000. Work and social policy with special reference to Indian conditions. Paper presented at UNRISD conference on social policy in a development context, Tamsvik, Sweden, 23–24 September.

Harriss-White, B. 2003. Inequality at work in the informal economy: key issues and illustrations. *International Labour Review*, 142(4), 459–69.

Hart, G. 1991. Engendering everyday resistance: gender, patronage and production politics in rural Malaysia. *The Journal of Peasant Studies*, 19(1), 93–121.

Hart, G. 1995. Gender and household dynamics: recent theories and their implications. *In:* M. Quirbir, ed. *Critical issues in Asian development: theories and policies*. Oxford: Oxford University Press, pp. 39–73.

Hart, G. 1996. The agrarian question and industrial dispersal in South Africa: agro-industrial linkages through Asian lenses. *The Journal of Peasant Studies*, 23(2), 245–77.

Hewitt de Alcántara, C. 1993. Introduction: markets in principle and practice. *In:* C. Hewitt de Alcántara, ed. *Real markets: social and political issues of food policy reform*. Geneva: Frank Cass in association with EADI and UNRISD, pp. 1–16.

Jackson, C. 1996. Rescuing gender from the poverty trap. *World Development*, 24(3), 489–504.

Jackson, C. 2003. Gender analysis of land: beyond land rights for women?. *Journal of Agrarian Change*, 3(4), 453–80.

Jackson, C. and R. Palmer-Jones. 1999. Rethinking gendered poverty and work. *Development and Change*, 30(3), 557–83.

Jackson, C. and N. Rao. 2009. Gender inequality and agrarian change in liberalising India. *In:* S. Razavi, ed. *The gendered impacts of liberalisation: towards 'embedded liberalism'?, Routledge/UNRISD research in gender and development*. New York, NY: Routledge, pp. 63–98.

Kabeer, N. 1994. *Reversed realities: gender hierarchies in development thought*. London: Verso.

Kandiyoti, D. 1988. Bargaining with patriarchy. *Gender and Society*, 2(3), 274–90.

Kandiyoti, D. 2003. The cry for land: agrarian reform, gender and land rights in Uzbekistan. *Journal of Agrarian Change*, 3(4), 225–57.

Kapadia, K. and J. Lerche. 1999. Introduction. *The Journal of Peasant Studies*, 26(2–3), 1–9.

Klasen, S. and C. Wink. 2003. Missing women: revisiting the debate. *Feminist Economics*, 9(2–3), 263–99.

Kynch, J. and M. McGuire. 1994. Food and human growth in Palanpur. *STICERD discussion paper no. 57*. London: London School of Economics and Political Science.

Lastarria-Cornhiel, S. 1997. Impact of privatisation on gender and property rights in Africa. *World Development*, 25(8), 1317–33.

Lerche, J. 1999. Politics of the poor: agricultural labourers and political transformation in Uttar Pradesh. *The Journal of Peasant Studies*, 26(2–3), 182–241.

Lockwood, M. 1992. Engendering adjustment or adjusting gender? Some new approaches to women and development in Africa. *Discussion paper no. 315*. Brighton: IDS.

Mackenzie, F. 1990. Gender and land rights in Murang'a District, Kenya. *The Journal of Peasant Studies*, 17(4), 609–43.

Mackintosh, M. 1990. Abstract markets and real needs. *In:* H. Bernstein, *et al.*, eds. *The food question: profits versus people?*. London: Earthscan, pp. 43–53.

Manji, A. 1998. Gender and the politics of the land reform processes in Tanzania. *Journal of Modern African Studies*, 36(4), 645–67.

Meena, R. 2008. General description of nurses and home-based care givers. *Research report 4: Tanzania, political and social economy of care project, Mimeo*. Geneva: UNRISD.

Miller, B. 1981. *The endangered sex: neglect of female children in North India*. Ithaca, NY: Cornell University Press.

Molyneux, M. 1979. Beyond the domestic labour debate. *New Left Review*, I(116).

Moore, H. 1994. Is there a crisis in the family? *Occasional paper no. 3 WSSD*. Geneva: UNRISD.

O'Laughlin, B. 1998. Missing men? The debate over rural poverty and women-headed households in Southern Africa. *The Journal of Peasant Studies*, 25(2), 1–48.

O'Laughlin, B. 2007. A bigger piece of a very small pie: intrahousehold resource allocation and poverty reduction in Africa. *Development and Change*, 38(1), 21–44.

O'Laughlin, B. 2008. Gender justice, land and the agrarian question in Southern Africa. *In:* H. Akram-Lodhi and C. Kay, eds. *Peasants and globalisation*. New York, NY: Routledge, pp. 190–213.

Omvedt, G. 1978. Women and the rural revolt in India. *The Journal of Peasant Studies*, 5(3), 370–403.

Palmer, I. 1991. *Gender and population in the adjustment of African economies: planning for change*. Geneva: ILO.

Pearson, R., A. Whitehead and K. Young. 1981. Introduction: the continuing subordination of women in the development process. *In:* K. Young, C. Wolkowitz and R. McCullagh, eds. *Of marriage and the market: women's subordination internationally and its lessons*. London: Routledge, pp. ix–xix.

Peng, D. and D.R. Phillips. 2004. Potential consequences of population ageing for social development in China. *In:* P. Lloyd-Sherlock, ed. *Living longer: ageing, development and social protection*. London: UNRISD/Zed, pp. 97–116.

Phillips, A. 2002. Does feminism need a conception of civil society?. *In:* S. Chambers and W. Kymlicka, eds. *Alternative conceptions of civil society*. Princeton, NJ: Princeton University Press, pp. 71–89.

Platteau, J.-P. 1995. Reforming land rights in sub-Saharan Africa: issues of efficiency and equity. *Discussion paper no. 60*. Geneva: UNRISD.

Polanyi, K. 1957. *The great transformation*. Boston, MA: Beacon Press.

Razavi, S. 1999. Gendered poverty and well-being: introduction. *Development and Change*, 30(3), 409–33.

Razavi, S. 2003. Introduction: agrarian change, gender and land rights. *Journal of Agrarian Change*, 3(1–2), 2–32.

Razavi, S. 2007. Liberalisation and the debates on women's access to land. *Third World Quarterly*, 28(8), 1479–500.

Razavi, S. and C. Miller. 1995. From WID to GAD: conceptual shifts in the women and development discourse. *Occasional paper for Beijing no. 1*. Geneva: UNRISD.

Rosenzweig, M.R. and T.P. Schultz. 1982. Market opportunities, genetic endowments and intrafamily resource distribution: child survival in rural India. *American Economic Review*, 72(4), 803–15.

Scoones, I. 2009. Livelihoods perspectives and rural development. *The Journal of Peasant Studies*, 36(1), 171–96.

Sen, A.K. 1990. Gender and cooperative conflicts. *In:* I. Tinker, ed. *Persistent inequalities*. Oxford: Oxford University Press, pp. 123–49.

Sender, J. and D. Jonson. 2004. Searching for a weapon of mass production in rural Africa: unconvincing arguments for land reform. *Journal of Agrarian Change*, 4(1–2), 142–64.

Sender, J., C. Oya and C. Cramer. 2006. Women working for wages: putting flesh on the bones of a rural labour market survey in Mozambique. *Journal of Southern African Studies*, 32(2), 313–33.

Sharma, M. 1985. Caste, class, and gender: production and reproduction in North India. *The Journal of Peasant Studies*, 12(4), 57–88.

Statistics South Africa 2007. *General household survey 2006*. Statistical Release PO318, July. Pretoria: Statistics South Africa.

Statistics South Africa 2008. *Income and expenditure of households 2005/6*. Statistical Release P0100. Pretoria: Statistics South Africa.

Tripp, A.M. 2004. Women's movements, customary law, and land rights in Africa: the case of Uganda. *African Studies Quarterly*, 7(4), 1–19.

Tsikata, D. 2003. Securing women's interests within land tenure reforms: recent debates in Tanzania. *Journal of Agrarian Change*, 3(1–2), 149–83.

Udry, C. and H. Alderman. 1995. Gender differentials in farm productivity: implications for household efficiency and agricultural policy. *Food Policy*, 20(5), 407–23.

Walker, C. 2003. 'Piety in the sky?' Gender policy and land reform in South Africa. *Journal of Agrarian Change*, 3(1–2), 113–48.

Walker, C. 2005. Women, gender policy and land reform in South Africa. *Politikon*, 32(2), 297–315.

White, C.P. 1986. Everyday resistance, socialist revolution and rural development: the Vietnamese case. *The Journal of Peasant Studies*, 13(2), 49–63.

Whitehead, A. 1981. 'I'm hungry mum': the politics of domestic budgeting. *In:* K. Young, C. Wolkowitz and R. McCullagh, eds. *Of marriage and the market*. London: Routledge.

Whitehead, A. 1990. Rural women and food production in sub-Saharan Africa. *In:* J. Dreze and A. Sen, eds. *The political economy of hunger. Volume I: entitlement and well-being.* Oxford: Clarendon Press, pp. 425–73.

Whitehead, A. 2001. Trade, trade liberalisation and rural poverty in low-income Africa: a gendered account. Background paper for the UNCTAD 2001 least developed countries report. Geneva: UNCTAD.

Whitehead, A. 2009. The gendered impacts of liberalisation politics on African agricultural economies and rural livelihoods. *In:* S. Razavi, ed. *The gendered impacts of liberalisation: towards 'embedded liberalism'?, Routledge/UNRISD research in gender and development.* New York, NY: Routledge, pp. 37–62.

Whitehead, A. and N. Kabeer. 2001. Living with uncertainty: gender, livelihoods and pro-poor growth in rural sub-Saharan Africa. *IDS working paper 134*. Brighton: Institute of Development Studies.

Whitehead, A. and D. Tsikata. 2003. Policy discourses on women's land rights in sub-Saharan Africa: the implications of the return to the customary. *Journal of Agrarian Change*, 3(1–2), 67–112.

World Bank 2001. *Engendering development through gender equality in rights, resources and voice*. New York, NY: Oxford University Press.

World Bank 2003. Land policies for growth and poverty reduction. *A World Bank policy research report*. Washington, DC: World Bank.

Everyday politics in peasant societies (and ours)

Benedict J. Tria Kerkvliet

Politics in peasant societies is mostly the everyday, quotidian sort. Hence, if one looks only for politics in conventional places and forms, much would be missed about villagers' political thought and actions as well as relationships between political life in rural communities and the political systems in which they are located. To make this case, the paper explains the meaning of everyday politics by distinguishing it from two other realms of politics and then discusses four types of everyday politics. The paper concludes by suggesting that, in addition to better understanding peasant societies, the concept of everyday politics makes us – researchers – further aware of, and realise the importance of, our own everyday political behaviour.

Introduction

The argument here is that everyday politics is a fruitful realm of study in peasant studies. Much of politics in peasant societies is of the quotidian sort. Hence, if one looks for politics only in the usual places and forms, as conventional political studies would direct, much would be missed about villagers' political thought and actions as well as relationships between political life in rural communities and the political systems in which they are located. To make this case, the paper explains the meaning of everyday politics by distinguishing it from two other realms of politics. Then the paper identifies four types of everyday politics, ranging from support for the status quo to resistance. In the course of this discussion, the paper shows that studying everyday politics in peasant societies helps us to discover and understand people's political views and analyse the impact of everyday politics on the other realms of politics. The paper concludes by suggesting that another benefit of examining everyday politics in peasant societies is that it makes us – researchers – aware of, and realise the importance of, our own everyday political behaviour.

Politics and conventional studies of it

Politics, according to one pithy definition, is about who gets what, when, and how (Lasswell 1958). To elaborate a bit, politics is about the control, allocation, production, and use of resources and the values and ideas underlying those activities.[1] Resources include land, water, money, power, education, among other

I thank Rosanne Rutten for her helpful suggestions.
[1]For this or similar conceptions of politics, see Ball (1993, 20–1), Leftwich (1984, 63–70), Miller (1993, 99–100), Stoker (1995, 5–7).

tangible and intangible things. Behavior regarding producing, distributing, and using resources can range from cooperation and collabouration to discussions and debates to bargains and compromises to conflicts and violence. Participants in such behaviour can include individuals, groups, organisations, and governments.

Defined this way, politics could be everywhere. Conventional political studies, however, usually limit investigation into the questions and processes implied in this view of politics to governments, states, and the organised efforts to influence what those two institutions do or to change them altogether. Hence, conventional studies pretty much restrict their examinations of the control, allocation, and use of resources and values underlying them to the activities of state authorities and agencies, political parties and their supporters, elections, organisations and individuals lobbying or otherwise trying to influence government officials and policies, and to movements, rebellions, and revolutions challenging existing governments and states or proposing different ones.

Citizens in many parts of the world also typically relegate politics to what government officials, politicians, lobbyists, and the like do. Journalists and other public commentators also usually talk and write this way, thereby deliberately or unintentionally encouraging and perpetuating this small arena for politics. Another strong tendency among people in the United States, Australia, the Philippines, to name just a few countries, is to link politics with unbecoming, nefarious, and other negative actions, usually involving government officials and agencies but sometimes referring to real or perceived unsavory actions in other aspects of life, such as when a disgruntled faculty member recently explained to me that the reason he did not get promoted was 'politics in the university's committee on promotions'.

This conventional view of politics has several shortcomings. One is the negative image just referred to, which marginalises the many positive and constructive aspects of political activities. Occasionally news accounts and politicians speak of 'politics' and 'political' in a positive manner, such as reporting that conflicts between Palestinians and Israelis require 'political solutions', not military ones. Far more prevalent, however, are off-putting characterisations of politics, obscuring the necessity, really, of politics. A society, any society, needs to devise ways to allocate resources. Those ways and the process of determining them are political. The allocation can be done badly, of course, but it can also be done well and somewhere in between.

Second, conventional political studies and commentary concentrates on a minute fraction of any country's population – primarily government officials, political parties, influential individuals, and activists in organisations trying to affect what government authorities do.[2] That number would be especially tiny in countries with political systems that allow citizens few opportunities to organise or speak openly about public issues. By excluding a huge percentage of populations in a country, the conventional view also puts politics in a realm far from most citizens – virtually relegating politics to places or arenas that few in society have any connection with or involvement in.[3]

---

[2] For an elaboration, which highlights the biases of viewing politics narrowly, see the stinging critique in Elshtain (1997, 12–35).

[3] A study of this view, which is also a marvelous investigation of Japanese women's civic and community service activities, finds that women in Tokyo see 'politics' as something far away and inaccessible to most people: 'Politics exists apart from even the housewives' most public actions – a fortress that cannot be broached from ordinary society'. Politics to them is essentially a private activity of a few unreachable, virtually unknowable people (LeBlanc 1999, 73, 84).

Third, such a restricted view of politics misses a great deal of what is politically significant. The allocation of important resources is rarely confined to governments and related organisations. Resources are distributed in corporations, factories, universities, religious groups, families, and other institutions. Conventional political analysis gives minimal attention to the political processes and significance for resource use, production, and distribution within these and other non-state entities. And what attention is given is usually only about the interaction between such entities and state or government officials and agencies.

Fourth, the conventional view privileges organisations that are explicitly political or seeking to influence politicians, governments, and state agencies. Yet, as emphasised already, political issues, processes, and outcomes occur far beyond these organisations. Moreover, people need not be organised to be political. Individuals and groups ponder and wrestle with problems and issues regarding the use, production, and distribution of resources among themselves, for their communities, as well as by or for their governments.[4]

Some of these problems with conventional understanding of politics have loomed large for me while doing research in the rural areas of the Philippines and Vietnam.

In the late 1970s, my wife and I lived for a year in San Ricardo, a village of about 230 households in Nueva Ecija, a Central Luzon province in the Philippines. During the 1980s and 1990s we returned several times for shorter stays. Nearly forty percent of the village's households relied primarily by farming, mostly rice, for their income; another forty percent had members who farmed but also worked in other sectors of the local and national economy as truck drivers, carpenters, maids, etc. The remaining households included non-agricultural workers, clerks, teachers, petty traders, small business people, and a few wealthy capitalists. I wanted to know about the political life of the village. More specifically, I wanted to understand what had happened in San Ricardo and surrounding villages following a large rebellion that many people in the area had directly participated in or had supported during the 1940s–50s. The causes of the rebellion, like most other violent uprisings, were complex, but central issues revolved around acrimonious relations between large landowners and their tenants and farm workers. I wanted to learn what had become of the organisations and people of that rebellion, what was the impact of the government's agrarian reform programs after the rebellion, how did those programs and the 'green revolution' in rice growing affect land tenure and villagers' lives, and what were relations like among villagers and between them and government authorities.

From a conventional perspective of politics, political life in San Ricardo was virtually morbid for most of the time between the early 1970s and mid 1980s. During that period the Philippines was under authoritarian rule, which President Ferdinand Marcos imposed in 1972. The regime allowed only government authorised organisations, in which few San Ricardo residents were active. The periodic elections during those years were ritualised affairs that permitted no opposition political parties. The village's local government had practically no money, few other resources, and little or no power. About all the village head and his council did was to mobilise residents to comply with the central government's programs to beautify villages and form security patrols.

---

[4]For clear arguments along this line, see Piven and Cloward (1977, 4–5), and Scott (1987, 418–23).

Yet, San Ricardo and its neighboring villages pulsated with discussions, debates, and other activity about access to land, grain, wages, and credit, about the availability of education and jobs, about better off families' obligations to their poor relatives and neighbors and vice versa, about the pros and cons of poor people being clients of local and distant wealthy people, about the adequacy or not of tenancy conditions, about how employers treated workers, about what is just and unjust, and many other issues about the control, allocation, and use of important resources and the values and ideas underlying those arrangements. People also discussed and debated among themselves various government policies and programs and the legitimacy or not of the entire Marcos regime. In short, political issues permeated everyday village life.[5]

Rarely was any of this political commotion manifested in public or in organised ways. Hence, looking for politics in conventional places and ways, one would have missed most of the political ferment in and around the village. And being unaware of that would seriously handicap anyone trying to understand the occasional times when political issues and activities suddenly became public and organised (a point I shall return to later).

In Vietnam, political analysis of a conventional sort would leave one confounded as to how to explain an amazing reversal in national policy. For years, beginning in the late 1950s, the Communist Party government in Vietnam – initially in the north then later in the south, after the war with the United States and the country's reunification – had enticed, cajoled, and sometimes forced peasant households to join cooperatives in which they had to farm rice and other crops collectively. Most cooperatives encompassed not only the households in one village but two, three, or even more villages. Each cooperative collectivised land as well as draft animals and labour. Officials clustered individuals into work teams and assigned each team specific farming and other tasks throughout the year. To Vietnam's national leadership, collective farming was a key institution to developing a socialist economy.

Rarely, however, were the nation's leaders pleased with how most cooperatives operated. Despite their many campaigns to strengthen the organisations and improve collective farming, problems persisted and, in the late 1970s–early 1980s, accelerated. Numerous collective farming cooperatives disintegrated. Land and draft animals reverted to individual households who resumed farming as family units, not as team members in a collective. By the mid 1980s, a large proportion of the cooperatives were phony; they had the appearance of being collective farms but actually individual households were doing much of the farming on land that had been divided up among them. Between December 1987 and April 1988, national government and Communist Party pronouncements officially marked the end of collective farming policy and endorsed family farming.

How did this remarkable reversal of a major program of the Communist Party government come about? That was the main question I researched, primarily in the Red River delta, the heartland of socialist Vietnam from the 1950s–80s, by talking to villagers, reading reports, interviewing policy makers, and using other sources. The policy transformation occurred without social upheaval, without violence, without a change in government, without even organised opposition. In other words, there was no apparent activity that conventional politics would recognise. Yet, national

---

[5]For elaboration, see Kerkvliet (1990).

authorities were pressured into giving up on collective farming and allowing family farming instead. To a significant degree, that pressure came from everyday practices of villagers in the Red River delta and other parts of northern Vietnam. To a considerable extent, those practices were political because they involved the distribution and control of vital resources. Those everyday political practices were often at odds with what collective farming required, what authorities wanted, and what national policy prescribed. Consequently, everyday politics put collective farming, authorities, and policy under enormous strain. A major reason why the government did a 180-degree policy turn in the late 1980s was the weakening and eventual collapse from within of the collective farming cooperatives. The policy change was less about dissolving collective farming and more about approving what was well underway in many places. Put simply, decollectivisation started locally, in the villages and largely by villagers; national policy followed behind.[6]

In both of the cases that I have quickly reviewed – San Ricardo village in Central Luzon, the Philippines, and collective farming cooperatives in Vietnam's Red River delta – the control, allocation, and use of resources – particularly land but also rice, draft animals, credit, jobs – were vital daily concerns to peasants and other villagers. So were the ideas and values underpinning how resources were produced, used, and distributed: family farming, collective farming, proper working conditions, appropriate tenancy conditions, justice. The concerns of ordinary people were not manifested in organisations that pressured government officials; they were not encapsulated by political parties or revolutionary movements; nor were people even saying much publicly and openly about these matters. Thus, to conventional political studies, they are invisible and do not matter. Even though they involve essential political questions – who gets what, when, and how – they do not count as politics, in the eyes of most political analyses, because they are not expressed in the manner and places conventional political studies expect.

Studies of peasant societies have helped to change the conventional perception of politics by demonstrating and analysing the significance of everyday politics for villagers. Before elaborating this point, I need to explain what everyday politics means.

Types of politics

To help to define everyday politics, it is useful to distinguish it from other types of politics. Two others are notable here; a more elaborate taxonomy is possible but unnecessary for this discussion. Official politics is one of those two. It involves authorities in organisations making, implementing, changing, contesting, and evading policies regarding resource allocations. The key words here are authorities and organisations. Authorities in organisations are the primary actors. The organisations include governments and states, but are not limited to those forms. Official politics also occurs, for example, in churches, universities, corporations, political parties, non-government organisations, labour unions, peasant associations, and revolutionary organisations. The people involved in official politics hold authoritative positions; they are authorised to make decisions or have a substantial hand in an organisation's decision making and implementation processes. Official political activities can range from public to private (even secret) spheres. For

---

[6]For more, see Kerkvliet (2005).

instance, government authorities might make budget decisions in an open, transparent manner, behind closed doors, or gradations in between. Official political activities also can range from formal to informal to illegal activities.[7] For example, authorities following the rules for collecting and spending membership dues is official politics; but so is authorities embezzling money belonging to the organisation.

Advocacy politics is another type of politics. It involves direct and concerted efforts to support, criticise, and oppose authorities, their policies and programs, or the entire way in which resources are produced and distributed within an organisation or a system of organisations. Also included is openly advocating alternative programs, procedures, and political systems. The key words here are direct and concerted. Advocates are straightforwardly, outwardly, and deliberately aiming their actions and views about political matters to authorities and organisations, which can be governments and states but need not be. Advocacy behaviour can extend from friendly, civil, and peaceful to hostile, rebellious, and violent. Advocates may be individuals, groups, or organisations. Usually advocacy politics is public, but some advocacy politics, such as that of a revolutionary organisation, may have clandestine features.

Conventional political studies are largely limited to aspects of official and advocacy politics. Typically, conventional political studies emphasise only a narrow range of organisations and authorities, especially those pertaining to governments, states, political parties, revolutionary movements, and groups or associations aimed at supporting, influencing, or opposing government officials and programs. Far fewer are studies of official and advocacy politics regarding other types of organisations, such as corporations, religious organisations, and unions.

Everyday politics involves people embracing, complying with, adjusting, and contesting norms and rules regarding authority over, production of, or allocation of resources and doing so in quiet, mundane, and subtle expressions and acts that are rarely organised or direct.[8] Key to everyday politics' differences from official and advocacy politics is it involves little or no organisation, is usually low profile and private behaviour, and is done by people who probably do not regard their actions as political. It can occur in organisations, but everyday politics itself is not organised. It can occur where people live and work. Often it is entwined with individuals and small groups' activities while making a living, raising their families, wrestling with daily problems, and interacting with others like themselves and with superiors and subordinates. Everyday politics also includes resource production and distribution practices within households and families and within small communities in ways that rely primarily on local people's own resources with little involvement from formal organisations.

---

[7]For some analysts, the meaning of informal politics may be broad enough to include what I call everyday politics. Most usage, however, restricts informal politics to types of behaviour among government leaders, bureaucrats, and other public officials. See the chapters in Dittmer *et al.* (2000).

[8]This synthesis of everyday politics is somewhat narrower than what John Hobson and Leonard Seabrooke (2007, 15–16) propose. For them, sizeable organisations can be among the actors in everyday politics. On the other hand, my synthesis is broader and different in other ways from Harry Boyte's recent discussion of everyday politics (2004). In an inspirational book, Boyte uses 'everyday politics' to advocate a renewal of democracy in the United States by de-professionalising 'politics', thereby reconnecting ordinary citizens to public life and getting people involved in doing public work for the good of their communities and nation.

As with most efforts in the social sciences to distinguish among similar social phenomena, the lines between these three types of politics can never be stark and bold. Overlapping and blurred boundaries are unavoidable. Such lack of precision, however, does not negate the value of drawing attention to the political significance of everyday practices.

Forms and significance of everyday politics

As the above conceptualisation of everyday politics suggests, it can take various forms, which I cluster under four headings: support, compliance, modifications and evasions, and resistance. The forms most studied in peasant societies are everyday resistance. Resistance refers to what people do that shows disgust, anger, indignation or opposition to what they regard as unjust, unfair, illegal claims on them by people in higher, more powerful class and status positions or institutions. Stated positively, through their resistance, subordinate people struggle to affirm their claims to what they believe they are entitled to based on values and rights recognised by a significant proportion of other people similar to them. While this definition may not have universal agreement among scholars studying peasant resistance, most agree on two essential points – intention and upward orientation. Resistance involves intentionally contesting claims by people in superordinate positions or intentionally advancing claims at odds with what superiors want. Acts at the expense of other people who are in the same or similar boat is not resistance. And behaviour by people in higher class or status positions to counter the actions of subordinates is not resistance. That is often repression or coercion. How subordinates resist can vary from organised and confrontational forms, such as peasant demonstrations and rebellions, to less elaborate but still direct and confrontational action, such as peasants boldly taking over land they claim belongs to them or petitioning authorities or other superiors to meet their demands; to subtle, indirect, and non-confrontational behaviour. The last is where everyday forms of resistance is found.

Everyday resistance involves little or no organisation; and the persons or institutions targeted by the resistance typically do not know, at least not immediately, what has been done at their expense (however meager it may be). The nasty, derogatory things peasants say or the jokes they crack about their landlords, employers, government officials, or the like behind their backs can be forms of everyday resistance. Other examples are peasants pilfering grain or tools from egregious landowners and employers; harvesting in the dead of night fruit or grain belonging to abusive officials; and surreptitiously setting fire to equipment, such as tractors or combines, that displaced people who are desperate for employment.

One of the first studies, at least in English, honing in on everyday forms of resistance draws out the political significance of peasants in colonial Burma and Java moving from place to place and doing other furtive things arising from their discontent while avoiding clashes with their oppressors (Adas 1981, also see Adas 1986). The most influential study has been James C. Scott's multi-layered analysis (1985) of a village in Malaysia. Scott's analysis brings out the political significance of verbal remarks, villagers' private characterisations of their superiors, people's recollections of past events, as well as pilfering, sabotage, and other everyday resistance. His research shows that daily life is rife with class struggle that only occasionally bursts into the open.

These studies have stimulated considerable research by scholars on everyday resistance in peasant societies in many other Southeast Asian countries; in China, India, and elsewhere in Asia; and in Africa and Latin America. The concept of everyday resistance also travels well when studying political behaviour and views of people in other sectors of society, not just peasants, who are in relatively weak and subordinate positions – office secretaries, factory workers, clerks, street vendors, and so on.[9]

Literature on everyday resistance in peasant societies contributes to debates about hegemony. A pronounced finding is that outward signs of peasants' quiescence or acceptance of impoverishment, exploitation, and the like cannot be taken at face value. Researchers find that often such appearances are facades hiding different, often contrary views and actions that grow from peasants' discontent and antipathy to how higher status, more powerful people and institutions treat them. Rather than accepting the status quo, peasants often harbor alternative visions, values and beliefs for how resources should be produced, distributed, and used.

Everyday resistance studies also help to understand better the emergence of peasant organisations, movements, and rebellions. Stated generally, peasants' everyday forms of resistance and the thinking underlying their hidden criticisms of prevailing political conditions can feed into confrontational forms of advocacy politics. A stark example occurred in San Ricardo. As mentioned earlier, discussions and debates about how land should be used and by whom are rife in everyday life in this Philippine village. One aspect of that political ferment in the 1970s and 1980s was landless villagers complaining among themselves about non-resident large landowners who left virtually unused over 100 hectares. To the villagers' way of thinking, they should be allowed to farm that area because they were poor, they needed land in order to make a living, and the landowners were wealthy and obviously had the means to help them by allowing them to at least rent the land. The villagers fantasised about unilaterally taking some of the land and farming it. They dared not do that, however, for fear the owners would immediately have police and soldiers arrest them, maybe even shoot them. But on a couple of occasions in the 1980s, seeing changed power configurations that appeared to advantage them and weaken the absent landowners, a number of poor villagers acted on their fantasies and began to farm several hectares of the vacant land.

Instances of everyday resistance being important precursors of open, confrontational, advocacy forms of politics have occurred in numerous rural societies. Among them are other land take-overs (Kerkvliet 1993). And there are many other forms. In China, to take recent examples from just one country, villagers' covert everyday resistance against corrupt local officials, taxes, government confiscation of land, and other adverse conditions in the mid 1980s developed into overt collective action in the late 1980s and 1990s. Angry villagers prepared and sent letters of complaints to authorities, demanded face-to-face meetings with officials, protested in front of local government offices, marched long distances to petition higher government authorities, blocked roads, occupied and even destroyed government buildings, and took officials hostage. In some parts of rural China, even rebel organisations have developed (Walker, K., 2008, 467, 469–70, also see O'Brien and Li 2006).

---

[9]See, for instance, Scott (1990).

Scholars have investigated the conditions under which everyday resistance might feed into open, organised, and confrontational resistance. One important reason appears to be changing political circumstances that favour peasants and disfavour individuals and agencies peasants have been surreptitiously resisting. Such altered circumstances help subordinates to 'cross the threshold of fear and insecurity' (Adnan 2007, 204, 214). In Bangladesh, a growing appreciation of their collective strength in numbers encouraged peasants to express their real views openly and support the candidates running for parliament whom they preferred rather than the ones supported by people to whom they were beholden (Adnan 2007, 214–15). Emboldening landless villagers in San Ricardo and elsewhere in the Philippines at various times in the 1980s to occupy land that owners had left vacant was support they thought they had from some key local authorities at a time when the national government was unstable. In China, a spate of central government pronouncements, laws, and other actions in the 1980s created, perhaps to some extent unintentionally, some opportunities and avenues for villagers to publicly object to abusive behaviour and taxes from local officials (Walker, K., 2008, 269–70, O'Brien and Li 2006, 26–49). Another condition is the emergence of leaders and groups who are able to 'frame' discontent and resistance in ways that enable peasants and agrarian workers to overcome or set aside reluctance and fear so as to begin to band together and confront collectively authorities and other powerful entities. Such framing typically involves joining local ideas and experiences regarding oppression and resistance with broader ones. The importance of this condition has been shown in recent studies of how peasant-based revolts and revolutions have developed.[10]

While considerable research has been done over last twenty-five years on everyday resistance in peasant societies, far less has been done on other forms of everyday politics. Like society generally, peasant societies are rife with relationships between people in unequal social, status, and class positions and between citizens and government or state authorities. Besides possible resistance, village life has everyday forms of support for and compliance with those relationships and their roles in the production, distribution, and use of resources. Village societies also teem with everyday forms of support and compliance with government authorities and prevailing political systems. The extent of intentionality marks the extent to which such everyday practices are support or compliance. Support involves deliberate, perhaps even enthusiastic endorsement of the system. Compliance is more a matter of going through the motions of support without much thought about it. Unlike everyday forms of resistance, everyday forms of support and/or compliance in rural societies of Asia, Africa, and Latin America can include daily practices of people in superordinate positions, such as landlords and employers, as well as subordinates, such as tenants and labourers.

One large realm of everyday political support and compliance in agrarian societies involves interpersonal relations within households and families, among neighbours and others in a village or community, between employers and employees, landowners and their labourers and tenants, etc. In San Ricardo and many other villages in the Philippines, creating and maintaining networks in order to have access

---

[10]For examples of studies in which this and other conditions are important explanations, see Collier (1997), Mason (2004), Rutten (1996), and Wood (2003). Also see Baud and Rutten (2004).

to land, labour, money, and emergency assistance is a big part of people's everyday politics. Families with little or no land, employment, or income regularly seek and cultivate favorable relations with people in the village or elsewhere who have land, businesses, money, and connections to still other individuals with these resources. And vice versa; people with means typically like to have a network of villagers on whom they can depend to provide services – plow and harvest their fields, tend their gardens, wash and iron their clothes, cook meals, serve guests during parties and holiday feasts, drive their vehicles, labour in their shops – without needing to pay them much nor even regularly employ them. These exchange networks and patron–client relationships, like everyday resistance, are largely private matters, unorganised, and go on pretty much day in and day out.[11] And like everyday resistance, they involve the production, distribution, and use of resources – they are political. But they are not resistance. They are the opposite; they reinforce class and status differences and help to perpetuate a political system in which inequalities, personal relationships and dependencies are endemic.[12]

Everyday support and compliance in a political system can also involve behaviour that is perhaps less about personal relationships and more about particular authorities, governments, and regimes. In a village in northern Thailand, Andrew Walker finds that discourses about government policies, elections, candidates, political party campaigns are a 'regular feature of day-to-day life. Electoral contests are embedded in local social relationships, and values that relate to the day-to-day politics of the village readily spill over into the electoral arena' (2008, 87). In central Thailand, peasants, labourers, shopkeepers, truck drivers, and other ordinary people regularly talk with relatives, friends, neighbours, customers, and others about local officials, government policies, public projects, and other such matters. Particularly important, argues Yoshinori Nishizaki (2005, 2008), is the pride people have developed in their province, Suphanburi, which now has roads, schools, hospitals, and other facilities envied by Thais living elsewhere. Much of the credit for these improvements, Suphanburi residents say, goes to Banharn Silpa-archa, a politician who, according to many scholars and journalists, is a corrupt, dirty, 'godfather' type politician. But a large proportion of Suphanburi people privately praise him because of what he has done to improve their province. Their positive view of him often feeds into advocacy politics, such as when people openly campaign for him during elections and publicly defend him against his critics.

More in the realm of compliance than support are daily activities that help to sustain authority and political systems. VaôÇclav Havel writes eloquently about this in his analysis of power and politics in communist-ruled Czechoslovakia. Using the example of a greengrocer who day after day 'places in his window, among the onions and carrots, the slogan, "Workers of the World, Unite!"', Havel talks about citizens doing such actions, often without much thought let alone conviction, 'because these things must be done if one is to get along in life. It is one of the thousands of details' to help have a 'relatively tranquil life "in harmony with society"' (1985, 27–8). Such actions, Havel argues, are signs of obedience, however modest or low level that may

---

[11]A valuable compendium of influential early research on exchange networks and patron–client relations in agrarian societies is Schmidt *et al.* (1977).

[12]I hasten to add that this is *not* to say people with patrons or seeking support from better off individuals and families are consequently incapable of resistance. Everyday resistance and everyday support or compliance are not mutually exclusive. Peasants and other subordinated people do both, either at different times or even simultaneously.

be. In Czechoslovakia and other countries with similar political systems aimed at maximising conformity, uniformity, and discipline, any behaviour that is exceptional – out of line with what authorities expect – is a transgression, a 'genuine denial of the system'. So, lest individuals be seen as transgressors and thereby risk punishment, they must 'behave as though' they believe, or at least 'tolerate ... in silence', what the system expects of them. In such behaviour, the greengrocer and others doing the routine, daily things expected of them by the system, they become players in the game, 'thus making it possible for the game to go on, for it to exist in the first place' (Havel 1985, 30, 31, 36).[13] Much the same can be said about virtually every other political system and set of authority–follower relationships.[14]

Between everyday compliance and everyday resistance are everyday modifications and evasions of what authorities expect or the political system presumes. These actions usually convey indifference to the rules and processes regarding production, distribution, and use of resources. They are typically things people do while trying to 'cut corners' so as to get by. Although they may approach becoming, or seem at first glance to be, forms of everyday resistance, they are not. They do not intentionally oppose superiors or advance claims at odds with superiors' interests. Besides lacking an intention to hurt or target authorities and other people in superordinate positions, these forms of everyday politics can include behaviour done at the expense of neighbours and other people in similar conditions as those acting or speaking. Examples are people privately bad mouthing fellow peasants and co-workers for their different ethnicities and religious beliefs, men putting down women and vice versa, or poor persons stealing from other impoverished individuals.

In the collective farming cooperatives of Vietnam's Red River delta during the 1960s–80s, villagers often 'cut corners', doing things that did not comply with what authorities expected of them but without intending to resist. For instance, unless a work team assigned to fertilise planted fields was closely supervised, members sometimes did the work sloppily, such as dumping excessive amounts of fertiliser in a few spots, so as to complete the task quickly and easily, rather than spreading it evenly, which would take more time and effort. Whether they did the job diligently or not, people reasoned, they received the same number of 'work points', so they took the easy way. Another example is peasants, when preparing manure from their pig pens to meet a quota each household had to contribute to collectively farmed fields, sometimes mixed it with banana stems and sand to add weight, thus satisfying their quota while retaining as much manure as possible for themselves to use on their own garden plots. Frequently people did such acts because they figured they would make better use of manure on their own land than the teams farming collective fields could do. It was a matter of not trusting fellow villagers' diligence rather than opposing officials or the government's collectivisation policy. During the late 1960s and 1970s, individual households in numerous cooperatives often secretly used as their own some portions of fields that were supposed to be farmed collectively. Sometimes such encroachments were acts of defiance against local and higher officials. Other times, however, people took land for themselves out of a conviction

---

[13]Of course, in the company of their families and close confidants, greengrocers and others could be lamenting their compliant actions, even cussing authorities or the entire political system – in short, engaging in a form of everyday resistance.
[14]Studies have shown that German citizens' efforts to come to terms with the Nazi regime, even if they did not particularly like it, helped to perpetuate it (Peukert 1987, 67–80, 109–10). Lisa Wedeen (1999) has a similar argument in her analysis of Syria's political system.

that they could farm it better individually than collectively. In numerous cooperatives, members also stealthily harvested rice from collective fields. In some instances, they did so to quietly oppose egregious and abusive officials – acts of everyday resistance. Other times, however, they did it as preemptive measures – to get grain that they presumed other members would steal because people did not trust one another or were continuing long-standing hostilities and rivalries between neighbouring villages even though they were all now in one collective farming cooperative and supposed to be working together.

A combination of everyday modifications and evasions and everyday resistance has been known to contribute to authorities adjusting policies or even making new ones. In other words, the actions can affect official politics. Take one humdrum example. A few years ago, a newly landscaped part of the campus of my university included paved walk ways that pedestrians and bicyclists were supposed to use. After a few months, however, a couple of unauthorised paths began to appear, cutting across the carefully planted and tended lawn. Apparently these paths were the creation of pedestrians and bicyclists who deviated from the designated walk ways. The paths are a political expression – they use a resource, land, out of line with how university authorities intended it to be used. Whether the paths' creators are resisters or not is hard to know. Persistent research could probably find out. Quite likely, many pedestrians and bicyclists simply find the paths more convenient, helping them get more rapidly from one place to another. Their action – using the unauthorised path – is everyday politics of the modification type. For others, however, using – maybe even initiating – the paths may be deliberate opposition – a form of everyday resistance – to the authorities who designed the designated walk ways and had them built. In any event, the paths exist, altering what authorities had intended for the land. When the university office in charge of campus grounds first noticed the infractions, it instructed grounds keepers to place small barriers at both ends of each path, hoping that would force people to use the authorised walk ways. But people soon started going around those obstacles to use the paths. Recently, those paths have been paved, making them even more convenient for users. University authorities apparently gave up trying to eliminate the deviant use of the grounds; instead, they incorporated everyday users' creations into the official landscape design.

A similar process happened to Vietnam's collective farming policy. At the instruction of national authorities, local officials created collective farming cooperatives encompassing virtually all peasants and agricultural land in northern Vietnam and, later, many villagers and much land in the southern half. Cooperative members were supposed to follow prescribed rules and procedures for every step in the farming process – plowing, sowing, weeding, irrigating, harvesting, etc. – and in related activities such as raising and caring for draft animals, acquiring and maintaining machinery and implements, and accumulating and distributing rice and other produce. Views and stances of peasants toward collective farming varied from place to place and over time. Generally speaking, a large proportion became unenthusiastic. But they were unable, given the Communist Party government's tight surveillance and discouragement of contrary views and its prohibitions against organised opposition, to wiggle out of the cooperatives or openly criticise them.

In some cooperatives, virtually all members complied with what collective farming procedures required. In many cooperatives, most members did as they were supposed to do some of the time but not all of the time. And over the years, in many parts of Vietnam, a large proportion of members in about seventy percent of the cooperatives were more often out of line than they were in line with prescribed rules and procedures for collective farming. The examples given earlier are only a few of the large number of ways people made 'paths' of their own rather than adhere to the ways officials prescribed. Officials themselves in numerous cooperatives also strayed from what they were supposed to do. The deviations, as examples above indicate, often arose from people modifying procedures so as to make work easier or because they felt entitled to stray from prescribed courses seeing that fellow cooperative members were doing so. Many deviations were more than that – they were forms of everyday resistance because people were angry with dishonest, self-serving, or mean cooperative managers and other officials or, especially by the late 1970s–early 1980s, had grown disgusted with the whole collective farming system.

On numerous occasions between the early 1960s and mid 1980s, authorities attempted to stop the deviations from prescribed production methods, land use, produce distribution methods, livestock handling, manure use, etc. They reorganised the cooperatives, reconfigured work teams, revamped work point systems, imposed penalties, punished corrupt officials, and took scores of other actions, which typically were effective only briefly. Short of using draconian, violent measures, which probably would not have achieved the desired results anyway, authorities ran out of options. Instead, they essentially gave in, incrementally at first – authorising some of the adjustments cooperative members had initiated – but then fully by endorsing household farming.[15] While peasants' everyday modifications of and resistance to collective farming were not the only reasons for this major policy change, they were significant influences.

Not only in Vietnam have everyday politics contributed to the government amending or changing policies. The communist government in neighbouring Laos suspended agricultural collectivisation in late 1979, after little more than a year of trying to implement the policy. A major reason was widespread opposition among peasants, not through confrontational resistance but in numerous indirect ways, most notably fleeing across the Mekong River into Thailand (Stuart-Fox 1996, 125–6, 168, also see Evans 1990). Communist Party leaders in China began in 1979–80 to back away from collectivisation and allow households to farm individually. Among the reasons were peasants in many places doing minimal required work for the collective and diverting labour and other resources away from collective production to their own household production. One analyst likened the phenomenon to an unorganised yet extensive 'invisible "sit-in"'.[16] In the mid 1950s, authorities in Poland and Yugoslavia gave up insisting on collective farming largely because peasants deviated significantly from how they were supposed to work and quietly refused to join, farmed less, and hid produce (Szymanski 1984, 27–35, Korbonski 1965, 5, 139–40, 164, 189, 194–5, 212–14, Bokovoy 1997, 116–17, 125–9, 132, Rusinow 1977, 36–40). In several other Eastern European countries and in the Soviet Union, collective farming persisted but in forms different from what authorities had

---

[15]Mostly I am drawing on my own research (Kerkvliet 2005), but other studies have indicated that local deviations created bottom-up pressures that influenced policy change: Ch Văn Lâm *et al.* (1992, 52, 78–9), Fforde (1989, 80–1, 85–6, 205), Lê Huy Phan (1985, 14).

[16]Zhou, Xueguang (1993, 67). Also see Kelliher (1992) and Zhou, K.X. (1996). China specialists disagree about the extent to which such bottom-up pressures influenced national policies (see Unger 2002, 95–115).

initially wanted. Peasants' everyday practices and resistance, sometimes together with their public statements, petitions, and other forms of confrontational opposition, forced officials to give more scope to family production and in other ways adjust the structure and operations of collective farming (Creed 1998, 3, 184–7, Kideckel 1977, 52–8, 1982, 321, 326, 328, Rev 1987, 343–4, Swain 1985, 25–6, 41–2, 60–6, Fitzpatrick 1994, 8–9, 62, 130–49, 242, Humphrey 1983, 305).

All of these examples involve collective farming. But the phenomenon of everyday modifications and everyday resistance contributing to authorities' rethinking programs and policies is possible, although of course not inevitable, in other situations in which political systems attempt to organise people's work, where they live, their families, their religious practices, etc. Such systems amplify the political aspect of small deviations from what officials expect. Initially the impact may be only local, but if those modest variations are persistent and widespread, they can have national implications.[17] In less stringent political systems, where considerable opportunities exist for people to organise and publicly express their concerns, preferences, and objections – in short, to engage in forms of advocacy politics – the impact of everyday modifications and resistance may actually be less remarkable.

Everyday politics and we

The 'we' here means all of us, but particularly readers of this and other academic journals. Everyday politics is both a field of study and a reality for us. Understanding and appreciating the significance of politics in everyday life of others helps, at least should help, to make us – academics, scholars, teachers, students, parents, supervisors, colleagues – aware and conscious of the political dimensions of our own daily practices and behaviours.

Everyday politics shows that politics is not a place or activity that people can opt into or out of. Journalists commonly talk about individuals 'going into' or 'leaving' politics. Social scientists often say much the same. Portraying politics as something we can choose to do or not, as a place or an occupation to go into or not, has the pernicious effect of allowing, even encouraging, anyone who is not a politician or self consciously engaged in political activity to somehow think s/he is above or beyond politics. That portrayal also encourages or endorses views that politics is something far removed from our own lives, something only politicians and political activists do – people who are distant from and usually unknown to us. It obscures the political significance and ramifications of things we, individually and collectively, regularly do that affects who gets what, when, and how. It allows us to use resources without reflecting much, if at all, about the distributional and accessibility dimensions of our use and the impact on others in society – near and far – as well as on the natural world and environment.

Being conscious of politics as an everyday experience makes us mindful of our relations with the people we live and work with and reflect on how our actions and relationships reinforce, amend, evade, or contest the political system(s) of which we are a part, however small or large the impacts may be. Such mindfulness and reflection may or may not prompt us to change our daily behaviour. But that state of mind prevents us from relegating politics to something other people do. Possibly it

---

[17]For engaging elaborations, see Rev (1987), Verdery (1991), and Creed (1998).

will help us to search for justifications for continuing, rather than altering, how we treat family members, neighbours, colleagues, students, and others with whom we interact daily; how our actions – and inactions – perpetuate the status quo; and how our daily consumption habits affect production, distribution, and use of resources far beyond our own lives.[18]

Awareness of everyday politics in our own lives can also prompt us to examine our behaviour in advocacy politics and official politics. Many of us have authoritative positions in universities and other organisations. Many of us are involved in advocacy politics of various sorts. We need to ask ourselves the extent to which the views and practices we project in those realms are consistent with our everyday political behaviour and views. Consistency may not be possible – maybe not even always desirable – but being self-conscious about the degree to which our political thoughts and actions in all three realms – official, advocacy, and everyday – are in sync is healthy.

References

Adas, M. 1981. From avoidance to confrontation: peasant protest in precolonial and colonial Southeast Asia. *Comparative Studies in Society and History*, 23(2), 217–47.
Adas, M. 1986. From footdragging to flight: the evasive history of peasant avoidance: protest in South and South-East Asia. *The Journal of Peasant Studies*, 13(2), 64–86.
Adnan, S. 2007. Departures from everyday resistance and flexible strategies of domination: the making and unmaking of a poor peasant mobilisation in Bangladesh. *Journal of Agrarian Change*, 7(2), 183–224.
Ball, A.R. 1993. *Modern politics and government*, fifth edition. London: Macmillan.
Baud, M. and R. Rutten, eds. 2004. *Popular intellectuals and social movements: framing protest in Asia, Africa, and Latin America*. Cambridge: Cambridge University Press.
Bokovoy, M.K. 1997. Peasants and partisans: politics of a Yugoslav countryside, 1945–1953. *In:* J.A. Irvine and C.S. Lilly, eds. *State-society relations in Yugoslavia, 1945–1992*. New York, NY: St. Martin's Press, pp. 115–36.
Boyte, H.C. 2004. *Everyday politics: reconnecting citizens and public life*. Philadelphia, PA: University of Pennsylvania Press.
Chử Văn Lâm 1992. *Hợp Tác Hóa Nông Nghiệp Việt Nam: Lịch Sử, Vấn Đề, Triển Vọng* [*Agricultural cooperativisation in Vietnam: history, issues, and prospects*]. Hanoi: Nxb Sự Thật.
Collier, C.J. 1997. Politics of insurrection in Davao, Philippines. PhD Dissertation. University of Hawai'i, Manoa.
Creed, G.W. 1998. *Domesticating revolution: from socialist reform to ambivalent transition in a Bulgarian village*. University Park, PA: The Pennsylvania State University Press.
Dittmer, L., *et al.*, eds. 2000. *Informal politics in East Asia*. Cambridge: Cambridge University Press.
Elshtain, J.B. 1997. *Real politics: at the center of everyday life*. Baltimore, MD: Johns Hopkins University Press.
Evans, G. 1990. *Lao peasants under socialism*. New Haven, CT: Yale University Press.
Fforde, A. 1989. *The agrarian question in North Vietnam, 1974–1979*. Armonk, NY: M.E. Sharpe.
Fitzpatrick, S. 1994. *Stalin's peasants: resistance and survival in the Russian village after collectivisation*. New York, NY: Oxford University Press.
Ginsborg, P. 2005. *The politics of everyday life: making choices, changing lives*. New Haven, CT: Yale University Press.
Havel, V. 1985. The power of the powerless. *In:* V. Havel et al. *The power of the powerless: citizens against the state in central-eastern Europe*. Armonk, NY: M.W. Sharpe, pp. 23–97.

[18]Paul Ginsborg (2005) makes an energetic argument for the political significance of our daily consumption patterns, family life, and use of time.

Hobson, J.M. and L. Seabrooke. 2007. Everyday IPE: revealing everyday forms of change in the world economy. *In:* J.M. Hobson and L. Seabrooke, eds. *Everyday politics of the world economy*. Cambridge: Cambridge University Press, pp. 1–23.

Humphrey, C. 1983. *Karl Marx collective: economy, society and religion in a Siberian collective farm*. Cambridge: Cambridge University Press.

Kelliher, D. 1992. *Peasant power in China*. New Haven, CT: Yale University Press.

Kerkvliet, B.J. Tria. 1990. *Everyday politics in the Philippines: class and status relations in a central Luzon village*. Berkeley, CA: University of California Press.

Kerkvliet, B.J. Tria. 1993. Claiming the land: take-overs by villagers in the Philippines with comparisons to Indonesia, Peru, Portugal, and Russia. *The Journal of Peasant Studies*, 20(3), 459–93.

Kerkvliet, B.J. Tria. 2005. *The power of everyday politics: how Vietnamese peasants transformed national policy*. Ithaca, NY: Cornell University Press.

Kideckel, D.A. 1977. The dialectic of rural development: cooperative farm goals and family strategies in a Romanian commune. *Journal of Rural Cooperation*, 5(1), 43–61.

Kideckel, D.A. 1982. The socialist transformation of agriculture in a Romanian commune, 1945–62. *American Ethnologist*, 9(2), 320–40.

Korbonski, A. 1965. *Politics of socialist agriculture in Poland: 1945–1960*. New York, NY: Columbia University Press.

Lasswell, H. 1958. *Politics: who gets what, when, how*. Cleveland, OH: World Publishers.

Lê Huy Phan 1985. Mấy Suy Nghĩ về Cơ Chế Quản Lyỉ Kinh Tế ở Nước Ta từ Trước đến nay và về Phương Hướng Đổi Mới Cơ Chế Đó [Thoughts regarding our current economic management mechanism and a way for renovating it]. *Nghiên Cứu Kinh Tế [Economic Research]*, 146(8), 10–22, 30.

LeBlanc, R.M. 1999. *Bicycle citizen: the political world of the Japanese housewife*. Berkeley, CA: University of California Press.

Leftwich, A. 1984. Politics: people, resources, and power. *In:* A. Leftwich, ed. *What is politics?*. Oxford: Basil Blackwell, pp. 62–84.

Mason, T.D. 2004. *Caught in the crossfire: revolutions, repression, and the rational peasant*. Boulder, CO: Rowman & Littlefield.

Miller, H. 1993. Everyday politics in public administration. *American Review of Public Administration*, 23(2), 99–117.

Nishizaki, Y. 2005. The moral origin of Thailand's provincial strongman: the case of Banharn Silpa-archa. *South East Asia Research*, 13(2), 184–234.

Nishizaki, Y. 2008. Suphanburi in the fast lane: roads, prestige, and domination in provincial Thailand. *The Journal of Asian Studies*, 67(2), 433–67.

O'Brien, K.J. and L. Li. 2006. *Rightful resistance in rural China*. Cambridge: Cambridge University Press.

Peukert, D.J.K. 1987. *Inside nazi Germany: conformity, opposition and racism in everyday life*, trans. R. Deveson. London: B.T. Batsford.

Piven, F.F. and R.A. Cloward. 1977. *Poor people's movements: why they succeed, how they fail*. New York, NY: Vintage.

Rev, I. 1987. The advantages of being atomised: how Hungarian peasants coped with collectivisation. *Dissent*, 34(3), 335–50.

Rusinow, D. 1977. *The Yugoslav experiment, 1948–1974*. Berkeley, CA: University of California Press.

Rutten, R. 1996. Popular support for the revolutionary movement CPP-NPA: experiences in a Hacienda in Negros Occidental, 1978–1995. *In:* P.N. Abinales, ed. *The revolution falters: the left in Philippine politics after 1986*. Ithaca, NY: Southeast Asia Program Series, Cornell University, pp. 110–53.

Schmidt, S.W., et al., eds. 1977. *Friends, followers, and factions: a reader in political clientelism*. Berkeley, CA: University of California Press.

Scott, J.C. 1985. *Weapons of the weak: everyday forms of peasant resistance*. New Haven, CT: Yale University Press.

Scott, J.C. 1987. Resistance without protest and without organisation: peasant opposition to the Islamic Zakat and the Christian Tithe. *Comparative Studies in Society and History*, 29(3), 417–52.

Scott, J.C. 1990. *Domination and the arts of resistance: hidden transcripts.* New Haven, CT: Yale University Press.

Stoker, G. 1995. Introduction. *In:* D. Marsch and G. Stoker, eds. *Theory and methods in political science.* London: Macmillan, pp. 1–18.

Stuart-Fox, M. 1996. *Buddhist kingdom, Marxist state: the making of modern Laos.* Bangkok: White Lotus.

Swain, N. 1985. *Collective farms which work?.* Cambridge: Cambridge University Press.

Szymanski, A. 1984. *Class struggle in socialist Poland.* New York, NY: Praeger.

Unger, J. 2002. *The transformation of rural China.* Armonk, NY: M.E. Sharpe.

Verdery, K. 1991. Theorising socialism: a prologue to the 'transition'. *American Ethnologist,* 18(3), 419–39.

Walker, A. 2008. The rural constitution and the everyday politics of elections in northern Thailand. *Journal of Contemporary Asia,* 38(1), 84–105.

Walker, K. 2008. From covert to overt: everyday peasant politics in China and the implications for transnational agrarian movements. *Journal of Agrarian Change,* 8(2–3), 462–88.

Wedeen, L. 1999. *Ambiguities of domination: politics, rhetoric, and symbols in contemporary Syria.* Chicago, IL: University of Chicago Press.

Wood, E.J. 2003. *Insurgent collective action and civil war in El Salvador.* Cambridge: Cambridge University Press.

Zhou, K.X. 1996. *How the farmers changed China.* Boulder, CO: Westview Press.

Zhou, X. 1993. Unorganised interests and collective action in communist China. *American Sociological Review,* 58(6), 54–73.

Synergies and tensions between rural social movements and professional researchers

Marc Edelman

This essay outlines approaches to analysing and managing relations between rural activists and academic researchers. It suggests (a) that contemporary social movements engage in knowledge production practices much like those of academic and NGO-affiliated researchers and (b) that the boundaries between activists and researchers are not always as sharp as is sometimes claimed. These blurred boundaries and shared practices can create synergies in activist–academic relations. The essay then examines tensions in the relationship, including activists' expectation that academic research will be immediately applicable to their struggles and researchers' expectation that movement participants will accommodate their needs. The final section discusses the pros and cons, from the perspective of each side, of several models of activist–researcher relations, ranging from 'militant' or 'engaged' research to the contractual agreement between a movement and those involved in research on it. It argues that one of the most useful contributions of academic researchers to social movements may be reporting patterns in the testimony of people in the movement's targeted constituency who are sympathetic to movement objectives but who feel alienated or marginalised by one or another aspect of movement discourse or practice.

There has been a certain timidity on the part of the professional, perhaps apathy. And on our part, a certain bitterness or resentment because of everything that has happened to us. There are people who have abused us … At times we feel that we're cows. The [researchers] give us a big milking and somebody else gets to drink the milk. You understand? – Sinforiano Cáceres Baca, Federación Nacional de Cooperativas, Nicaragua, interview with the author, Managua, 4 July 1994.[1]

I presented a shorter, very different version of this essay at a January 2006 conference on 'Land, Poverty, Social Justice and Development: Social Movements Perspectives' at the Institute of Social Studies in The Hague. Many activists and colleagues at that forum provided comments and pointed critiques, as did Brenda Biddle, Jun Borras, Philippe Bourgois, Jeff Boyer, Annette Desmarais, Jonathan Fox, Lesley Gill, Angelique Haugerud, and Jeff Rubin. I am very grateful to all for their input, which helped immeasurably in improving the argument. The deficiencies of the essay are, of course, my responsibility.
[1]'Ha habido cierta timidez de parte del profesional, tal vez cierta apatía. Y cierto resquemor de parte nuestra por todo que nos ha pasado. Hay gente que nos ha abusado … Nos sentimos como vacas a veces. Y nos pegan una gran ordeñada y otro se bebe la leche. ¿Entendés?'

Introduction

Can and should rural social movements and professional or academic researchers work with each other and, if so, under what conditions and in pursuit of what objectives? In what ways are professional or academic researchers and movement researchers similar and in what ways are they different? What types of collaboration and cooperation might be fruitful? When do the relations between social movements and academic or professional researchers become problematical? What are some possible models or ways of specifying or negotiating mutually beneficial relationships? Who gets to 'drink the milk' and could the 'milking' metaphor be transcended?

This essay attempts to sketch some approaches to these issues. It starts with an analytical distinction between three categories of people: movement activists, academic researchers in universities and similar institutions, and professional researchers in other kinds of institutions, such as non-governmental organisations (NGOs). It then argues, however, that the distinction is partly, though not entirely, a heuristic one and that the lines between activist researchers and other researchers are in practice often blurred.[2] To make matters worse, or at least more complicated, another useful heuristic that breaks down under even minimal scrutiny is central to the way the problem here is framed. That is, the distinction between activists and researchers (of all kinds) rests to a large extent on a spurious distinction between 'doing' and 'thinking'. While such distinctions are dubious in practice, they nonetheless retain some limited analytical value inasmuch as activists and professional researchers (of both academic and other varieties) often occupy different social roles and institutional spaces and emphasise different kinds of social action.

The essay does not pretend to provide definitive answers, but aims instead at stimulating reflection, debate and mutual understanding. It draws on a reading of materials produced by movement and professional and academic researchers, on many conversations over the years, and on my own experience as a researcher sympathetic to and yet critical of some of the movements I have studied. I should state at the outset that I do not see the approaches of movement and academic or other professional researchers as incompatible or even necessarily as all that distinct. This does not mean, however, that the relations between them or their knowledge production practices are entirely unproblematic. Indeed, when professional (academic and non-academic) researchers and activists enter into relations tensions may always be present, in greater or lesser degrees and sometimes in subterranean forms, unrecognised by one or both parties. This does not necessarily mean, however, that the relation cannot be fruitful for each. It is also important to note at the beginning that (1) the essay focuses mainly on peasant and farmer movements, particularly transnational ones, even though many of the issues raised are likely relevant as well for other kinds of activist projects and for the researchers in and out of universities who accompany, study and write about and partner with them; and (2) the discussion of approaches to engaged research, of activists' concerns about researchers and of the history of professional researchers' acting for and against movements draws heavily on

---

[2]Moreover, as Silber (2007, 15) suggests, it is important to examine the 'move to theorise a created temporal and spatial community of engagement and suffering as activism'.

examples from anthropology. In part, this results from the author's own disciplinary location, but more importantly it reflects anthropology's role in the social scientific division of labour as the field assigned to 'the savage slot' (Trouillot 1991) of less-developed countries, the rural and urban poor, and peasant and indigenous peoples.

Academics and activists: blurred boundaries

Contemporary social movements engage in knowledge production practices much like those of academic and NGO researchers and the boundaries between activists and researchers are not always as sharp as is sometimes claimed. These blurred boundaries and shared practices can create synergies in activist–researcher relations. There are nonetheless some critical differences that may give rise to tensions. While tensions between social movements and NGOs are notorious and widespread, they generally involve questions of representation and competition over access to resources and decision-making fora (Borras 2004, 2008, Edelman 2003, Desmarais 2007, 21–6). Tensions between activists and academics, on the other hand, tend to revolve more narrowly around the research process and the purpose and methods of knowledge production and dissemination.

The complex challenges facing today's social movements have required activists to become researchers. In many cases, researchers in the social movements (and among their NGO allies) employ methods, technical language, and publication practices that resemble those of academics. Examples are numerous, but include many fine reports on biotechnology, global trade and Canadian agriculture written and published by the National Farmers Union (Canada).[3] Leading figures (or, in some cases, former participants) in the transnational peasant and small farmer organisations have also written rigorous and perceptive 'insider' histories of their movements, at times in academic journals (Bové 2001, Bové and Dufour 2001, Desmarais 2002, 2007, Holt-Giménez 2006, Stédile 2002, 2007, Borras 2004, 2008). In other cases, non-farm intellectuals and farm activists have collaborated closely in producing political and historical analysis (Alegría and Nicholson 2002, Stédile and Fernandes 1999). A few of these non-farm intellectuals have been integrally involved in formulating strategy, publicising movement platforms and activities, and carrying out research and training directly geared to movements' needs (du Plessis 2008, Monsalve Suárez *et al.* 2008, Rosset 2003).[4] All of these are potential synergies between social movement activists and non-farm researchers that could be put into practice more systematically.[5]

Some movement activists view academics as coming from an alien world. They draw sharp distinctions between activists and academics (and sometimes between movement organisations and NGOs, even though these lines are frequently more

---

[3]Many of these are available at http://www.nfu.ca/.

[4]Importantly, however, most of these are associated with NGOs and not academic institutions. The implications of this distinction will be examined more below.

[5]Fox (2006, 28–29) points to the US environmental justice movement as exemplifying the potential of 'partnerships between engaged researchers and grassroots organisations. In the US debate numbers and quantitative analysis were the key battleground for revealing the racial and class imbalance in exposure to toxic hazards. Alternative numbers empowered alternative ideas, turning them into mainstream common sense'; see also Hale (2007a, 21).

blurred than certain activists like to acknowledge).[6] Yet many leading activists also have considerable academic experience and credentials. They do not always 'wear these on their sleeve', since they are participants in and leaders of organisations that represent – or seek to represent – people who typically have much less formal education and sometimes none at all.[7] For a few movement activists involved in farming, the off-farm employment that permits them to survive as agriculturalists includes holding academic positions available only to those who have obtained a post-graduate degree.[8]

The movements have also produced, though their own training programs or those carried out in conjunction with various NGOs, a significant cadre of grassroots intellectuals.[9] Elsewhere, for example, I have written about the development in Central America of a large group of highly sophisticated peasant activists, trained in diverse areas such as trade policy, cooperative administration and agroecology (Edelman 1998, also see Rappaport 2005). These could be viewed as one impressive indicator of movement success, even when peasant movements in the Central American region have suffered major reverses in other aspects of their work (Edelman 2008).[10]

To summarise briefly, then, some important synergies between social movements and academics could involve exchanges of knowledge and contacts, joint strategy discussions, publicising organisations' platforms and activities and analysing their histories, and engaging in collaborative research and training.

Sources of tension

Relationships between academics and social movements are, not surprisingly, sometimes characterised by tensions. These include activists' expectation that academic research will be immediately applicable to their struggles and academics' expectation that movement participants will accommodate their needs. Moreover, professional intellectuals, and perhaps especially those who work in academic institutions, are often deeply invested in the search for detail and complexity and for comprehensiveness and 'truth', even when they recognise the illusory, relative and unattainable nature of the latter two objectives. Some of the professional intellectuals' best work, like that of good investigative journalists, involves probing

---

[6]On the blurred boundaries between social movements and NGOs, see Alvarez (1998), Bickham Méndez (2007), and Edelman (2008).
[7]Brazilian MST leader João Pedro Stédile, for instance, reports that 'probably the best period of [his] life' was when he was able to study in Mexico for two years in the 1970s. He reported meeting major figures in the Brazilian and Latin American left who were living in exile there, such as Francisco Julião, who had led the Ligas Camponesas (Peasant Leagues) in the early 1960s, and outstanding intellectuals such as Rui Mauro Marini, Vânia Bambirra, Teotônio dos Santos and Jacques Chonchol (Stédile 2002, 78–9).
[8]This is the case, for example, with at least one major figure in the National Farmers Union of Canada.
[9]Less commonly, as in Central America in the 1980s and 1990s, peasant organisations have provided the major impetus for creating NGOs to serve movement objectives.
[10]This process also points to the inadequacy of Gramsci's frequently cited concept of 'organic intellectuals'. As a good Marxist, Gramsci assumed that such intellectuals would come from the working class and, if they did emerge from the peasantry, would never remain loyal to peasant interests (Gramsci 1971, 6). Despite Gramsci's doubts about peasants' political reliability, in some countries, notably Bolivia, his language of 'organic intellectuals' has been widely adopted by militant activists of rural origin (Zamorano Villarreal 2009).

beneath the surface, questioning appearances and asking uncomfortable questions both of their movement interlocutors and of data that they may have obtained elsewhere (conversely, some of the worst work fails to do precisely this). The uncomfortable questions alone may generate frictions, but more fundamental is the activists' investment in presenting overly coherent 'official narratives' about their movements and in making representation claims that may or may not have a solid basis.[11] At times academic researchers and other professional intellectuals knowingly or unknowingly collude in producing and propagating those narratives and in 'airbrushing' (or, to be more up-to-date, 'photo-shopping') out dimensions of activists' biographies and of social movement practice that conflict with or complicate the 'official' picture or line. Whether or not this cosmetic approach, which in its more extreme manifestations critics sometimes characterise as 'self-censorship', 'uncritical adulation' or even 'cheerleading', really serves the needs of social movements is an important question, about which I will shortly have more to say.[12]

Academic researchers and social movement activists, even if they have similar knowledge production practices, sometimes seek to produce knowledge for different objectives.[13] Or, even if the objectives of each are similar, they rank the same objectives differently. To be more concrete, a movement researcher and a university-based researcher might each wish to write a book that examines recent agrarian struggles in country X. Each might genuinely want the results of the research to serve the needs of contemporary and future movements for change. The university-based scholar, however, would likely be at least somewhat more interested in writing the book for an audience of professional intellectuals, addressing arcane academic debates, and publishing with a prestigious academic press.[14] The movement-based author of such a book, on the other hand, would be more likely to seek a politically

---

[11]On the latter point, see Borras *et al.* (2008, 182–89) and Edelman (2008, 231). Speed (2007, 215) suggests that 'the multiple tensions and contradictions that exist between anthropologists and those we work with' need to be viewed as 'productive tensions that we might strive to benefit from analytically rather than seek to avoid'.

[12]Warren (2006, 221) asks about 'self-censorship', 'Does it mean that whole domains of social life have been, in effect, off the table for richer ethnographic analysis?'. Bevington and Dixon (2005, 191) argue that 'uncritical adulation' does not provide a movement 'with any useful information and does not aid the movement in identifying and addressing problems which may hinder its effectiveness'. The 'cheerleading' concept is often invoked in off-the-record conversations among researchers who study social movements and rarely appears in print. Scholars not identified with (or even hostile to) the movements are those most likely to employ the term in publications (e.g., Wickham-Crowley 1991, 4).

[13]Martínez (2007, 191) compares the researcher–activist relation to the rural Haitian *konbit* or communal reciprocal work party: 'activist anthropology puts people to work alongside each other, each side maintaining a distinct project, the anthropologist hoping to harvest academic publications even as he helps activists cultivate political or organisational gains. As in peasant agriculture, the goal of activist anthropology is not generating maximum output but generating sustainable and equitably shared gain'.

[14]Academics also, of course, are expected to publish not just books, but articles in specialised journals. In certain countries (such as the United States) and in some academic disciplines, professional advancement is strongly correlated with publishing in 'disciplinary' journals, particularly those affiliated with the main professional associations. More innovative work on social movements, especially that which challenges positivist paradigms or which manifests even mild political engagement, tends to be relegated to smaller, less prestigious but frequently more stimulating publications or to those journals that nobody reads but which exist mainly to credential scholars in 'second tier' institutions and to make profits for academic publishers.

progressive publisher that will guarantee a large print run, wide distribution and perhaps a low price for the work. She or he would also possibly put more energy into a collective discussion of the research findings and into translating any published work into the main language of country X, if it was not first written in that language.

The point is not that academic researchers are selfishly pursuing riches or hoping to inflate their curricula vitae at the expense of, or using knowledge generated by, the social movements. Virtually all university researchers who study or accompany social movements are profoundly sympathetic to the activists' goals (the exceptions are usually those who study right-wing extremist and religious fundamentalist movements).[15] Academics have political projects too, and those who study or partner with social movements tend to do so because they see in activism the realisation of some of their goals and hopes, movement toward a more just world and toward the kind of society in which they would like to live. Almost all of the scholarly books that academics write generate trivial amounts of royalties (though what is 'trivial' may appear different in the Global South and very occasionally, of course, academic books actually become bestsellers).[16] Rather, it is the institutional situation of university-based researchers that powerfully shapes the extent to which different objectives seem important to them. Especially for early-career academics in major universities, the possibilities of continuing to be able to work in their chosen fields depend mightily on the kinds of journals and presses that have published their work (and on the language in which they publish). It is not that they are generally persecuted for publishing in other kinds of outlets, or for translating their work into other languages, it is just that these activities typically have to be something 'extra', carried out alongside and in addition to the more traditional – I am tempted to say, more soporific and dry – academic work that secures their careers (which, in the United States at least, must almost always be in English). This 'extra' effort, of course, entails risks and has costs in time and career advancement that need to be anticipated by the professionally vulnerable, early-career academic and that, in the interests of real transparency, also ought to be brought to the attention of and acknowledged by her or his social movement interlocutors or collaborators. My observations here may reflect the particular characteristics of US academia (and particularly the more exalted 'Research I' institutions) where 'engaged' or 'action' research and acting as a public intellectual are less accepted than in Europe or Latin America (see Greenwood 2007, 322).[17] But the rules of promotion and tenure in

---

[15]For an excellent synopsis of the relevant literature and issues involved in the latter sort of study, see Blee (2007). Some researchers, while not sympathetic to rightist movements' goals, nonetheless strive to identify 'the human dimension' of even pathologically violent participants (see Cívico 2006).

[16]In the offices of some very distinguished scholars, I have actually seen royalty checks framed on the wall as ornaments, the amounts so laughably tiny as to make them not worth cashing.

[17]The 'Research I' category of universities, developed by the Carnegie Foundation (which discontinued use of the term in the late 1990s) refers to US doctoral-level institutions that place a heavy emphasis on research and obtain large amounts of federal grants. Criteria for promotion and tenure in Research II and other institutions may include a greater emphasis on teaching and service and less on research and publication. It is possible that researchers in these less prestigious institutions and particularly in fields that in the Research I universities avidly defend their disciplinary boundaries, may actually have greater leeway in pursuing unconventional career paths, integrating activism into their scholarly work, and publishing with other than the supposedly most important journals and presses.

European and Latin American universities do not vary much from North American ones, which is the important thing here in terms of explaining academics' priorities.

The time frames of academic and movement-based researchers are also different. The movement-based researcher typically wants research results now, to serve immediate political needs. Many social movements function in a permanent crisis-response mode as they attempt to adapt to fast changing political events. They do not enjoy the luxury of long-term reflection that academics aspire to have (even if this rarely exists in reality). Moreover, the organisation of university work, with its summer and sabbatical research leaves, 'is incompatible with any form of activism' or at least makes the academic largely unavailable to 'external stakeholders' between fieldwork periods (Greenwood 2007, 333–34).[18] Academics, unlike journalists, are socialised in the universities to write slowly. Sometimes they have also unlearned the ability to write clearly and succinctly.[19] Nor do academics always have access to great audiences or powerful media, as some activists seem to think. It is exceedingly difficult, for example, to place an opinion column in a major daily newspaper in the United States and very few scholars manage to do it more than occasionally, if at all.[20] Academic journals are notoriously slow in publishing research reports and I am convinced that very few people actually read most of them.[21] These differences of pace, style, perception, and audience between activists and academics may be another source of tension.

Among social movement activists the perception sometimes exists that university-based researchers control huge economic resources. This is rarely the case. University jobs, especially in the public universities, tend to be poorly paid in most countries, at least in comparison with those available in other sectors of the economy to individuals with comparable advanced training. The grants that university-based researchers receive are frequently insufficient to cover their costs. Young, graduate student researchers commonly live in quite precarious circumstances. Academic and NGO-based researchers can hopefully be an *intellectual* or *political* resource for social movement activists, able to connect them to knowledge, information and policymakers. Researchers may also be able to facilitate activists' connections to

---

[18]Warren (2006, 221) points out that 'most of [academic anthropologists] are part-time observers'. Many nonetheless position themselves in heroic and self-aggrandising ways in their written accounts as a way of establishing narrative authority and fail 'to acknowledge scholarly networks and lines of transnational solidarity that provide the basis upon which innovative findings and activism are constructed'.

[19]Conversely, many have been trained to employ deliberately obscure yet high-prestige, jargon-laced prose styles accessible only to a small, initiated group of similarly specialised colleagues. It is not always a simple matter to unlearn these rhetorics of (academic) power in the interests of communicating about or supporting a broader political project.

[20]But see González (2004b) and Besteman and Gusterson (2005) for forceful interventions by academics in, respectively, major media and public policy debates.

[21]Credentialing, rather than knowledge diffusion, is often the main function of most academic journals. This is rarely evident to non-academics, however, and is rarely acknowledged by academics. Opinions differ as to whether activists actually read social movement theory produced by professional scholars (as opposed to case study histories of movements). Shukaitis and Graeber (2007) point to intense engagement with certain varieties of 'high' theory by (mostly European) anarchist and autonomist activists. Bevington and Dixon (2005, 189), writing mainly about US political science and sociology's social movement studies, ask, 'what does it say if the social movement theory being produced now is not seen as helpful by those persons who are directly involved in the very processes that this theory is supposed to illuminate?'

certain funding agencies. But academics themselves are unlikely to be a direct source of money resources or to have significant funds to contribute to the movements.

Another problem that arises from the academic–activist relation is the activists' fear that the academic might be gathering intelligence or functioning as an agent provocateur.[22] This reflects activists' well-founded anxieties about omnipresent imperial or state power, even as it is also suggestive of their sometimes exaggerated belief in their own political significance. The involvement of supposedly neutral academic researchers in intelligence work dates back to World War I and the first decades of professional social science, when the Mayanist archaeologist Sylvanus Morley reported from Mexico's Yucatán Peninsula to the US Office of Naval Intelligence on Germany's sympathisers in the region and its shipping in the Caribbean (Sullivan 1991, Harris and Sadler 2003). In the inter-war years, British anthropologists, in particular, worked among colonised peoples and, while their work tended 'to obscure the systematic character of colonial domination' (Asad 1973, 109), their advice to the authorities was rarely sought and, when offered, was almost always ignored (James 1973, 48–9). During World War II, numerous anthropologists lent their skills to the struggle against Nazism and Japanese imperialism and to the administration of local populations, particularly on Pacific islands that had been retaken by the United States (Price 2008, Wolf and Jorgensen 1970). In the post-war era, anthropologists and other social scientists were involved in research and in intelligence and military work that came to be viewed as ethically and morally questionable in a range of Cold War hot spots, including Guatemala, Thailand and Vietnam, and Chile (Berger 1995, Gusterson 2003, Horowitz 1967, 'Newbold' 1957, Price 2002, Watkin 1992, Wolf and Jorgensen 1970).[23] More

---

[22]Price points out that the situation is actually more complex, since scholars may be witting or unwitting participants in research that is directly funded or sought by intelligence agencies or in independent research that is later used by such agencies. 'The following four scenarios are possible: Witting-Direct, Witting-Indirect, Unwitting-Direct, Unwitting-Indirect' (2002, 17). He concludes that 'most of anthropology's interactions with intelligence agencies have probably been unwitting and indirect, with anthropological work being harvested by intelligence agencies as it enters the public realm through conferences and publications' (2002, 21). Many social scientists have been reluctant to examine scholars' past links to intelligence agencies, arguing that to do so will result in reduced possibilities for field research. Price argues forcefully for scrutinising such ties, since 'we all risk reduced field opportunities as these largely unexamined historical interactions become documented' (2002, 17).

[23]Richard N. Adams' survey research with political prisoners in Guatemala, following the 1954 CIA-directed coup, was published in 1957 under the pseudonym 'Stokes Newbold', a composite of his own middle name and that of Manning Nash, who collaborated in the research. In a 1998 reminiscence, Adams noted that since his survey did not provide much evidence that the rural population had been influenced by Communist proselytising, as the State Department had alleged, the US Embassy rapidly forgot about it. Ironically, some left-leaning social scientists subsequently found Adams' survey of considerable value in examining the social class origins and attitudes of Guatemalan rural activists of the 1950s (Grandin 2004, 226n54, Wasserstrom 1975, 464). Adams also indicated that he used a pseudonym at the insistence of his employer, the Pan American Sanitary Bureau, and that 'I never hid the fact that I was the author, but it was some years before it became widely known' (Adams 1998, 15, 20n9). During the rest of his long career, Adams developed a pronounced concern about research ethics (Adams 1967) and a strongly critical stance regarding US policy in Guatemala and the Guatemalan military's abysmal human rights record, as did June Nash, who also participated in the survey.

Wolf and Jorgensen (1970, 32) cite a US counterinsurgency specialist in Thailand, who was interviewed by a *New York Times* reporter: 'The old formula for successful counterinsurgency

recently, US social scientists have deployed to Afghanistan and at least one played a key role in authoring the US Army's new counterinsurgency manual (González 2007, Rohde 2007).[24]

Clearly, then, the activists' fears regarding the researcher's possible covert activities or loyalties are not based solely on febrile imaginings. Nonetheless, three important points deserve consideration. First, it is worth remembering, especially if suspicion falls on a foreign researcher, that most countries' secret services almost always employ nationals of the country in which they are operating, rather than their own nationals, to do most of the actual spying.[25] Second, it is only a tiny minority of outside researchers who are compromised by ties to intelligence agencies and false accusations of such links have occasionally resulted in tragedy.[26] Third, in various world regions outside (and frequently foreign) researchers are among those who have produced many of the most compelling exposés of powerful institutions, structural violence, and militarisation, as well as the most trenchant critiques of the deficiencies of mainstream punditry, scholarship and policymaking.[27]

Even if the academic researcher is entirely beyond reproach, activists are also concerned that the data gathered or the reports published might find their way into the wrong hands or strengthen the analytical capabilities of their antagonists. If these concerns arise, they need to be addressed explicitly and clear agreements need to be reached about how to assure that no harm results from the researcher's activities. Much academic research – probably most – does not do any direct or indirect harm and it does not do much direct good either, since, as I noted above, hardly anybody usually reads it.[28] But activists can and ought to do background research on

---

used to be 10 troops for every guerrilla ... Now the formula is ten anthropologists for each guerrilla' (Braestrup 1967).

[24]The manual, known as *FM 3-24*, includes a chapter by 'Montgomery McFate', a pseudonym for Mitzy Carlough, who received a PhD in anthropology from Yale.

[25]The writings of Philip Agee, who worked for the US Central Intelligence Agency in Latin America for many years before abandoning the CIA and authoring an exposé, provide abundant evidence of this practice. Agee remarked, however, that it was harder to recruit agents in countries with a higher standard of living and a developed welfare state than in less-developed countries. 'Uruguayan communists simply are not as destitute and harassed as their colleagues in poorer countries and thus are less susceptible to recruitment on mercenary terms' (1975, 339).

[26]Price (2002, 17) mentions the case of Raymond Kennedy, a US scholar who had worked in the Office of Strategic Services during World War II and then joined the State Department. An opponent of Dutch colonialism in Indonesia, he resigned from the State Department in protest against US policies. Four years later he was executed in Java by anti-colonial fighters who mistakenly believed he was a CIA agent.

Similar suspicions fall on non-researchers as well, undoubtedly in larger numbers and with tragic results, especially during revolutionary armed movements. See, for example, Bourgois's description of internecine violence among the Salvadoran guerrillas in the 1980s (2001).

[27]Examples are many, but include Besteman and Gusterson (2005), Gill (2004), González (2004a), Lutz (2004), McCaffrey (2002), and Vine (2007).

However, the ethical and legal dimensions of foreign activists involving themselves in the politics of other nations are rarely considered in the literature on engaged research. Some, such as Juris (2008) and Scheper-Hughes (1995), apparently consider such involvement a matter of internationalist or ethical commitment and completely unproblematic. Speed (2007, 11) is among the few who acknowledge the problem and note its impact on her research; she nonetheless attributes her difficulties in Mexico to a government 'xenophobia campaign'.

[28]Vargas notes 'that scholars, especially those in the beginning of their career, benefit from their involvement with grassroots organisations in ways glaringly disproportionate to what we can offer them' (2007, 164–65, also 178).

academic researchers if they have doubts about them. Other academics might even be able to help with this, though this of course runs the risk of creating unfounded paranoia and destroying cooperative relationships (see Price 2002, 17).

Some academic researchers expect that the people they accompany or study will or ought to accommodate their needs for time-consuming conversations or other contributions (searching for documents, photocopying, making introductions, chauffeuring them to meetings, and so on).[29] This expectation, perhaps the outcome of the exalted status that some university professors enjoy in their home institutions and societies, understandably irritates movement activists. Like overworked university faculty members, activists have many demands to which they must respond. The benefits of meeting the researcher's needs may be unclear or abstract, minimal or nonexistent, or realisable only far in the future. Not all human relationships, fortunately, require a clear quid pro quo, but ongoing ones do usually require some sort of reciprocity. Attending the academic researcher can become yet another burden and source of stress for the activist. How might each party to this relation see the other more clearly? How might the expectations and perceptions each has of the other be made more explicit? What can each party offer the other?

Some models of activist–academic relations

It may be useful, in addressing the question of activist–academic relations, to distinguish two types of researchers that enter into relations with rural social movements: first, the researcher who has knowledge or connections sought by the activists, such as an expert understanding of agronomy, trade policy, intellectual property, web design, foundations, or legislative processes and lobbying; and second, the researcher who seeks to study and write about the rural social movements.[30] I have already suggested that the first kind of academic or NGO-affiliated expert may be an important intellectual or political resource for the movements. The second type is potentially more complicated. This is primarily because the category may include researchers who fall in different places along a wide continuum of epistemologies and political commitment. These may range from a positivist stance of 'neutrality' and 'objectivity' to a 'militant' position that conceives of the researcher's role as a publicist and/or uncritical supporter of the movement under study.[31] In between these two poles are various degrees of 'engagement' or 'commitment' and at least the possibility of a critical interrogation of both activists' and scholars' activities and knowledge production practices. Also important, of course, is the specific set of issues that the researcher intends to examine. Some situations of extreme political urgency or polarisation – massive human rights violations, structural violence,

---

[29]Nabudere (2007, 79), writing on Uganda, remarks on the 'deep mistrust on the part of people who had developed "research fatigue" from constant harassment by hordes of researchers since colonialism had first knocked on their doors. They had seen researchers come and go while their own conditions had steadily worsened. This suggested to them, with some justification, that the researchers were part of their problem.'
[30]Fox (2006, 30) remarks that 'one kind of contribution that scholars can offer to social actors is to wade through, decipher, and boil down the mind-bending quantities of arcane and hard-to-access information that is produced by mainstream institutions'.
[31]Given the hegemony of positivist approaches in some disciplines, it may be important to recall that the very origins of modern social science were thoroughly activist and its practitioners were often deeply involved in what might today be called social movements and in attempting to modify policy (see Calhoun 2007, Greenwood 2007).

famines or epidemics, for example – may require that researchers 'commit' in more sustained, profound ways. This may not be simply a moral or ethical imperative or a decision that is up to researchers. The research subjects with whom they work may very likely push them in this direction.

Nonetheless, there is no single model of activist–academic relationship that will address all of the tensions, possible synergies, or questions. Some models of relationship will accomplish some things better than others. For this reason, both parties to such relationships might do well to think in terms of creativity and variety rather than in terms of a single desirable model or pattern.[32]

'Militant' or 'engaged' research constitutes a range of approaches, although there is not, of course, consensus about the practices and ideas that these rubrics might include or about how much they might overlap. At the more 'militant' end of the spectrum, the academic or NGO-affiliated researcher places herself or himself at the service of an organisation, takes direction from that organisation, works as a sort of publicist, and reports findings that advance the organisation's agenda.[33] Admirable as this commitment sometimes might be, the 'militant' model has some shortcomings, whether seen from the position of the movement or from academia.[34] The work of intellectuals closely identified with particular organisations or political tendencies often has less credibility in the larger society and especially in the media, academia and among policymakers than does the work of intellectuals who are sympathetic to the movement but maintain some critical distance and independence. In the role of advocate or publicist, the formally independent voice is likely to be heard more widely and to be taken more seriously than the one seen as 'militant' and compromised. Moreover, as Jennifer Bickham-Méndez (2007, 144) argues, 'Decision makers might assume research presented by academics to be more rigorous and reliable than that put forth by campaigning NGOs, lending a level of legitimacy and credibility to social justice struggles.'

---

[32]Three colleagues who commented on an earlier version of this essay pointed out that activist–researcher synergies are most likely to occur when both parties develop strong feelings of trust in each other as a result of daily interactions around small matters. Activists naturally observe outside researchers at least as much as the other way around, and they make decisions about the extent of their collaboration on the basis of how they evaluate the researcher's integrity, sincerity and decency, among other things.

[33]Efforts to theorise these ideas and practices include Scheper-Hughes (1995), the essays in Harrison (1991) and in Hale (2007b), and the mostly anarchist-influenced essays in Shukaitis *et al.* (2007). Juris (2008, 20, 319) calls for researchers to practice 'proactive solidarity' and remarks that, 'Militant ethnography ... refers to ethnographic research that is not only politically engaged but also collaborative, thus breaking down the divide between research and object.' This formulation, apart from its emphasis on ethnography to the exclusion of other research practices, obscures the extent to which collaboration between researchers and subjects may be of varied forms and intensity. In other words, it elides discussion of a collaborative engagement that might not be militant in the sense of subordinating the researcher to a larger political project, but which still might serve objectives shared by the researcher and the movement.

[34]At times, the 'militant' stance, presented as an unproblematic matter of pre-existing ethical-political principles, verges on a troubling naivete, as when one prominent US anthropologist – newly arrived in South Africa – identified herself in a squatter camp as 'a member of the ANC [African National Congress]' (Scheper-Hughes 1995, 414). At least one critic has charged, in relation to this case, that the 'militant' position rests on 'an amalgamation of sociobiological and religious ideas' and substitutes an outsider's politics for research itself (Trencher 1998, 122).

An additional problem derives from the frequent gaps between leaders and grassroots activists and the notorious factionalism (and at times, corruption) that pervade social movements (Edelman 2001, 310–11). To which individuals or groups is the 'militant' activist committed? What does that commitment imply about the analytical weight accorded dissenting voices and alternative claims? Even more troubling can be the problem of what to do 'when faced with dirty laundry' (Fox 2006, 35), that is, when the researcher runs across anti-democratic practices or instances of malfeasance. Jonathan Fox argues compellingly for an approach of 'first do no harm'. Simple whistle blowing to one or another movement sector (or to donor organisations) may, as Fox indicates, constitute unacceptable, impolitic and counterproductive external intervention. But, as he also suggests, doing nothing or pretending that nothing is wrong may not help the movement and represents another kind of external intervention, albeit one characterised by inaction.

The degeneration of the United Farm Workers (UFW) since its founding by César Chávez in California in the 1960s is a striking case of what can occur when knowledgeable outsiders (and insiders) eschew speaking out or intervention in the face of questionable practices (Cooper 2005, Pawel 2006a, 2006b, 2006c, 2006d, Weiser 2004). In its early years, the UFW led dramatic organising campaigns and consumer boycotts and the charismatic Chávez, with his principled dedication to non-violent direct action, was widely considered a Mexican-American version of the great US civil rights leader, Martin Luther King, Jr. In the late 1970s, faced with growing tensions among his top lieutenants, Chávez implemented self-awareness encounter sessions (called 'The Game') in the UFW modelled on the programs of Synanon, a cult-like authoritarian drug rehabilitation organisation, and 'drifted toward a more autocratic management style' (Ferriss *et al.* 1998, 212, Pawel 2006c). By 2005, less than 2 percent of California's agricultural labour force belonged to any union and a mere 5,000 workers had UFW contracts – one-tenth of the organisation's peak strength.[35] Following the death of Chávez in 1993, the union largely abandoned its original mission of organising farm labourers and his heirs poured their efforts

> into a web of tax-exempt organisations that exploit his legacy and invoke the harsh lives of farmworkers to raise millions of dollars in public and private money. The money does little to improve the lives of California farmworkers, who still struggle with the most basic health and housing needs and try to get by on seasonal, minimum-wage jobs ... [The] tax-exempt organisations ... do business with one another, enrich friends and family, and focus on projects far from the fields. (Pawel 2006a)

As the UFW began to unravel in the late 1970s, few activists stood up to Chávez or resisted his increasingly erratic leadership (Pawel 2006c) and when his family members began to 'enrich' (Pawel 2006a) themselves and one another it was

---

[35]In part this resulted from anti-labour policies in the state capital and from growers' increasing reliance on undocumented Mexican workers, many of whom came from indigenous communities and spoke little Spanish. The UFW lacks organisers who speak indigenous languages. One indigenous worker-activist quoted in a press report declared: 'We hear a lot about the achievements of Cesar Chavez. But we can't see any of them. Where are they? Truth is, the UFW has no strength here, not among our people. We remember how, when the Mixtecs first began to organise, Cesar called us "communists". That's okay, he's gone. We need our own organisations now that speak to our heart, our own union' (Cooper 2005).

investigative journalists, not internal dissenters or 'militant' researchers, who encouraged greater scrutiny.[36]

Another role, not incompatible with 'militant' research though not necessarily linked to it either, is being an 'engaged' public intellectual and witness. The public intellectual or 'citizen-scholar' (González 2004a, 3) speaks out in the print and electronic media, debates the spokespeople of the opposition, exposes hidden wrongdoing, provides expert testimony in courts and legislatures, discusses lobbying and other strategy with activists, teaches students and colleagues, and brings to a larger public the issues raised by grassroots movements.[37] The public intellectual works fundamentally to bring about political and cultural shifts in society. Often his or her contributions are not especially dramatic, but they are grains of sand that hopefully contribute to eroding the legitimacy of existing power structures, exposing abuses, strengthening the legitimacy of grassroots movements, and informing broader publics about alternative visions, other experiences and strategies for change.

Most social movements claim to be representatives of particular sectors of the population (women, farmers, indigenous people, immigrants, and so on), but it is nearly always a small minority of the group that the movement claims to represent that actually participates in it (Borras *et al.* 2008, 182–86). Activists are often baffled about why people fail to join them in larger numbers. One of the most significant contributions of academic researchers to social movements may be reporting patterns in the testimony of people in the movement's targeted group who are sympathetic to movement objectives but who feel alienated or marginalised by one or another aspect of movement discourse or practice (Burdick 1998, see also Bevington and Dixon 2005). Movement activists and leaders are often unable to identify problematical organisational patterns that for them have become 'naturalised' and that outside researchers easily detect, such as ethnic or regional imbalances and exclusions. Importantly, the 'militant' researcher, closely identified with the movement, is unlikely to do these things as well as a more independent researcher to whom people will speak candidly.[38]

Reporting the testimony of the alienated or uninvolved is one important role, but asking challenging questions more generally is another.[39] Both are delicate matters, since they potentially dispute movements' 'official stories', activists' self-concepts and organisations' strategic visions. In the world of transnational activism, for example, national organisations that participated in cross-border alliances at times withdrew in order to struggle at the national level or to work with alternative transnational

---

[36]Family members of UFW leader Dolores Huerta are also prominent in the same organisations. One of the first journalistic exposés of the UFW affiliates wryly remarked that 'some would call this nepotism' (Weiser 2004).
[37]Examples of mass media interventions of this sort are collected in González (2004b).
[38]One strand of activist research may simply involve generating critical knowledge, i.e., 'an effort to understand how things could be different and why existing frameworks of knowledge do not recognise all the actual possibilities' (Calhoun 2007, xxiv-xxv).
[39]In a brilliant essay on 'Research as an experiment in equality', Portelli (1991, 44) suggests that the research interaction itself is both laden with political content and an opportunity for political work. 'There is no need to stoop to propaganda in order to use the *fact* itself of the interview as an opportunity to stimulate others, as well as ourselves, to a higher degree of self-scrutiny and self awareness; to help them grow more aware of the relevance and meaning of their culture and knowledge; and to raise the question of the senselessness and injustice of the inequality between them and us.'

networks (Borras 2008).[40] Because sympathetic researchers tend to assume that transnational strategies and organisational forms will always be the most effective, they rarely pose this assumption as a question even though activists frequently mention as a problem the conflicting demands of local-, national- and transnational-level activities. The balance of what is gained *and* what is lost in moving from a national to a transnational level (or back) is among the critical issues that activists need to consider as a matter of long-term survival and researchers need to analyse either as part of a more genuine solidary relation with the activists or as a matter of political-intellectual honesty.[41]

This researcher contribution to the movements – posing difficult questions and especially reporting the testimony of the disaffected – may be more possible when the scale of the research is smaller and place-based rather than when it has a wider, multi-sited focus (it is also easier when organisations are flourishing than when they are in decline). In my own case, I felt that I was more effective in this way when I did extended field research in the countryside in Costa Rica in the 1980s and 1990s, and when I knew many campesinos in and out of the movements, than when I began to study transnational farmer networks and was constantly moving from place to place and had more superficial, shorter-term interactions almost exclusively with movement participants and few or none with non-participants.[42] The issue of non-participants in targeted constituencies also raises the broader question of what Fox terms 'the *directionality* of the researcher's goals – are we drawing from the movement in order to project analysis *outward*, or are we drawing from the external environment in order to project analysis *inward*?' (2006, 30, emphasis in original).

The last model of activist–academic relations that I will mention involves a formal contract specifying shared objectives and what each party is expected to do and over what period of time. Some indigenous peoples have long required that outsiders hoping to carry out research among them be vetted by a council of experts. The Kuna in Panama, for example, require written contracts with researchers and demand that all studies carried out in their *comarca* or district be published in Spanish or Kuna and provided to the group's own archives. The statutes governing their territory specify as a crime 'the free giving of Kuna-related data and documents

---

[40]In 2008, for example, several Mexican organisations that had participated in the transnational Vía Campesina network wrote an open letter explaining their decision not to attend the organisation's Fifth International Conference in Mozambique. They charged that the North American regional coordinating group of Vía Campesina engaged in 'verticalist and antidemocratic' practices and spread 'disinformation' and was more interested in 'international activities' than in supporting local and regional initiatives (AMUCSS, ANEC, CNPA, and FDC 2008, see also Edelman 2005, 37–9).
[41]In an analysis of the impact of transnational consumer boycotts, 'stateless regulation' and NGO monitoring of corporations, Seidman writes, 'In an era when most national governments seem weaker than footloose multinational corporations, the international human rights movement and past examples of transnational consumer-based pressure on corporations seem to offer promising new directions for transnational campaigns … I interrogate this promise, hoping not to undermine efforts by transnational activists to find new approaches to organising workers, but to provoke discussion: in the effort to create new support for workers' struggles, why do so many activists neglect or bypass local institutions designed to protect citizens, and what might be gained or lost as a result?' (2007, 15–16).
[42]I do not mean to suggest that I have found it easy to raise challenging questions of activists. On the contrary, at times it can be decidedly uncomfortable for both the researcher and the activist (see Edelman 1999, 34–6).

to any institution without an equitable and reciprocal enrichment' (CGK and CISAI 2003).

Would this type of contractual relationship work in the rather different context of relations between social movements and researchers who wish to study and write about them? The Vía Campesina and some of its constituent organisations have been moving in the direction of establishing such a written protocol for their relations with researchers and academic and non-governmental organisations, loosely modelled on similar agreements that indigenous peoples in Canada have found useful. The goals of this more formalised relationship would include assuring that the research is of use to the movement, helping the movement to understand what the research is about, and – in some cases – shaping or determining the research agenda.[43] While these seem like reasonable objectives, particularly from the point of view of the activists, they also entail potential difficulties. The first of these relates to what I have called above activists' penchant for presenting overly coherent 'official narratives' or 'stories' about their movements and in making representation claims that may not have a solid basis. To what extent would the written contract limit researchers in the kinds of testimony they could report or in asking delicate questions? The contractual agreement should ideally include a genuine movement commitment to openly entertain uncomfortable questions or findings and to refrain from personalising any discomfort experienced in the process. Activists need to ask whether the refusal to countenance challenging questions or consider problematical findings really serves the movement's interests. The second difficulty concerns the potential of the written contract to become a kind of gate-keeping or an ideological litmus test. Gate-keeping of this sort would not only preclude facing uncomfortable research findings, but it could also introduce a burdensome formality and bureaucracy into the relation that consumes activists' (and researchers') time and energy.

Despite these caveats, however, it is essential that researchers recognise that their very presence entails costs for the movements, even if there are also clear benefits of collaboration. The activists' desire for a written contract that spells out the parameters of the relationship is thus eminently understandable. So too is the possibility that close collaboration can identify shared objectives and powerful synergies.[44] Researchers in private universities in the Global North, for example, are often able to fund activists' international speaking tours and other activities. Contractual agreements could also contemplate the disposition of royalties from books or other research products, even if these are unlikely in most cases to be very substantial.

What can or should movements ask or insist on from the academic researcher who expresses an interest in doing research with and writing about them? Really anything they would insist on in another political or personal relationship, including any or all of the following things: transparency about funding sources and objectives and about how, if at all, the movement might benefit from having a researcher in its midst; frank dialogue and exchanges of opinion about issues of mutual concern; prompt, clear and succinct reporting of research results in a form accessible to movement participants; co-authorship of publications or visual or other media, if

---

[43] Annette Desmarais, personal communication, 15 September 2008.
[44] Juris (2008) and the essays in Shukaitis *et al.* (2007) provide a number of compelling examples from the perspectives of 'militant', anarchist-oriented researchers.

both parties so desire; and collaboration in day-to-day movement tasks. As regards the latter, in the 1980s and 1990s when I was involved in work in Central America, for example, I sometimes rented jeeps to get myself and activist friends to assemblies in remote areas. I participated in some memorable joint forums with Costa Rican academics, NGO researchers and peasant leaders that sometimes degenerated into heated arguments from which everybody nonetheless learned a great deal. I occasionally translated correspondence and grant proposals, and copyedited the text of the English edition of the ASOCODE newsletter.[45] Several times I hosted ASOCODE representatives on their visits to UN agencies, foundations, and church and university groups in New York. These were small forms of solidarity that I would have offered to anyone, but which I felt especially pleased to share with those throughout Central America who had treated me with the utmost graciousness and hospitality and who expected little or nothing in return for helping me to understand their struggles and write about their lives and aspirations.

Concluding remarks

The epigraph of this essay quotes a long-time campesino activist-intellectual lamenting that too often researchers 'milk' their subjects, who then don't get to drink any of the milk. Activists (and many researchers too) have been rightly critical of this extractive, exploitative and unidirectional model of movement–researcher relations. Yet developing different, more horisontal and collaborative kinds of relations and research practices is not entirely straightforward either. Movements frame issues and make claims in ways that may not withstand close examination, even by sympathetic observers, and they may resist hearing information that contradicts the stories they tell about themselves (and possibly ostracise the people who bring that information). Researchers' professional priorities may not coincide and may conflict with those of the movements they study. NGO-affiliated intellectuals, while less constrained by the career imperatives that often shape academics' research and dissemination practices, sometimes assume that they may speak in the name of those on whose behalf they claim to work. All parties to this complicated, multi-faceted relation are involved in generating new knowledge. This can be a shared practice, though too frequently differences arise over what knowledge to produce, how to produce it, who should produce it, and what to do with it and who 'owns' it once it is produced. Some of these tensions become explicit, while many others remain uncommunicated and sometimes even unconscious.

What can be done to realise the potential synergies between professional researchers and social movements and to ensure on both sides greater clarity about shared (and divergent) objectives and more realistic expectations and understandings of limitations and possibilities? This essay has analysed several models of activist–researcher relations and argued that different approaches are suitable for different goals and that no single approach is able to address or resolve all or even most of the tensions in the relationship. Indeed, I have suggested that some tensions are very likely inherent aspects of any process of collaboration between social movements and researchers. What, then, of a process in which the activist–researcher distinction

---

[45]ASOCODE was the Asociación Centroamericana de Organizaciones Campesinas para la Cooperación y el Desarrollo (Association of Central American Peasant Organisations for Cooperation and Development). Its rise and demise are outlined in Edelman (2008).

is erased? This 'militant', 'engaged' or 'committed' stance also entails strains and, in its more extreme versions, a narrowed vision that may turn out to be of little help – or indeed, harmful – to the movements themselves. Rather than reifying the 'militant' or researcher as a single thing or practice, the essay has argued that 'engagement' and 'commitment' are best understood as existing along a continuum that has many dimensions. Sometimes the most enduring contribution of the researcher to the social movement may be in challenging its activists' assumptions with fresh data and an outsider's insights. Not to do this, as suggested above, can involve an abdication of responsibility that flies in the face of genuine engagement.

References

Adams, R.N. 1967. Ethics and the social anthropologist in Latin America. *American Behavioral Scientist*, 10(10), 16–21.
Adams, R.N. 1998. Ricocheting through a half century of revolution. Kalman Silvert Award address, Latin American Studies Association. *LASA Forum*, 29(3), 14–20.
Agee, P. 1975. *Inside the company: CIA diary*. New York, NY: Stonehill.
Alegría, R. and P. Nicholson. 2002. Nous espérons que ce livre aidera à faire connaître les luttes paysannes et à construire des ponts avec d'autres mouvements sociaux. *In:* M. Mazoyer, *et al. Vía Campesina: une alternative paysanne à la mondialisation néolibérale*. Geneva: CETIM, pp. 5–6.
Alvarez, S. 1998. Latin American feminisms 'go global': trends of the 1990s and challenges for the new millennium. *In:* E. Dagnino, S.E. Alvarez and A. Escobar, eds. *Cultures of politics/ politics of cultures: re-visioning Latin American social movements*. Boulder, CO: Westview, pp. 293–324.
AMUCSS, ANEC, CNPA, and FDC [Asociación Mexicana de Uniones de Crédito del Sector Social, Asociación Nacional de Empresas Comercializadoras de Productores del Campo, Coordinadora Nacional Plan de Ayala, Frente Democrático Campesino de Chihuahua] 2008. Por qué no podremos asistir a la 5ª Conferencia Internacional de La Vía Campesina. 21 October. Email document from ANEC.
Asad, T. 1973. Two European images of non-European rule. *In:* T. Asad, ed. *Anthropology and the colonial encounter*. Atlantic Highlands, NJ: Humanities Press, pp. 103–18.
Berger, M. 1995. *Under northern eyes: Latin American studies and US hegemony in the Americas, 1898–1990*. Bloomington, IN: Indiana University Press.
Besteman, C. and H. Gusterson, eds. 2005. *Why America's top pundits are wrong: anthropologists talk back*. Berkeley, CA: University of California Press.
Bevington, D. and C. Dixon. 2005. Movement-relevant theory: rethinking social movement scholarship and activism. *Social Movement Studies*, 4(3), 185–208.
Bickham Méndez, J. 2007. Globalizing scholar activism: opportunities and dilemmas through a feminist lens. *In:* C.R. Hale, ed. *Engaging contradictions: theory, politics, and methods of activist scholarship*. Berkeley, CA: University of California Press, pp. 136–63.
Blee, K.M. 2007. Ethnographies of the far right. *Journal of Contemporary Ethnography*, 36(2), 119–28.
Borras, S.M. Jr. 2004. La Vía Campesina: an evolving transnational social movement. *TNI briefing series*, No. 6:32. Amsterdam: Transnational Institute. Available from: http:// www.tni.org/reports/newpol/campesina.pdf [Accessed 20 January 2005].
Borras, S.M. Jr. 2008. La Vía Campesina and its global campaign for agrarian reform. *Journal of Agrarian Change*, 8(2–3), 258–89.
Borras, S.M. Jr., M. Edelman and C. Kay. 2008. Transnational agrarian movements: origins and politics, campaigns and impact. *Journal of Agrarian Change*, 8(2–3), 169–204.
Bourgois, P. 2001. The power of violence in war and peace: post-Cold War lessons from El Salvador. *Ethnography*, 2(1), 5–34.
Bové, J. 2001. A farmers' international?. *New Left Review*, 12, 89–101.
Bové, J. and F. Dufour. 2001. *The world is not for sale: farmers against junk food*. London: Verso.
Braestrup, P. 1967. Researchers aid Thai rebel fight. *The New York Times*, 20 March, p. 7.

Burdick, J. 1998. *Blessed Anastácia: women, race, and popular Christianity in Brazil.* New York, NY: Routledge.

Calhoun, C. 2007. Foreward. *In:* C.R. Hale, ed. 2007. *Engaging contradictions: theory, politics, and methods of activist scholarship.* Berkeley, CA: University of California Press, pp. xiii–xxvi.

CGK and CISAI [Congreso General Kuna and Centro Interdipartimentale di Studi Sull'America Indígena, Università di Siena] 2003. Acuerdo Bilateral Entre el Congreso General de la Cultura Kuna y el Centro Interdipartimentale di Studi Sull'america Indígena dell'Università degli Studi di Siena. Available from: http://www.congresogeneralkuna.org/acuerdos.htm [Accessed 5 October 2008].

Cívico, A. 2006. Portrait of a paramilitary: putting a human face on the Colombian conflict. *In:* V. Sanford and A. Angel-Ajani, eds. *Engaged observer: anthropology, advocacy, and activism.* New Brunswick, NJ: Rutgers University Press, pp. 131–46.

Cooper, M. 2005. Sour grapes: California's farm workers' endless struggle 40 years later. *LA Weekly*, 11 August. Available from: http://www.laweekly.com/2005–08–11/news/sour-grapes/ [Accessed 15 October 2008].

Desmarais, A.-A. 2002. The Vía Campesina: consolidating an international peasant and farm movement. *The Journal of Peasant Studies*, 29(2), 91–124.

Desmarais, A.-A. 2007. *The Vía Campesina: globalisation and the power of peasants.* Halifax, NS: Fernwood Publishing.

du Plessis, A. 2008. *New global contract: values in conflict: how trade and finance rules curtail our rights.* Minneapolis, MN: Institute for Agriculture and Trade Policy. Available from: http://www.iatp.org/iatp/publications.cfm?accountID=451&refID=103539 [Accessed 30 September 2008].

Edelman, M. 1998. Transnational peasant politics in Central America. *Latin American Research Review*, 33(3), 49–86.

Edelman, M. 1999. *Peasants against globalization: rural social movements in Costa Rica.* Stanford, CA: Stanford University Press.

Edelman, M. 2001. Social movements: changing paradigms and forms of politics. *Annual Review of Anthropology*, 30, 285–317.

Edelman, M. 2003. Transnational peasant and farmer movements and networks. *In:* M. Kaldor, H. Anheier and M. Glasius, eds. *Global civil society 2003.* London: Oxford University Press, pp. 185–220. Available from: http://www.lse.ac.uk/Depts/global/yearbook03chapters.htm [Accessed 15 January 2009].

Edelman, M. 2005. When networks don't work: the rise and fall and rise of civil society initiatives in Central America. *In:* J. Nash, ed. *Social movements: an anthropological reader.* London: Blackwell, pp. 29–45.

Edelman, M. 2008. Transnational organising in agrarian Central America: histories, challenges, prospects. *Journal of Agrarian Change*, 8(2–3), 229–57.

Ferriss, S., R. Sandoval and D. Hembree. 1998. *The fight in the fields: Cesar Chavez and the farmworkers movement.* New York, NY: Harvest/HBJ Book.

Fox, J. 2006. Lessons from action–research partnerships. LASA/Oxfam America 2004 Martin Diskin Memorial Lecture. *Development in Practice*, 16(1), 27–38.

Gill, L. 2004. *The school of the Americas: military training and political violence in the Americas.* Durham, NC: Duke University Press.

González, R.J. 2004a. Introduction. *In:* R.J. González, ed. *Anthropologists in the public sphere: speaking out on war, peace, and American power.* Austin, TX: University of Texas Press, pp. 1–20.

González, R.J., ed. 2004b. *Anthropologists in the public sphere: speaking out on war, peace, and American power.* Austin, TX: University of Texas Press.

González, R.J. 2007. Towards mercenary anthropology? The new US army counterinsurgency manual FM 3–24 and the military anthropology complex. *Anthropology Today*, 23(3), 14–19.

Gramsci, A. 1971. *Selections from the prison notebooks.* New York, NY: International Publishers.

Grandin, G. 2004. *The last colonial massacre: Latin America in the Cold War.* Chicago, IL: University of Chicago Press.

Greenwood, D.J. 2007. Theoretical research, applied research, and action research: the deinstitutionalization of activist research. *In:* C.R. Hale, ed. *Engaging contradictions: theory, politics, and methods of activist scholarship.* Berkeley, CA: University of California Press, pp. 319–40.

Gusterson, H. 2003. Anthropology and the military – 1968, 2003, and beyond?. *Anthropology Today*, 19(3), 25–6.

Hale, C.R. 2007a. Introduction. *In:* C.R. Hale, ed. *Engaging contradictions: theory, politics, and methods of activist scholarship*. Berkeley, CA: University of California Press, pp. 1–28.

Hale, C.R., ed. 2007b. *Engaging contradictions: theory, politics, and methods of activist scholarship*. Berkeley, CA: University of California Press.

Harris, C.H. III and L.R. Sadler. 2003. *The archaeologist was a spy: Sylvanus G. Morley and the office of naval intelligence*. Albuquerque, NM: University of New Mexico Press.

Harrison, F., ed. 1991. *Decolonising anthropology: moving further toward an anthropology for liberation*. Washington, DC: American Anthropological Association, Association of Black Anthropologists.

Holt-Giménez, E. 2006. *Campesino a Campesino: voices from Latin America's farmer to farmer movement for sustainable agriculture*. Oakland, CA: Food First Books.

Horowitz, I.L., ed. 1967. The rise and fall of Project Camelot. *In:* I.L. Horowitz, ed. *The rise and fall of Project Camelot*. Cambridge, MA: MIT Press, pp. 3–44.

James, W. 1973. The anthropologist as reluctant imperialist. *In:* T. Asad, ed. *Anthropology and the colonial encounter*. Atlantic Highlands, NJ: Humanities Press, pp. 41–69.

Juris, J.S. 2008. *Networking futures: the movements against corporate globalisation*. Durham, NC: Duke University Press.

Lutz, C. 2004. Militarisation. *In:* D. Nugent and J. Vincent, eds. *A companion to the anthropology of politics*. Oxford: Blackwell, pp. 318–31.

Martínez, S. 2007. Making violence visible: an activist anthropological approach to women's rights investigation. *In:* C.R. Hale, ed. *Engaging contradictions: theory, politics, and methods of activist scholarship*. Berkeley, CA: University of California Press, pp. 183–207.

McCaffrey, K.T. 2002. *Military power and popular protest: the US Navy in Vieques, Puerto Rico*. New Brunswick, NJ: Rutgers University Press.

Monsalve, S., *et al.* 2008. *Agrofuels in Brazil*. Heidelberg: FIAN International. Available from: http://www.fian.org/resources/documents/others/agrofuels-in-brazil/pdf [Accessed 30 September 2008].

Nabudere, D.W. 2007. Research, activism, and knowledge production. *In:* C.R. Hale, ed. *Engaging contradictions: theory, politics, and methods of activist scholarship*. Berkeley, CA: University of California Press, pp. 62–87.

'Newbold, Stokes' [Richard Newbold Adams and Manning Stokes Nash] 1957. Receptivity to communist fomented agitation in rural Guatemala. *Economic Development and Cultural Change*, 5(4), 338–61.

Pawel, M. 2006a. UFW: a broken contract. Farmworkers reap little as union strays from its roots. *Los Angeles Times*, 8 January. Available from: http://www.latimes.com/news/local/la-me-ufw8jan08,0,6620187.story?coll=la-home-headlines [Accessed 16 October 2008].

Pawel, M. 2006b. UFW: a broken contract. Linked charities bank on the Chavez name. *Los Angeles Times*, 9 January. Available from: http://www.latimes.com/news/local/la-me-nonprofits9jan09,0,378433.story [Accessed 16 October 2008].

Pawel, M. 2006c. UFW: a broken contract. Decisions of long ago shape the union today. *Los Angeles Times*, 10 January. Available from: http://www.latimes.com/news/local/la-me-history10jan10,0,3382590.story [Accessed 16 October 2008].

Pawel, M. 2006d. UFW: a broken contract. Former Chavez ally took his own path. *Los Angeles Times*, 11 January. Available from: http://www.latimes.com/news/local/la-me-medina11jan11,0,1905063.story [Accessed 16 October 2008].

Portelli, A. 1991. *The death of Luigi Trastulli and other stories: form and meaning in oral history*. Albany, NY: SUNY Press.

Price, D.H. 2002. Interlopers and invited guests: on anthropology's witting and unwitting links to intelligence agencies. *Anthropology Today*, 18(6), 16–21.

Price, D.H. 2008. *Anthropological intelligence: the deployment and neglect of American anthropology in the Second World War*. Durham, NC: Duke University Press.

Rappaport, J. 2005. *Intercultural utopias: public intellectuals, cultural experimentation, and ethnic pluralism in Colombia*. Durham, NC: Duke University Press.

Rohde, D. 2007. Army enlists anthropology in war zones. *The New York Times*, 5 October. Available from: http://www.nytimes.com/2007/10/05/world/asia/05afghan.html [Accessed 7 November 2007].

Rosset, P. 2003. Food sovereignty: global rallying cry of farmer movements. *Food First Backgrounder*, 9(4), 1–4. Available from: http://www.foodfirst.org/pubs/backgrdrs/2003/f03v9n4.html [Accessed 3 June 2004].

Scheper-Hughes, N. 1995. Objectivity and militancy: a debate. The primacy of the ethical: propositions for a militant anthropology. *Current Anthropology*, 36(3), 409–40.

Seidman, G.W. 2007. *Beyond the boycott: labour rights, human rights, and transnational activism*. New York, NY: Russell Sage Foundation.

Shukaitis, S. and D. Graeber. 2007. Introduction. *In:* S. Shukaitis, D. Graeber and E. Biddle, eds. *Constituent imagination: militant investigations/collective theorisation*. Oakland, CA: AK Press, pp. 11–34.

Shukaitis, S., D. Graeber and E. Biddle, eds. 2007. *Constituent imagination: militant investigations/collective theorisation*. Oakland, CA: AK Press.

Silber, I.C. 2007. Local capacity building in 'dysfunctional' times: internationals, revolutionaries, and activism in postwar El Salvador. *Women's Studies Quarterly*, 35(3–4), 167–83.

Speed, S. 2007. Forged in dialogue: toward a critically engaged activist research. *In:* C.R. Hale, ed. *Engaging contradictions: theory, politics, and methods of activist scholarship*. Berkeley, CA: University of California Press, pp. 213–36.

Stédile, J.P. 2002. Landless battalions: the Sem Terra movement of Brazil. *New Left Review*, 15, 76–104.

Stédile, J.P. 2007. The class struggles in Brazil: the perspective of the MST. *In:* L. Panitch and C. Leys, eds. *Global flashpoints: reactions to imperialism and neoliberalism – socialist register 2008*. London: Merlin Press, pp. 193–216.

Stédile, J.P. and B.M. Fernandes. 1999. *Brava Gente: a Trajetória do MST e a Luta pela Terra no Brasil*. São Paulo: Fundação Perseu Abramo.

Sullivan, P. 1991. *Unfinished conversations: Mayas and foreigners between two wars*. Berkeley, CA: University of California Press.

Trencher, S.R. 1998. Righteous anthropology. *Society in Transition*, 29(3–4), 118–29.

Trouillot, M.-R. 1991. Anthropology and the savage slot: the poetics and politics of otherness. *In:* R. Fox, ed. *Recapturing anthropology. Working in the present*. Santa Fe, NM: School of American Research Press, pp. 18–44.

Vargas, J.H.C. 2007. Activist scholarship: limits and possibilities in times of black genocide. *In:* C.R. Hale, ed. *Engaging contradictions: theory, politics, and methods of activist scholarship*. Berkeley, CA: University of California Press, pp. 164–82.

Vine, D. 2007. Island of injustice: the US has a moral duty to the people of Diego Garcia. *The Washington Post*, 2 January. Available from: http://www.washingtonpost.com/wp-dyn/content/article/2007/01/01/AR2007010100698.html?referrer=emailarticle [Accessed 30 September 2008].

Warren, K. 2006. Perils and promises of engaged anthropology: historical transitions and ethnographic dilemmas. *In:* V. Sanford and A. Angel-Ajani, eds. *Engaged observer: anthropology, advocacy, and activism*. New Brunswick, NJ: Rutgers University Press, pp. 213–27.

Wasserstrom, R. 1975. Revolution in Guatemala: peasants and politics under the Arbenz government. *Comparative Studies in Society and History*, 17(4), 443–78.

Watkin, E. 1992. *Anthropology goes to war: professional ethics and counterinsurgency in Thailand*. Madison, WI: University of Wisconsin, Center for Southeast Asian Studies.

Weiser, M. 2004. UFW, affiliates are a family affair. *The Bakersfield Californian*, 9 May. Available from: http://www.fels.org/Staff/Union/Inside%20UFW-May%202004-7.pdf [Accessed 14 October 2004].

Wickham-Crowley, T.P. 1991. *Exploring revolution: essays on Latin American insurgency and revolutionary theory*. Armonk, NY: M.E. Sharpe.

Wolf, E.R. and J.G. Jorgensen. 1970. Anthropology on the warpath in Thailand. *The New York Review of Books*, 15(9), 26–35.

Zamorano Villarreal, G. 2009. Reimagining the state: politics, video and indigenous struggles in Bolivia. PhD Dissertation, City University of New York.

Index

Adnan, S. 6
advocacy politics 5, 6, 220
African National Congress (ANC) 208
Agarwal, B. 191, 196, 200, 207
Agarwala, R.: and Herring, R. 13
agrarian populism 63-6; current
    challenges 69-73; and neo-populism
    63-6
agricultural production: growth 3
agricultural technology 4-5
agricultural trade: growth 3
agriculture and development strategies
    10-11, 94-128; agrarian reform 111;
    dual economy model 97; East Asia
    108-12, 121; industrialisation 96-8;
    industrialisation sequence 111-12;
    industrialisation and urban bias 106;
    industry-agriculture relations 106-8;
    industry-agriculture synergies 110-11,
    113-15, 120; international factors
    105-6; Latin America 108-12, 121;
    poverty causes/persistence 104-5;
    primacy of agriculture 100, 107;
    rural-urban linkages 113-15, 121;
    Soviet industrialisation debate 98-100,
    121; unequal exchange 105-6; urban
    bias thesis 96, 100, 101-5; World Bank
    and rural poverty 115-16; World
    Development Report (2008) 115-20
Agriculture and Human Values 146
Akram-Lodhi, A.H. 115, 118; and Kay, C.
    3-4, 9
Alderman, H.: and Udry, C. 192
Amanor, K.S. 118-19

Araghi, F. 134, 144, 145
Arce, A. 172

Batterbury, S.: and Bebbington, A. 175-6
Baviskar, A. 18
Bebbington, A. 164, 166; and Batterbury,
    S. 175-6
Becker, G. 188-9
Bennholdt-Thomsen, V. 193, 194
Bernstein, H. 1, 4, 9, 10, 13, 50-75;
    et al 174
Bezemer, D.: and Headey, D. 105, 115
Bickham-Mendez, J. 242
biofuels 5
Black Death 33-4
Bloch, M. 29-30, 40
Bourgholtzer, F. 53
Brenner, R. 28-9, 41
Brezhnev, L. 84, 89
Brundtland report (1987) 162, 163
Bukharin, N. 82, 98, 99
Burch, D.: and Lawrence, G. 148
Byres, T.J. 10, 19-20, 28-49, 102, 103,
    105, 106

Campbell, H. 149, 150
capitalist agrarian transformation:
    England 28-30, 31-6; France 29,
    36-41; Prussia 29, 41-6
Carney, D. 166
Carson, R. 150
Chambers, R.: and Conway, G. 161, 163
Chávez, C. 243-4
Chayanov, A.V. 13; and agrarian populism